DRIVEN WILD

How the Fight against Automobiles

Launched the Modern Wilderness Movement

PAUL S. SUTTER

Foreword by William Cronon

UNIVERSITY OF WASHINGTON PRESS

Seattle and London

TO JULIE, HENRY, AND WYATT

Driven Wild by Paul Sutter has been published with the assistance of a grant
from the Weyerhaeuser Environmental Books Endowment, established by
the Weyerhaeuser Company Foundation, members of the Weyerhaeuser
family, and Janet and Jack Creighton.

ISBN 0-295-98219-5
(cl.; alk. paper)

The paper used in this publication is acid-free and recycled from
10 percent post-consumer and at least 50 percent pre-consumer waste.
It meets the minimum requirements of American National Standard
for Information Sciences-Permanence of Paper for Printed Library
Materials, ANSI Z39.48-1984. ♻ ∞

CONTENTS

FOREWORD: WHY WORRY ABOUT ROADS

William Cronon

Among the benchmark environmental events of the twentieth century was the U.S. government's decision in 1964 to protect from development a growing acreage of public lands by legally designating them as "wilderness." The process began in a few obscure places in remote corners of the country: the Gila National Forest in New Mexico, Trapper's Lake in Colorado, and the Boundary Waters in northern Minnesota. By the 1930s, a new national organization—the Wilderness Society—had been created with the explicit mission of protecting wild places on the public lands and securing legislation that would guarantee that they remain forever wild. Although less well known by the public than it deserves to be, the Wilderness Society played an essential role in drafting and lobbying for the legislation that created the national wilderness system as we know it today. The 1964 Wilderness Act that eventually resulted from this effort remains among the most important environmental laws ever passed in the United States.

Why should Americans be so interested in protecting wilderness? This question has long been at the heart of American environmental history. It has many answers. One is the romantic sublime: the belief since the late eighteenth century that certain natural sites and phenomena—the mountain top, the chasm, the waterfall, the storm, the rainbow—are the places on earth where God is most immanent and where we are most likely to experience the deity at firsthand. Another is the frontier: the longstanding conviction among many Americans that their nation was forged by the pioneer encounter with wilderness. One of the founding myths of American nationalism is articulated in Frederick Jackson Turner's famous (and infamous) frontier thesis, which argues (problematically, but in the popular imagination still compellingly) that American character and American democracy are both the products of a frontier encounter with wilderness. There can be little doubt that the sublime and the frontier played key roles in the early movement to set aside national parks in places like Yosemite and Yellowstone. Our affection for such parks, based on their natural beauty but also on the romantic and nationalist symbolism we still find in them, continues to this day.

But neither the sublime nor the frontier can adequately explain one curious feature of the 1964 Wilderness Act. The authors of the Act, in their effort to protect lands "where the earth and its community of life are untrammeled by man," included in Section 4(c) a "Prohibition of Certain Uses" that were to be explicitly outlawed in wilderness. The Prohibition declares that there shall be "no permanent road within any wilderness area," and furthermore that there shall be "no temporary road, no use of motor vehicles, motorized equipment or motorboats, no landing of aircraft, no other form of mechanical transport, and no structure or installation within any such area." Despite all the possible activities that threaten the integrity of wilderness and that could easily have been named in this crucial section, the Act's overwhelming concern is to outlaw motorized vehicles and roads from the lands it seeks to protect. At a time when automobiles were very nearly the defining core of what was proudly described as "the American way of life," and when the nation was in the midst of constructing an Interstate Highway System that was among the wonders of the modern world, this hostility toward cars and roads in the 1964 Act seems at least intriguing, if not downright puzzling. Where did it come from? What can it tell us about the origins of wilderness protection in the United States? And what might be its lessons for today?

Paul Sutter's signal contribution in *Driven Wild* is to answer these and many other questions by arguing that one cannot understand the role of wilderness in modern American culture without recognizing its crucial relationship to roadlessness. Sutter demonstrates that the movement to protect wild land reflected a growing belief among many conservationists that the modern forces of capitalism, industrialism, urbanism, and mass consumer culture were gradually eroding not just the ecology of North America, but crucial American values as well. For them, wilderness stood for something deeply sacred that was in danger of being lost, so that the movement to protect it was about saving not just wild nature, but ourselves as well.

To shed new light on the ideas and values that underpinned the early days of this movement, Sutter adopts as especially appealing strategy. By looking at four of the men who founded and led the Wilderness Society in the years surrounding its creation, he weaves together biography and history to demonstrate that people's motives for protecting wilderness were surprisingly diverse. The different backgrounds of these four men help account for the richness and sophistication that characterized wilderness advocacy by the mid-twentieth century, and make it even more intriguing that they were so much in agreement about the dangers posed by roads and cars.

Of the four figures Sutter studies in depth, some will be more familiar than others. Robert Marshall, for instance, will be known to anyone who is even casually familiar with the early history of wilderness preservation in the United States. No one was more responsible in the 1930s for persuading the U.S. Forest Service and the Bureau of Indian Affairs to set aside wilderness areas on the lands they managed. Marshall was an extraordinary figure, tireless in his travels and lobbying efforts on behalf of wilderness, and his tragic early death at the age of thirty-eight only makes more remarkable the number of acres that he was personally responsible for protecting. Equally well known among the founders of the Wilderness Society was Aldo Leopold, the eminent wildlife ecologist and nature writer who served as the chief theorist of wilderness protection in the 1930s and 1940s. It was Leopold who argued that wilderness provided a crucial ecological baseline against which more humanized environments could be compared, and he also produced some of the most lyrical celebrations of wilderness and wilderness values in all of American literature. His *A Sand County Almanac* remains essential reading for anyone interested in wild land protection today.

Much less well known among the subjects of *Driven Wild* is Robert Sterling Yard, who was involved after 1916 in the early development of the National Park System before deciding that the parks were doing an inadequate job of protecting wild lands. Yard played an essential role in the day-to-day management of the Wilderness Society during its early years, was among the most important publicists for wilderness, and helps us understand why a growing number of advocates believed that the National Park Service should not be solely responsible for protecting such places. Finally, Benton MacKaye, a regional planner, was as interested in protecting rural communities as he was in protecting wilderness. One of his most valuable contributions to the movement was his suggestion in the 1920s that a trail be constructed along the crest of the Appalachian Mountains all the way from Georgia to Maine. That suggestion led over the next two decades to the construction of the famed Appalachian Trail, as potent a symbol of wilderness as anything in the eastern United States.

Each of these men brought quite different experiences and ideas to the task of defending wilderness. The founders of the Wilderness Society were neither starry-eyed romantics blessed with a singular revelation of how wilderness should be protected nor conservative reactionaries fleeing into the back country because they could not make their peace with modernity. Instead, Sutter shows how much they were men of their era, fully in dialogue with the intellectual ferment and politics of the 1920s and 1930s. Marshall and MacKaye, for instance, leaned toward socialism and had little doubt that the best way to protect wilderness (and also people) in the modern world was to rely on a strong activist state as a bulwark against the ravages of capitalism. Leopold, on the other hand, was far more dubious that government could be relied on to do all that was needed to protect wild nature, and was therefore just as interested in working to preserve wilderness on private property as on public. Coming to their work with such divergent politics, these men nonetheless mounted a defense of wilderness that will probably feel surprisingly contemporary to readers even three-quarters of a century later.

Part of what makes their ideas feel so modern is precisely their focus on motorized recreation. Without the evidence that Sutter unearths in these pages, it would be all too easy to imagine that the founders of the Wilderness Society sought to protect wild land from a familiar set of corporate villains: ranchers whose animals wreak havoc in riparian areas, grasslands, and forests; loggers intent on cutting down the last old-growth forests to exploit their commodity values; and mining corporations devastating entire landscapes by stripping away whole mountains and leaving polluted wastelands in their stead. Marshall, Leopold, Yard, and MacKaye surely recognized these threats to wilderness. But as Sutter brilliantly demonstrates, none of these was the reason they felt such urgency to devise legal protections for wilderness in the 1930s.

No, the founders of the Wilderness Society saw the gravest danger coming from an entirely different direction: ordinary middle-class tourists seeking to visit wild places in their automobiles. It was to serve such tourists and build a powerful political constituency that the first director of the National Park Service, Stephen Mather, embarked on a massive promotional campaign to encourage Americans to visit their "national playgrounds." To enable park visitors to experience these natural wonders for themselves, even through the windows of their automobiles, Mather and his successors initiated some of the most heroic feats of highway engineering of the twentieth century. By pushing strips of concrete and asphalt ever deeper into some of the nation's most remote and scenic land-scapes, the Park Service created some of the most exhilarating driving experiences to be found anywhere on the planet.

The combined craft of engineers and landscape architects built roads that have become so famous that they are almost synonymous with the parks in which they are located: Trail Ridge Road in Rocky Mountain National Park, Going to the Sun Highway in Glacier National Park, and, perhaps most famous of all, Skyline Drive in Shenandoah National Park. All followed high-elevation, ridge-top routes that no ordinary highway would ever traverse, assuring the visitor that these roads would themselves be among the highlights of the park experience.

For people like the founders of the Wilderness Society, the commitment of the National Park Service to automobile-based tourism made it an adversary of the very places it was legally supposed to protect. If wilderness was to be a sanctuary in the modern world for nature untrammeled by humanity, if it was to remain one of the last places on earth where the primitive conditions of the frontier could be experienced at first hand, then the intrusions of automobiles and highways must be resisted at all costs. The effects of logging and mining might be managed to keep them localized and under control, and cattle grazing, if properly regulated, might even serve as a reminder of the frontier experience. But once a road had been pushed into the heart of a wilderness area, there was almost no stopping the forces of development that would erode and finally destroy it.

This is why, despite the many possible ways of defining wilderness and despite many other possible threats, the legal definition of wilderness as written into American law by the 1964 Act depends more than anything else on the touchstone quality of roadlessness. When one considers the numerous battles during the past half century to protect wilderness from logging or grazing or mining or dam-building or oil-drilling, the decision by the Wilderness Society's founders to focus primarily on preventing the construction of new roads may seem oddly parochial and narrow-minded. But one of Paul Sutter's greatest contributions in this fine book is to demonstrate just how wise and prescient their emphasis on roadlessness truly was.

For Sutter, the most remarkable feature of early wilderness advocacy was the effort to redefine and reposition wilderness so that it stood in creative tension with mass consumer culture and modernity itself. Although motorized recreation might seem to give every car-owner easy access to "nature," when judged against the standard of roadless wilderness the effect of highways and other tourist amenities was to build a wall between travelers and the very nature they sought to visit. By making access to wilderness too easy, the road-builders turned it into just another consumer good, its worth measured in mere dollars instead of honest physical exertion and deeper spiritual renewal. The result was to alter and devalue wild landscapes so fundamentally that their special qualities were lost even for those who still made the pilgrimage on foot or horseback. If the value of wilderness was not merely ecological but moral as well, then one of its most precious features was an experience of non-human nature in which the mediating distractions of modern technology were kept to a minimum. To let cars and other motorized vehicles intrude into wilderness was to make this most non-human of places too much like the cities and other humanized landscapes that tourists were seeking to escape in the first place.

Defending wilderness, in other words, did not mean rejecting modernity; rather, it meant preserving alternatives to modernity so as not to forget the larger contexts and counterpoints that continue to give our human world some of its most important meanings. What Marshall, Leopold, Yard, and MacKaye realized before most other Americans is that roads literally paved the way for all other threats to wilderness, so that by stopping them one might hope to fend off others as well. Without roads, natural ecosystems might more successfully resist other human modifications, and older, humbler forms of travel might persist without being replaced by noisy, exhaust-spewing gasoline engines. Without roads, logging, mining, dam-building, and oil-drilling would all become much harder if not impossible. And without roads, wilderness might continue to offer a place where we might better understand our human selves by standing in the presence of the non-human Other. Focusing on roadlessness, in other words, was a brilliantly simple and concrete way to embody and defend in law the less tangible qualities that must somehow be protected if wilderness is to survive in the modern world. The Wilderness Society's founders, in focusing with such single-minded conviction on cars and roads, knew precisely what they were up to all along.

ACKNOWLEDGMENTS

This book began as a Ph.D. dissertation at the University of Kansas, where I was privileged to have Donald Worster as my advisor. Don's wisdom and patience were, in equal measure, critical to the successful completion of this book and to my maturation as a historian. His influence is everywhere in these pages. Beyond that, I am indebted to him for his passion and strong sense of moral purpose. For as long as I write environmental history, his will be the voice of my scholarly conscience. Finally, I thank Don for giving so much to his graduate students and for building and presiding over a vibrant community of inquiry in Lawrence and its hinterland.

I am also grateful to the other members of my dissertation committee: John Clark, Peter Mancall, Bill Tuttle, and Rick Prum. Rick was a good-natured, last-minute recruit who averted a crisis and served admirably under the circumstances. Bill gave the dissertation a great read and provided valuable suggestions for connecting my study to the broader contours of modern U.S. history. Beyond his assistance with my dissertation, Peter was a source of great encouragement throughout my graduate career. I profited tremendously from John's wide-ranging intellect and his renowned candor. Sadly, John passed away before I could send him a copy of this book. Finally, I owe a word of thanks to the History Department faculty at the University of Kansas for their support and general excellence.

I was fortunate to have an exemplary group of peers in graduate school. I thank Mike French and Matt Logan for their friendship and for keeping me in my place (which was, more often than not, the Free State Brewery). They were both, a few plumbing-related incidents aside, ideal housemates and intellectual sparring partners. Brian Black, James Pritchard, and Adam Rome were newcomers to Lawrence, as was I, in the fall of 1991, and ever since I have been scrambling to match the high scholarly standards they have set for our cohort. Each has become a great friend. I am also grateful to, among others, Jay Antle, Kevin Armitage, Karl Brooks, Kip Curtis, Sterling Evans, Rusty Monhollon, Amy Schwartz, and Frank Zelko for their friendship and camaraderie.

After completing my Ph.D., I had the luxury of a three-year postdoctoral fellowship sponsored by the Committee on the History of Technology and the Environment at the University of Virginia, during which time I completed major revisions on this manuscript. I thank the members of the History Department and the Division of Technology, Culture, and Communication who made my stay a productive one. I am particularly indebted to Ed Russell, Brian Balogh, Bernie Carlson, and Jack Brown. I also had the good fortune of getting to know Matt Dalbey and Dan Philippon while I lived in the shadow of the Blue Ridge Mountains.

I want to thank my new colleagues in the History Department at the University of Georgia, where I put the finishing touches on this book. I feel lucky (yet again) to have landed in a terrific place to live and work.

There are so many others who have shaped this book. I recall vividly, and embarrassingly, a visit I made to the Wilderness Society offices in Washington, D.C., at an early stage of this project. There a gentleman, who I assumed was a low-level staffer, ably assisted me with what historical materials they had. His name was T. H. Watkins. Only later did I realize that Tom Watkins was an accomplished historian, environmental writer, and biographer—and a great champion of American wilderness. Tom generously read and commented on early drafts of a couple of chapters. His untimely death was a great loss to the entire environmental community.

Larry Anderson shared with me his immense knowledge of Benton MacKaye's life and thought and provided valuable comments on my MacKaye chapter. Curt Meine went above and beyond the call of duty, on a cold January morning no less, to drive into Madison, have breakfast with me, and give me a tour of the Aldo Leopold Papers. Curt has subsequently been an invaluable guide to all things Leopold. Peggy Shaffer commented on an early version of my chapter on Robert Sterling Yard, and she kindly shared her important work on the "See America First" movement—and some images from the era as well. More recently, Phil Terrie read over my Bob Marshall chapter, shared with me his thoughts on Marshall's life and wilderness advocacy, and provided me with a copy of Marshall's FBI file. Animated conversations with Neil Maher have helped me to understand the complex contours of interwar environmental politics, and, in the process, Neil has become a valued friend.

Nancy Scott Jackson has been a font of sage advice on the publishing process. I am grateful to have as colleagues a group of scholars who make the yearly meetings of the American Society for Environmental History such a joy.

A number of other people have read and commented on the manuscript in whole or in part. Mark Harvey read it not once but (if the word on the street is correct) twice. His comments, and our conversations, have helped me immensely. Brian Balogh gave the manuscript a careful reading and helped me to think through the political aspects of my research. Ed Russell read portions and provided valuable advice. The following people also read and commented on chapters or portions thereof: Brian Black, Matt Klingle, Curt Meine, Tim Silver, and Bryant Simon. Over the years, I have presented material from this book at various conferences and workshops (and job talks!), and I thank everyone who has provided formal and informal comments along the way.

I could not have completed this research without the help of archivists at the following archives and libraries: the Appalachian Trail Conference, the Bancroft Library, the Dartmouth College Library, the Denver Public Library, the Forest History Society, the Franklin Delano Roosevelt Library, the Hoover Library, the Kansas Collection at the University of Kansas, the Knox County Public Library, the Library of Congress, the National Archives and Records Administration, the National Park Service Harper's Ferry Center, and the University of Wisconsin Archives. Several institutions gave me crucial research support: The Forest History Society awarded me with a Bell Fellowship to complete research there; the Roosevelt Library provided a research travel grant; a Hoover Scholarship enabled my research at the Hoover Library; and a Lila Atkinson Creighton Fellowship from the History Department at the University of Kansas provided funds for a summer of research.

Portions of this manuscript appeared previously as journal articles, and I appreciate receiving permission to reprint them here. An earlier version of the first half of chapter three appeared in the *Western Historical Quarterly* (Summer 1998), and an earlier version of the first half of chapter five appeared in *Environmental History* (October 1999). David Rich Lewis, Hal Rothman, and the anonymous reviewers gave me valuable feedback on content and style that not only improved those pieces but also helped me to sharpen the entire manuscript.

From the moment I approached Bill Cronon, he has been an enthusiastic supporter of this project. Working with Bill, one gets used to receiving email responses from exotic locales (mostly airports, actually) that begin with the phrase, "I don't have time for a full reply now, but . . ." This preamble is invariably followed by a lengthy and substantive response that, by most people's standards, would qualify as full and then some. Bill has given tremendous time and energy to working with me on this book, and the final product is much stronger as a result. Julidta Tarver and the entire staff of the University of Washington Press gave the manuscript great care at all stages of the production process, and both Bill and Lita showed a deft touch in dealing with this nervous first-time author. They make a great team.

My family has supported me throughout the long process of creating this book. As importantly, they have been a wonderful haven from it. My in-laws, Bob and Sara Rothschild, graciously allowed their daughter to marry a struggling graduate student with an uncertain future, and then helped to make that future more certain in innumerable ways. My parents, Dick and Joan Sutter, have sustained me with their love (and sometimes their money), and they have shown enough confidence in me to carry me through those times when I lost confidence in myself. My greatest debt is to my wife, Julie, who has not known me when I was not working on, or thinking about, this book. She and our two sons, Henry and Wyatt, have made my life better in ways as yet uncaptured by language, and I dedicate this book to them.

Driven Wild

CHAPTER 1

The Problem of the Wilderness ·

In October 1934, the American Forestry Association (AFA) held its annual meeting in Knoxville, Tennessee. Among those on the program was a young forester, then working for the Bureau of Indian Affairs (BIA), named Bob Marshall. Marshall had distinguished himself as a strident critic of the timber industry and federal forestry policy. His 1933 book, *The People's Forests,* made a forceful case for socializing the nation's industrial timberlands. Yet among certain attendees of the AFA conference, Marshall was better known for a 1930 article, "The Problem of the Wilderness," in which he called for the "organization of spirited people who will fight for the freedom of the wilderness."[1]

Benton MacKaye, a forester and regional planner who was living in Knoxville and working for the Tennessee Valley Authority (TVA) at the time of the AFA meeting, had read and been moved by Marshall's plea. Indeed, MacKaye was confronting his own problem of the wilderness. In 1921, he had proposed a visionary plan for "an Appalachian Trail." Although his trail was nearing completion by 1934, it was threatened by a series of federally funded skyline drives being planned for and built along the Appalachian ridgeline.[2] MacKaye and a number of his supporters were busy organizing a protest against these incursions, and they were eager to talk with Marshall about their efforts.

They had their opportunity when, on October 19, Marshall joined MacKaye, Harvey Broome, and Bernard and Miriam Frank for an all-day field trip to a Civilian Conservation Corps (CCC) camp outside of Knoxville. The AFA had arranged the trip to give conference-goers a sense of the profound changes occurring in the upper Tennessee Valley. Broome knew the region well. He was a Knoxville lawyer and a leading member of the Smoky Mountains Hiking Club, one of the groups most important to the construction of the Appalachian Trail (AT) in the South. Bernard

Frank, newer to the region, was a watershed management expert on the TVA's forestry staff and, as Broome would later recall, "a genius at reading the landscape."[3] As the group drove north toward Norris Dam in the Franks' car, they discussed forming the sort of organization that Marshall had proposed in 1930. In fact, they had broached the idea during a brief visit Marshall had made to Knoxville two months earlier, and in the interim someone—probably MacKaye—had drafted a constitution that became the focus of discussion during the drive. As the conversation became more animated, the group decided to pull over and get out of the car. They clambered up an embankment by the side of the road—"between Knoxville and Lafollette somewhere near Coal Creek," Broome would later remember—and there they agreed upon the principles of what was to become the Wilderness Society, the first national organization dedicated solely to the preservation of wilderness. It was in just such a setting that the founders felt most keenly what Marshall had called "the problem of the wilderness."[4]

The Wilderness Society's roadside creation was rich with symbols of the founders' motivating concerns. Foremost among those concerns were the road and the car. The group had come together to define a new preservationist ideal because of a common feeling that the automobile and road building threatened what was left of wild America. Wilderness, as they defined it, would keep large portions of the landscape free of these forces. And yet, despite their flight from the Franks' car, a gesture evocative of their agenda, they could not escape the fact that, literally as well as figuratively, the automobile and improved roads had brought them together that day. The very conditions that had prompted their collective concern for protecting wilderness had also enabled their concern. That paradox gave wilderness its modern meaning.

The larger setting was also of symbolic import: the roadside caucus occurred in a region being transformed by New Deal capital and labor. The unprecedented federal mobilization of resources in the name of conservation was a promising development to these advocates, most of whom had long argued for a greater (and often more radical) federal commitment to environmental protection. Yet New Deal conservation work projects, particularly by emphasizing road building and recreational development, threatened wilderness as these activists defined it. Indeed, the New Deal represented the climax of a two-decade-long effort to modernize the public lands for motorized recreation. These New Deal developments precipitated the founding of the Wilderness Society.

As the rest of the AFA caravan whirred by, the roadside conspirators proceeded to draft a letter of invitation to join the Wilderness Society, which they agreed should go to six other potential founders: Harold Anderson, Robert Sterling Yard, Aldo Leopold, Ernest Oberholtzer, John Collier, and John Campbell Merriam. Their aim was to keep the group small and focused on defending an ideal that they feared might be compromised or misconstrued. "We want no straddlers," Marshall succinctly insisted in a note attached to each invitation, and they got none.[5] What they did get was a group of advocates whose varied backgrounds revealed the modern wilderness idea's complex pedigree.

Both Harold Anderson and Robert Sterling Yard had been privy to organizational conversations prior to the AFA meeting, and their inclusion among the founders was thus assumed. Anderson was a Washington, D.C., accountant, a prominent member of the Potomac Appalachian Trail Club, and a friend and supporter of MacKaye's. Some months earlier, he had urged the formation of an organization to fight skyline drives along the AT, and to counter the failure of the Appalachian Trail Conference (ATC), the confederation of hiking clubs responsible for the trail's completion, to take action to oppose such schemes. Anderson wanted an organization composed of ATC malcontents who would fight for the integrity of the AT, but Marshall convinced him of the need for a group with an expanded scope.[6] Yard was a national parks watchdog who, as the motive force behind the National Parks Association (NPA) since its inception in 1919, had fought for the maintenance of park standards. He had entered park politics in the mid-teens as the publicity man for his friend Stephen Mather, the first director of the National Park Service, but he soon soured on the Service and its developmentalist tendencies. He was being squeezed out of the NPA for his public criticisms of the Park Service and was more than happy to devote his energies to a new organization.

To give the organization a stronger national standing, the group also invited Aldo Leopold and Ernest Oberholtzer to join as founding members. Leopold was, in 1934, a newly appointed professor of game management at the University of Wisconsin. In the early 1920s, while working for the Forest Service in the Southwest, he had been the first to push for wilderness protection within the national forests, and during the mid-1920s he wrote extensively about the wilderness idea. While Leopold had not been as active a voice in wilderness debates in the years leading up to the 1934 AFA meeting, Marshall still thought of him as "the Commanding General of the Wilderness Battle."[7] Although not entirely comfortable with this sobriquet, Leopold was eager to serve as a foot soldier. Oberholtzer was an advocate for the preservation of the Quetico-Superior lake country in northern Minnesota and southwestern Ontario. During the previous decade, he had done battle against various schemes to develop the region for both its natural resources and its tourist amenities. He headed the Quetico-Superior Council, a group that worked to protect the unique wilderness of water that became the Boundary Waters Canoe Area, and Franklin Roosevelt had just appointed him to chair the Quetico-Superior Committee, a body charged with creating a transnational preserve in the region.[8] After some initial hesitation, Oberholtzer signed on with the Wilderness Society as well.

Only two of the proposed founders declined. One was John Collier, a long-time advocate for Native American rights recently named by Franklin Roosevelt to head the BIA. Collier, who was in the midst of orchestrating what would become known as the Indian New Deal, was Marshall's boss at the time. Although he expressed enthusiasm, he decided not to join the Wilderness Society as a founder. It is not clear why he declined, though he was burdened with other responsibilities and may have worried about mixing such advocacy with high-level government service.[9] Nonetheless, the decision to invite Collier, and Collier's serious interest in the group, hint at the complex relationship between the modern wilderness idea and interwar Native American policy. The other refusal came from John C. Merriam, a paleontologist, head of the Carnegie Institution and an expert on the aesthetics of "primitive" nature. Merriam was an active member of the NPA whose advocacy, like Yard's, was informed by an older tradition of scenic preservation most at home in the national parks lobby. Indeed, it was likely Yard who, much impressed by the way Merriam had brought science to bear on explanations of scenic magnificence, urged that Merriam be included. Merriam was enthusiastic about the group's aims, but he begged off because of too many claims on his time.[10]

Five of the eight founding members—Anderson, Broome, MacKaye, Marshall, and Yard—met again on January 20 and 21, 1935, at the Cosmos Club in Washington, D.C., to formally organize the Wilderness Society and to give definition to the modern wilderness idea: the notion that the federal government ought to preserve large expanses of roadless and otherwise undeveloped nature in a system of designated wilderness areas.[11] This gathering heralded the beginning of a long political fight for federal wilderness legislation, a fight that climaxed with the passage of the Wilderness Act of 1964. But the meeting was also the culmination of individual efforts over the previous quarter century to make sense of what preserving nature meant in an automotive era. Historians have long seen the founding of the Wilderness Society as a watershed event, but few have recognized the origins of modern wilderness sentiment. This book is about those origins—the various streams of thought that came together to launch a new idea and, equally important, the context in which that confluence occurred.

During the last decade or so, no concept has been more hotly contested within the American environmental community than wilderness. Although my major goal in this study is a historical one—to explain the interwar rise of modern wilderness advocacy on its own terms, not through a presentist lens—this book, itself almost a decade in the making, has become inextricably linked to this contest over wilderness. The current debate has changed this study in dramatic ways, as I explain below, and I would be lying if I said that I had no ambition to shape the debate in turn. Nonetheless, to the extent that I have an agenda, it is less to argue for or against wilderness (though my sympathies are fairly transparent) than it is to suggest that scholars and activists have ignored or misconstrued a foundational chapter in the history of the wilderness idea—a chapter that has tremendous relevance for contemporary environmental politics. Much of the wilderness debate now being played out has its roots in interwar wilderness advocacy, though those roots have gone unrecognized.

When I began my inquiry into the origins of modern wilderness, the wilderness debate was in its infancy, and there were only faint glimmerings of the new wilderness historiography that has been so important to that debate. As a result, my initial approach was fairly traditional. Historians had long studied the centrality of the wilderness idea in American history, from its importation as a filter for viewing the colonial landscape to its role as a shibboleth of the postwar environmental movement, and I was fascinated by the same questions that preoccupied many of these scholars: How was it that a nation founded upon an antipathy for the wilderness had come to cherish and protect it? What had produced this intellectual and cultural sea change?

There were already some good answers to this particular problem of the wilderness when I began my research. Some suggested that the change was simply a matter of abundance and scarcity. Early American settlers had been too close to the wilderness to appreciate it. Overwhelmed by wild nature and its great power over their lives, they sought its transformation. But by the time Americans had successfully subdued a large part of the continent, they began to feel the absence of wilderness as a physical and cultural loss. As wilderness became scarce, its value shot up.[12] Other scholars, less keen on this model of supply and demand, thought that the attitudinal transformation had more to do with an increasingly sophisticated ethical approach to the natural world. Where we once had treated nature as a mere instrument, we came in time to appreciate that the nonhuman world was worthy of moral consideration. The appearance of wilderness advocacy, in this interpretation, signaled an appreciation of the rights of nature, the rise of a biocentric ethic, and a foreshadowing of deep ecology.[13] Still other scholars suggested that the change was the political product of major demographic shifts. As Americans became affluent and educated consumers whose urban and suburban lives were disconnected from a direct economic relationship with the land, they pined for the sorts of recreational and aesthetic amenities that wild nature provided.[14]

All of these interpretations struck me—and continue to strike me—as sound but limited. The abundance and scarcity argument is true enough but not particularly sensitive to shifting meanings. The ethical argument, though edifying, is too beholden to a neat idealism that conforms more to the logic of philosophy than the messiness of history. And while the demographic argument does a satisfying job explaining the growth of political support for wilderness preservation, it is too faceless and deterministic to explain the intellectual development of the modern wilderness idea.

Sensing these inadequacies, I decided to pursue a closer reading of America's historical reassessment of wilderness. Doing so required chronological focus, and so I set out to find a crucial moment in this transition, a fulcrum upon which the balance seemed to tip in favor of wilderness preservation. The founding of the Wilderness Society fit the bill. Although preservationist groups such as the Sierra Club predated the Wilderness Society, their efforts focused on national parks and scenic preservation. While making a case for the value of parks and natural scenery was crucial to later arguments for wilderness, national parks and wilderness areas were not one and the same thing—politically or aesthetically. Modern wilderness politics began with the founding of the Wilderness Society, and I expected that the key aesthetic distinctions that allowed preservationists to move beyond the scenic would be located there as well. Moreover, while there was some excellent scholarship on a number of the Wilderness Society's founders—Aldo Leopold and Bob Marshall in particular—no one had yet provided a detailed look at the Society's origins. Indeed, this specific hole in the scholarship was a manifestation of a general neglect of interwar environmental thought and politics.

Finding the pivotal moment that I was after amid relatively unexplored terrain, I set out to add my own piece of the wilderness puzzle. I began with a couple of basic hypotheses that conformed to conventional wisdom. First, I anticipated that I would find in modern wilderness advocacy a set of strong arguments against resource exploitation. Wilderness, I assumed, was an idea defined in opposition to the forces of production, and to a brand of utilitarian conservation that sought to make those forces more efficient. More specifically, I expected to discover that the wilderness idea was the result of an aesthetic shift within the preservationist community, with the insights of the young discipline of ecology playing a starring role. I would explain in detail how a group of important preservationists rejected the static and human-centered aesthetic of scenic beauty—an esthetic that defined the park-making process but failed to provide a preservationist impetus in the absence of spectacular scenery—for a dynamic and nature-centered wilderness ideal that proved more powerful in opposing resource development.

The sources told a different story. While I did find some evidence to support my initial hypotheses, it was mere background noise compared with the decibel level of another set of concerns voiced in the first issue of *The Living Wilderness,* the Wilderness Society's magazine. In a cover article describing their mission, the founders proclaimed:

> Ten years of warfare in Congress have saved the National Park System from water power and irrigation, but left the primitive decimated elsewhere. What little of it is left is passing before a popular craze and an administrative fashion. The craze is to build all of the highways possible everywhere while billions may yet be borrowed from the unlucky future. The fashion is to barber and manicure wild America as smartly as the modern girl. Our duty is clear.[15]

This call to arms seemed odd. Where were the denunciations of industrial offenders? Where was the repudiation of the instrumental utilitarian worldview? Where was ecology's influence? Such concerns were barely visible. Instead, the founders collectively bemoaned the "craze" for road building that was swiftly opening up the nation's few remaining wild landscapes, and they criticized emergency conservation initiatives that prioritized the recreational development and beautification of the public domain, largely for recreational motorists, at the expense of wilderness conditions. Almost every contribution to that first issue of *The Living Wilderness* was about the automobile, roads, and the federal government's willingness to countenance, even encourage, the modernization and mechanization of roadless areas. The founders of the Wilderness Society, I realized, had been *driven wild*.

There were two important implications to this realization, one substantive and the other methodological. First, I recognized the causative importance of road building and the nascent American car culture to the emergence of modern wilderness advocacy. This relationship between the automobile and the making of modern wilderness is my overarching theme and thesis. Methodologically, accepting that the founders were driven wild meant embracing an approach to the intellectual history of the wilderness idea that emphasizes material and cultural context over detached idealism. Context *drove* the creation of modern wilderness.

Such a contextualist approach challenges a notion often at the core of traditional wilderness narratives: that the history of preservationist sentiment in the United States has evolved from lower to higher forms of appreciation. Rather than assuming that our ideas about how to preserve nature simply become more refined over time, I suggest that each era reworks its ideas to fit and reflect contemporary circumstances. The modern wilderness idea was the product of such a process. It was not the result of enlightened minds decoding an idea's internal logic; wilderness was not a pure, platonic form that had flickered away for eons, waiting to be correctly deciphered and appreciated. It was a product of intellectual engagement with specific circumstances.

The founding of the Wilderness Society was a crucial moment in the history of American environmental thought and politics not because it embodied a collective epiphany that wilderness was the ultimate expression of preservationist sentiment, but because it involved the pragmatic act of giving a name to certain qualities that were disappearing from the landscape because of road building and the automobile. The value of wilderness was not so much reassessed by the founders of the Wilderness Society as it was reinvented.

As I have developed these arguments about the centrality of the automobile and the importance of context, I have had to confront a spate of new scholarship challenging the legitimacy of wilderness as a preservationist ideal. And while the new wilderness historiography of the last decade has forced me to reframe my analysis, it also has sharpened my understanding of the interwar generation of advocates by exposing deeper and more powerful currents in their advocacy. To appreciate those currents, and their relevance to the present debate, it is first necessary to understand the recent criticisms of wilderness. Although they vary and are interrelated, they can be broken down into four categories: ecological, ethnocentric, social, and cultural.

The ecological critique of wilderness is premised on shifting scientific understandings of how nature works. Where ecologists once saw order, harmony, equilibrium, and purpose in the natural world, many now see stochasticity, competition, and pervasive disturbance. Utilizing the insights of this new ecology, one group of critics has suggested that the complexity of natural processes invariably complicates attempts to preserve wilderness. To preserve wilderness, ecologists tell us, is *not* to keep nature in a timeless equilibrium. Rather, it is to draw boundaries around a world in flux. Indeed, scholars and activists, including staunch wilderness defenders, now recognize that containing wild nature within discrete parks and wilderness areas can create profound ecological and management problems. Moreover, the new ecology has complicated the use of wilderness as a standard against which to measure human-induced change. If nature is a moving target characterized by disturbance, patchiness, and randomness, then setting normative standards—saying that a given area shall be protected in a certain state—becomes difficult. Finally, ecological studies increasingly show that the human imprint is ubiquitous. If by wilderness we mean a "pristine" landscape absent all human influence, we now know that few such areas exist. Wilderness, these critics suggest, is too general and value-laden a term to describe, and perhaps even protect, a complex ecological reality.[16]

Other scholars have shown that Euro-Americans used the wilderness idea to dispossess Native Americans who had competing and prior claims to the landscape. By implying that certain areas were empty of human settlement, and thus of legitimate legal claims to landownership, those who wielded the wilderness idea justified theft. This was true not only of early European settlers who vilified the American wilderness (and its inhabitants), but also of more recent advocacy for the protection of national parks and other natural areas.[17] Moreover, by calling an area wilderness and preserving it as such, these critics point out, preservationists have aided and abetted a pristine myth that continues to obscure a deep history of Native American land use. This line of criticism has developed in tandem with a growing body of scholarship on the relationship between colonialism, the developing world, and the appropriateness of exporting wilderness as a guiding ideal for international nature preservation efforts.[18]

A third group of critics has used the force of social history to argue that the wilderness idea reflects considerable class bias. These scholars, extending the wilderness-as-dispossession argument, have pointed out that the preservation of nature and protection of wildlife often have served the recreational and aesthetic interests of an urban leisure class while closing off subsistence options to local, and often marginal, populations. Indeed, there have been a number of cases in which poor residents were removed from areas that were essentially rural, sometimes forcibly, to create national parks. In this critique, the state is often portrayed as a crucial ally in the preservation of areas that privilege the recreational desires of the wealthy while limiting subsistence access to the marginal and landless. Wilderness preservation, these critics suggest, often has functioned as a form of recreational enclosure, a modern (and statist) analogue to the expropriation of common lands in early modern Europe that destroyed traditions of communal use.[19]

Other critics have made use of the insights of cultural history and theory to critique the wilderness idea. Rather than accept the traditional assumption that the rise of preservationist sentiment was a progressive intellectual development, these critics have taken a more sophisticated look at the connection between preservationist ideals, recreational interest in nature, tourism, and consumerism. Preservationists, according to this analysis, know and value the natural world almost solely through their leisure. Middle- and upper-class Americans have constructed a natural ideal—wilderness—and preserved it in discrete and distant areas of pure nature. Wilderness, these critics suggest, is a consumer construct that has enthralled environmentalists while the entrenched illnesses of our living and working relationships with nature go unexamined. Moreover, new scholarship on the history of landscape design has pointed to the role of human artifice in what appear to be natural spaces. Wilderness, in other words, is not only a cultural construct but sometimes an architectural one as well.[20]

Alone and in combination, these critiques have loosened the wilderness idea from its moorings, forcing scholars and environmentalists alike to take a hard look at their commitments. Wilderness areas, we have learned, cannot always contain what we hope to preserve ecologically; the wilderness idea has often come tainted with social, ethnic, and racial biases; and we can no longer ignore the fact that wilderness conditions often have been *produced,* constructed materially as well as intellectually. Perhaps most important, we now recognize the historical development of preservationist sentiment as part of a broader shift from a producer to a consumer economy and culture. This is all to the good. But these critiques have tended to caricature the wilderness idea and wilderness advocates. Rather than attending to the complex history of wilderness advocacy, wilderness critics have conveniently or inadvertently lumped wilderness advocates together, intimating that all hold to an idea of wilderness that is by turns ecologically naïve, dispossessive, class-biased, consumerist, and hopelessly separated from concerns for social justice. These critics have abstracted the wilderness idea from politics, reified it, and built by logic and selectivity a profile of advocacy that misses complexity, contingency, and context. In pointing out the weaknesses of the wilderness idea, they have made them paramount.

Wilderness critics have all but ignored the thought of interwar wilderness advocates, preferring to dwell in Progressive and postwar terrain. This has been a mistake, for the founders of the Wilderness Society formulated the modern wilderness idea in ways that challenge the logic of these recent critiques. Indeed, interwar advocacy highlights an irony that has lurked unrecognized amid the recent wilderness debate: some of the very arguments that critics are using today to *challenge* the appropriateness of wilderness preservation were developed by interwar advocates in making a case *for* wilderness.

A brief comparison of each of the bodies of wilderness criticism with the ideas of the Wilderness Society's founders suggests some important lessons about the origins of modern wilderness. The ecological critique equates modern wilderness advocacy with the preservation of "pristine" nature. In other words, this critique generally assumes that modern wilderness was and is an *ecological* idea, though one based in an outdated scientific model of succession and climax. This model was the product of the first generation of American ecologists, among them Frederick Clements and Henry Cowles, who argued that ecological processes were marked by an orderly march toward equilibrium, and that, in the case of disturbance, recovery occurred in a series of predictable successional steps toward a stable and harmonious climax. Although ecologists and naturalists still speak of succession and climax, most have rejected the teleological and normative aspects of successional theory—that succession is purposeful and that climax states are the way nature ought to be. In the process, many have questioned the feasibility of what they see as the modern wilderness idea's main aim: the protection of stable and mature states of nature.[21]

The founders of the Wilderness Society did see wilderness areas as places meant to preserve pristine nature—"virgin," "primitive," or "primeval" were the more frequently used terms—but they almost always spoke of such an ideal in relative rather than absolute terms. Ecological critiques of wilderness, in other words, have tended to overstate the extent to which wilderness was, and is, an ideal of ecological purity. Indeed, ecological concerns were not a central causative agent or a major component in the founders' definition of modern wilderness. This is not to say that the founders were ignorant of ecological theory or the workings of nature; they appreciated the complexities of wilderness preservation and understood quite well what wilderness did and did not preserve ecologically. But more important to the founders was the contrast between the modern, mechanized world of the early twentieth century and the few remaining large areas in the United States where nature dominated. Wilderness was as much about "wildness," the absence of human control, as it was about pristine ecological conditions. The modern wilderness idea was shaped more by a collective uneasiness with the enormity of change at a given historical moment than it was by the emergence of a new scientific way of looking at nature.

This question of the role of ecology in the advent of modern wilderness advocacy highlights another anomaly. Many of the founders of the Wilderness Society had the scientific training and field experience to think through the ecological complexities of wilderness preservation because they were professional foresters. Yet we generally do not associate foresters with preservationist sentiment precisely because their scientific training and silvicultural goals have tended to imbue them with a managerial approach to conservation that is— or has seemed—antithetical to the wilderness idea. How, then, does one explain that the interwar era's most important wilderness advocates were foresters? It might be tempting to see their advocacy as a rejection of the utilitarian mentality, but such an explanation does not hold water. While the trained foresters among the founders were critical of the single-minded commodity focus of utilitarian forestry, none of them rejected its basic premise that resources should be managed wisely and scientifically for the public good. Instead, they crafted a wilderness idea consonant with their utilitarian commitments, an idea focused on shaping the behavior not of resource producers but of recreational consumers. Understanding how foresters became wilderness advocates thus involves rethinking the traditional historiographical division between preservationists and conservationists—and the role that ecological knowledge played in defining that split. It also involves a more subtle reading of the tensions between producer and consumer mentalities as they shaped modern environmental thought.

The charge that wilderness has been a dispossessive ideal is perhaps the most troubling indictment. There certainly have been cases where preservationist efforts have come at the expense of certain social and ethnic groups. But, again, when I applied such critiques to the Society's founders, I found a poor fit. While the founders defined wilderness areas as recreational preserves that were uninhabited and generally free from the workings of the commercial economy, they made important exceptions. A number of them suggested that wilderness preservation could be compatible with subsistence or local resource use. A few even proposed wilderness designation for Native American lands as a strategy for preserving traditional economic relations with nature; they saw wilderness preservation as a way of keeping native populations on the land at a time when the forces working against preservation seemed more likely to dispossess. Such proposals were problematic, but they do suggest that the modern wilderness idea was not necessarily hostile to subsistence resource use and human habitation. Indeed, some of the founders were among their generation's most important thinkers on the relation between work and nature, mixing their wilderness advocacy with social and political radicalism that made them particularly sensitive to charges of dispossession.

Of all the categories of wilderness criticism, I am most heartily in agreement with the cultural critique. I too am fascinated and troubled by the recreational relationships that Americans have crafted with nature during the twentieth century, and I too believe that a full explication of how Americans came to embrace modern wilderness must involve a serious reckoning with consumerism. But, contrary to the upshot of the cultural critique, I do not think that the trouble is with wilderness. Modern wilderness, as the founders conceptualized it, was certainly a recreational ideal, but its more important function was as a recreational critique. The founders' preoccupation with the automobile and roads was part of their broader discomfort with consumerism, tourism, mechanization, advertising, landscape architecture, and the various other forces that remade outdoor recreation during the interwar period. What made modern wilderness distinct, separate from the national park ideal, was the *critique* of consumerism that was central to it. For the founders of the Wilderness Society, modern wilderness advocacy sprang from a sense that as roads and the automobiles carved up the nation's remaining wild spaces, the American desire to retreat to nature, traditionally a critical gesture, was becoming part of the culture's accommodation to the modern social and economic order. No feature of interwar advocacy is more relevant to the current debate than this one.

Driven Wild is thus both an attempt to correct traditional narratives of wilderness history and an effort to temper and redirect recent wilderness criticisms. I focus on the decades leading up to the founding of the Wilderness Society, with the founding itself as the narrative climax. While I occasionally go as far back as the earliest years of the twentieth century, the mid-teens mark a convenient starting point for tracing the growth of modern wilderness sentiment. The battle over the Hetch Hetchy Valley in Yosemite National Park, which has long symbolized the peak of the Progressive Era conflict between utilitarian conservationists and preservationists, in fact signaled a turning point in American environmental politics. By 1916, there was a National Park Service to protect the parks against such intrusions, to promote their use, and to develop them for a growing number of motor tourists. The mid-teens also saw a rapid increase in the recreational use of the national forests, and the beginning of Forest Service efforts to provide for recreation on their holdings. In 1916, Congress allocated the first significant funds for a federal road-building program that would grow throughout the interwar period.[22] At the same time, Henry Ford refined assembly line production, which made the automobile affordable to middle-class Americans.[23] With Europe closed to tourist travel by World War I, motor tourists, prodded by "See America First" boosterism, headed west to witness what their own country had to offer. Together, these developments marked the dawn of a new era in conservation politics, one in which questions about recreational development often overshadowed tensions between preservation and use.

The main chapters of this study, chapters two through six, trace the developments that led the founders to fear the disappearance of wilderness. After a survey of outdoor recreational trends during the interwar era, I provide close readings of how and why four of the founders of the Wilderness Society came to advocate wilderness preservation. While each of the founders saw the problem of the wilderness in distinct ways, all of them hit upon wilderness only after realizing that older preservationist ideals did not fit a new era. In charting the paths that these men took to reach collective and organized wilderness advocacy, I spend considerable time surveying the political, social, cultural, institutional, and physical terrain over which they traveled.

For a variety of reasons—including the availability of sources, my sense of who the most influential and representative founders were, and considerations of length and redundancy—I have focused my analysis on only four of the eight founders: Aldo Leopold, Robert Sterling Yard, Benton MacKaye, and Bob Marshall. At various points, I contemplated additional chapters that would have added even greater detail to the story. A chapter on Harold Anderson and Harvey Broome, for instance, could have traced their evolving wilderness advocacy as it related to the growth of urban hiking clubs, and a chapter on Ernest Oberholtzer's advocacy could have provided a close-grained case study of preservationist politics in the Boundary Waters Canoe Area. But, in the end, I was able to say most of what I wanted to say by looking at just four of the founders.

Because I have kept a tight focus on the interwar years, I neglect the much longer and deeper tradition of American wilderness thought that predates World War I, a tradition that must be seen as prologue to the story I tell. Other scholars have treated this subject in detail, and I will not duplicate their efforts here.[24] Moreover, I deal only briefly—in the Epilogue—with another important story, the rancorous postwar political battle that led to the creation of a statutory wilderness system. Although others have told parts of that story, the history of the Wilderness Act of 1964 awaits definitive treatment.[25] But one point about the postwar era does bear emphasizing here: there were fundamental differences between the wilderness politics of the interwar and postwar eras. Those differences have not only masked the historical particulars of interwar advocacy, in turn contributing to the scholarly neglect of the era, but they have also obscured some important continuities that make the present problem of the wilderness much more like the interwar problem than we realize. After World War II, as dam builders and timber cutters ran amok, the politics of wilderness preservation fell back into the traditional mold of utilitarian conservation versus preservation that had defined the Progressive

Era. It was easy to think that little had changed since Hetch Hetchy. But a lot had. Roads and automobiles continued to proliferate to an unprecedented degree; visitation to the national parks and forests grew dramatically, raising profound management dilemmas; and the car culture continued to shape the experiences most Americans had, and continue to have, in "nature." That roadlessness remains the defining characteristic of modern wilderness suggests the continuing relevance of the interwar generation of wilderness advocates to the current wilderness debate.

In the last decade or so, as the wilderness debate has raged, protectors of America's wild lands increasingly find their attention drawn to the confounding results of a peculiar condition with roots in the interwar period: our love for wild nature is intimately connected with our affection for the automobile and other forms of mechanized transport. This is evident in proposals to limit the number of automobiles in some of the most heavily visited national parks, and in campaigns to control the impacts of off-road vehicles (ORVs) on other federal lands. As we confront these seemingly new problems of the wilderness, we would do well to remember the tradition of thinking critically about these issues that is at the core of American wilderness advocacy. But, beyond that, each of the founders of the Wilderness Society insisted that Americans needed to think critically about the connections between roads, cars, and environmental sentiment more generally. That message rings particularly true today. If the only nature we can recognize is the nature that we drive hours, or even days, to see—and then we see it mostly through our windshields—preserving wilderness may be the least of our worries.

CHAPTER 2

Knowing Nature through Leisure:
Outdoor Recreation during the Interwar Years

"One of the significant trends in modern recreation," Jesse Steiner wrote in *Americans at Play* (1933), his contribution to the Recent Social Trends studies commissioned by President Herbert Hoover, "is the increasing demand for great open spaces set apart for the enjoyment of those outdoor diversions which have become so eagerly sought as a means of escape from the noise and confusion of urban life." As Steiner suggested throughout his study, widespread American participation in outdoor recreation was one of the signature developments of the interwar period, and he rightly saw a connection between outdoor recreation, the increasing popularity of commercial amusements, and "the rising tide of sports and games." But the boom in outdoor recreation was not just the product of more time for leisure. "With *the improvement of means of travel*," Steiner continued, "people are finding it possible to go even further afield in their search for recreation and readily travel long distances during week-ends and vacations to places of scenic interest where their favorite forms of outdoor life may be enjoyed [italics added]."[1] One important result, he intimated, was the incorporation of remote natural areas into the recreational orbits of modern Americans. This unprecedented access to wild nature resulted in some important landscape changes during the interwar period—changes that would preoccupy wilderness advocates.

The interwar boom in outdoor recreation also reshaped American conservation politics and played a complex role in the evolution of preservationist sentiment. Historians traditionally have assumed an easy congruence between the growth of American interest in outdoor recreation and a deeper appreciation of wild nature. This interest certainly led Americans to demand, in Steiner's words, "great open spaces set apart" for their leisure, thus lending the cause of preservation broader political support. But this simple association between recreational interest in nature and more sophisticated preservationist sentiment obscures political and cultural factors that complicated the equation.

In political terms, the preservationist community was far from united during this period, and their differences were crucial in shaping preservationist ideas and policies. We tend to assume that the idea of wilderness emerged as a recreational alternative to the worked landscape, and that wilderness was meant to be a haven from the forces of production. But to stop there, as most scholars have, is to miss a much more important level on which wilderness advocacy operated. In the years between the two world wars, the modern wilderness idea emerged as an alternative to landscapes of modernized leisure and play, and it was preeminently a product of the discordant internal politics of outdoor recreation.

More recently, as cultural historians have turned a critical eye toward the growth of American interest in recreational nature, the very notion of "appreciating nature" has become more complex terrain. Many have argued that the popularity of outdoor recreation and the growth of preservationist sentiment were less progressive intellectual developments than they were consumer trends, part and parcel of the cultural tectonics of the last century or so. Such arguments have performed the crucial service of historicizing and contextualizing Americans' recreational affection for nature. Indeed, one purpose of this chapter is to show that knowing nature through leisure was a historical development to which the interwar era was crucial. But I hope, as well, to lay the groundwork for showing that wilderness advocacy arose in reaction to the landscape changes, the political changes, and the cultural and technological changes that shaped and accompanied the growing popularity of outdoor recreation. Modern wilderness was more a response to than a product of the ways in which Americans were coming to know nature through leisure during the interwar years.[2]

Outdoor Recreation Before World War I

Traditional interpretations of when and how Americans came to know nature through leisure have focused on the years between 1880 and 1920, and have stressed urbanization and industrialization as overarching interpretive structures.[3] During these decades, Americans produced and consumed a voluminous literature on natural and wild themes; they built vacation homes and camps; they initiated a wide variety of programs in scouting and woodcraft; they developed a distinctive hunting culture and ethos; they adopted nature study as a prominent hobby; and they embraced the "strenuous life" that could be found only in the "great outdoors." The era also produced a country life movement and early stirrings of the suburban exodus that would transform America later in the century. Authors such as John Burroughs, John Muir, and Jack London celebrated the wilderness experience and packaged it for American readers.[4] In the Progressive Era, wild nature was a potent symbol and a foil for the dramatic urban-industrial changes overtaking society.

This Progressive Era revaluation of nature was also a product of important cultural changes that accompanied urbanization and industrialization. American culture was undergoing a dramatic shift away from an ethos that stressed the virtues of work, savings, and delayed gratification, and toward a more therapeutic worldview that sanctioned consumption and self-realization. Rejecting Victorian decorum and recoiling from modernity itself, many Americans turned to a "cult of experience," a craving for "real life" that manifested itself in a variety of impulses, including the era's back-to-nature movement.[5] To many Americans, nature, once a raw material to be transformed by ceaseless labor, became a place of relaxation, therapeutic recreation, and moral regeneration. For many, nature offered psychic accommodation to a changing world. More than a reaction to urban-industrial life, this interest in nature had the early stirrings of consumerism at its core.

Yet despite the influence of an emerging consumer culture, back-to-nature sentiment during the late 1800s and early 1900s had a strong element of protest to it. For many Americans, nature stood as a symbolic haven from the ethos of industrial capitalism. Leading American figures, from radical reformers to members of the capitalist establishment, shared this reaction to grasping commercialism and turned to recreation in nature, and the preservation of recreational nature, as a critical gesture. For some, this gesture was a limited one; they simply felt that there should be some checks on a socioeconomic order which for the most part they embraced. For others, advocacy for nature was part of a more sweeping critique. John Muir became the nation's foremost preservationist during the Progressive Era not only by celebrating nature's scenic wonders and spiritual powers but also by blasting the culture's pecuniary monomania and aesthetic irreverence.[6] For quite a few Progressive Era Americans, nature promised to restore civic virtue and counter the atomizing trends of the commercial world. Thus, while many Progressive Era Americans went back to nature as consumers, nature was also a place where Americans confronted their anxieties about consumption.

Other cultural anxieties fed America's growing fascination with recreational nature in the years before World War I. All sorts of outdoor recreation organizations and movements emerged, most of them dedicated to the character-building qualities of appropriately structured leisure in nature. Leading intellectuals such as psychologist G. Stanley Hall lauded retreats to the natural world as a way of recapturing earlier, more primitive stages in human historical and moral development. Such outings regenerated spirits apparently sapped by the artificialities and comforts of modern life.[7] Often these themes mingled with fears of Anglo-Saxon racial degeneration, as contemporaries worried that the martial vigilance required to survive in a hostile environment was being blunted by the relative ease of civilized life. These themes were also played out in gendered language. As America's urban culture struck many as more "feminized," leisure in nature provided an arena in which one might restore "manliness."[8] Finally, recreational nature fit neatly into a "culture of authenticity" that emerged in the early twentieth century. In a world increasingly dominated by artifice and ephemerality, by mass-produced consumer goods and rapid technological change, nature seemed real, a tangible and timeless source of meaning and moral authority.[9]

As fin-de-frontier Americans began to think about nature's role differently, so they rethought the broad category of leisure. Before the imposition of time discipline on industrial and institutional workplaces, labor and leisure were less segmented categories. Whether tangible leisure time increased significantly in the late 1800s and early 1900s is hard to say. "What did occur," Daniel Rodgers has suggested, "was an increasing segregation of work and play into distinct categories in place of the old interfusion of work and free time."[10] Such a separation, which was both temporal and spatial, was vital to the era's constitution of outdoor recreation. A vision of the nation's collective labor engaged in transforming raw nature had long provided many Americans with a metaphor for their national identity. By the turn of the century, however, that myth seemed in crisis. With growing urban populations, degraded work routines, and declining access to frontier land, Americans looked to leisure in nature as a new forge of national character.

Recreation in nature was but a facet of the nation's new interest in leisure activities. Commercial amusements were perhaps the most popular way of "spending" one's leisure time in the decades straddling the turn of the century. But concerns arose that such amusements threatened to degrade leisure as industrial mechanization had degraded work.[11] In a culture more attentive to leisure, and with leisure increasingly a function of free time rather than social class, there occurred an important debate about the fine line between leisure and idleness.[12] Moral reformers who had traditionally viewed work as a source of meaning and virtue came to associate those qualities with tightly controlled leisure activities. As the vacation habit spread, reformers thought that the influence of nature could make leisure an uplifting experience; some even speculated that giving the working classes recreational access to nature could quell labor unrest. Within this broad shift from a producer to a consumer society, then, nature took the edge off the perils of leisure.

If the apparent absence of nature from the lives of many Progressive Era Americans made their hearts grow fonder for it, one result was an idealized and nostalgic affection. Americans not only gained an appreciative perspective on nature in their retreat from a direct and unmediated economic relationship with it, but they also experienced nature in new ways. One important change was the very sense that nature was somehow absent from much of the landscape, that it had been banished to the margins in a fit of urban-industrial conquest. Moreover, in idealizing this absent nature, Americans forgot the drudgery of making a living from the land and projected upon nature their dreams of escape from certain aspects of their industrial existences. The nature they longed to go back to was usually one they had never known in the first place.

The years from 1880 to 1920 were thus crucial to the culture's idealization of nature as a space for leisure. In many ways, the intellectual and cultural groundwork for a mass embrace of outdoor recreation was laid during this period. But most Americans still lacked the time, the money, and the tools to heed the call of the wild. For Americans of ordinary means, outdoor recreation remained largely a local phenomenon and a relatively tame affair before World War I.

Defining the Outdoor Recreation Age

While there were important cultural continuities from the Progressive Era, three broad changes marked the interwar years as a distinct period in the history of outdoor recreation. The first was the rapid proliferation of the automobile. Although the automobile had been a symbolic fixture of American life for more than a decade by the mid-teens, ownership was only just spilling beyond an elite group of Americans. In 1900 there were about 8,000 registered automobiles in the United States, and though that number rose to more than 100,000 by 1906, it represented a small socioeconomic segment. It was Henry Ford's Model T, introduced in 1908 but not widely affordable until assembly-line production began in the mid-teens, that "democratized" automobile ownership. By 1913 there were more than one million registered automobiles nationally; by 1922, more than ten million. When the stock market crashed in 1929, there were 23 million automobiles on American roads. In 1910 there had been only one automobile for every 265 Americans; by 1929 the ratio was about one in five.[13]

Contemporary commentators were quick to note the dramatic changes that came with the automobile's proliferation. In *Middletown*—their now-famous 1929 survey of social change in Muncie, Indiana—Robert and Helen Lynd began the chapter "Inventions Remaking Leisure" by recounting their conversation with "a lifelong resident and shrewd observer of the Middle West" who questioned the need for an extensive sociological inquiry into contemporary American culture. "'Why on earth do you need to study what's changing in this country?'" the observer had asked them. "'I can tell you what's happening in just four letters: A-U-T-O.'"[14]

Although the automobile was changing many aspects of American life, the Lynds were right to see it as "an invention remaking leisure," for that was its primary impact by the mid-1920s. The automobile, according to the Lynds, had been "influential in spreading the 'vacation habit.'" Although many among Muncie's working-class residents still did not receive paid vacations, about half of them owned automobiles, most purchased on credit. "[O]wnership of an automobile," the Lynds concluded, "has now reached the point of being an accepted essential of normal living."[15] More important, the automobile was "extending the radius of those who are allowed vacations with pay and putting short trips within the reach of some for whom such vacations are still 'not in the dictionary.'"[16] What the Lynds found in Muncie was true generally: the automobile gave a broad swath of the American populace unprecedented access to exurban nature.

The second major change during the interwar years involved government and its willingness to sponsor both road building and recreational development. During the Progressive Era, despite some interest in recreational nature and its preservation, the federal government prioritized resource conservation.[17] But beginning in the mid-teens, the government took a different turn. The creation of a National Park Service in 1916 marked an explicit recognition that the national parks deserved the same sort of institutional support and clear policy voice enjoyed by the national forests. Another piece of legislation in 1916, the first Federal Aid Highway Act, signaled the federal entrance into road building and its recognition of an automobile infrastructure as a public good. Although this act and a series of later acts by the same name were not explicitly recreational, they had important implications for outdoor recreation— not least of which was their generous funding for roads on public lands.

The federal government also began to take outdoor recreation much more seriously as a subject of national import. In the mid-1920s, a high-profile National Conference on Outdoor Recreation considered the novel concept of adopting a national recreational policy, and during the New Deal, the Roosevelt administration devoted significant resources to outfitting the public domain for recreational use. From the mid-teens through the 1930s, the federal government emerged as an important force in the growth of outdoor recreation and in the development of a public recreational landscape.

State and local governments also catered to the rising popularity of outdoor recreation. Municipalities, particularly those in the Midwest and West, actively sought motor tourists by developing municipal autocamps. Counties and states took advantage of an infusion of federal road-building funds to develop and mark routes to attract tourists. With the encouragement of both local and federal authorities, states began developing state parks and forests in the 1920s, often with the motor tourist specifically in mind. In almost all of these cases, regional nongovernmental interests, like chambers of commerce and automobile associations, shaped the actions of local and state governments. Thus, while federal sponsorship of outdoor recreation was strong during the interwar period, much of it was built upon state and local efforts and was congruent with the "associative" tenor of the era.[18]

The nation's experience in World War I also led to increased interest in outdoor recreation and spurred government into greater support for recreational development. As one Forest Service official observed, the war had "introduced a large number of the male population of the country to outdoor life and physical exercise in the open."[19] Aside from such anecdotal evidence, it is difficult to correlate service in the armed forces during World War I with participation in outdoor recreation after the war, but the two processes did have important parallels. Soldiers did plenty of backpacking, camping, and living in the open, and their training stressed self-reliance, fitness, and the ability to survive in a hostile environment. Even the equipment American soldiers used to wage war in Europe—their packs, tents, and stoves—became recreational equipment after the war, through surplus sales and mimicry by recreational outfitters. But perhaps the most important connection between war and outdoor recreation was a more abstract one: both tested the physical and moral fitness of the nation. A number of contemporaries argued that outdoor recreation helped to maintain the nation's military preparedness, and when Franklin Roosevelt launched his "moral equivalent of war" against the Depression, he relied heavily on the belief that contact with nature provided national moral uplift.[20] The federal government thus justified its involvement in outdoor recreation by claiming that it had such civic benefits.

A third important way that outdoor recreation changed during the interwar years was in its relationship to the maturing consumer culture of the era. Historians have long recognized that the interwar period was a distinct one in the emergence of an American culture of consumption premised on the mass production and standardization of goods, the development of national systems of marketing and distribution, the rise of modern advertising, and the emergence of commercial mass media such as radio and motion pictures. Higher wages, the extension of credit, and a broadening affluence allowed a larger portion of the populace to join the consumer ranks in the years after World War I. More than that, the interwar period saw the solidification of a much-emulated middle-class ideal, and the rise of what Richard Wightman Fox has called "a new consciousness of the centrality of consumption in American life."[21] Americans increasingly came to see themselves as consumers, and to ground their identity in consumption, after World War I.

Outdoor recreation became more consumer oriented during this period, though it is important to be clear about what that meant. There are a number of ways of seeing outdoor recreation as a consumer activity. First, one can simply equate consumption with leisure. According to this logic, insofar as outdoor recreation is a leisure activity, one that does not involve economic production, it is a consumer activity. The division between production and consumption here is essentially temporal. One can also make distinctions between commercial and noncommercial forms of outdoor recreation. In other words, outdoor recreation becomes a consumer activity—or perhaps more of a consumer activity—when it involves considerable spending, and when entrepreneurs successfully set up commercial ventures to cater to and profit from that spending. Money rather than time is the distinguishing factor here. A third level on which one can see outdoor recreation as a consumer activity involves intellectual orientation. As a number of scholars have suggested, consumerism taught Americans to see the world in more possessive and materialistic ways, not only in the acquisition of goods but also in the accumulation of experiences.[22] In this sense, outdoor recreation becomes a consumer activity when it is characterized by consumer habits of mind—when Americans start seeing recreational nature as an experiential commodity.

By all three definitions, outdoor recreation became more intimately connected with consumerism during the interwar years. Certainly Americans had more leisure time, and with the automobile they were more likely to head out into nature to enjoy it. More strikingly, outdoor recreation became a decidedly commercial phenomenon after World War I. American expenditures on recreation during the decade increased by 300 percent. Among other effects, this created anxiety among those who saw nature as a bulwark against commercialism.[23] Finally, with the growth of both a car culture and a consumer culture, Americans turned to recreational nature with a new set of acquisitive habits of mind. Nowhere were these changes more apparent than in the rise of mass nature tourism.

Nature and the Logic of Tourism

Understanding the historical development of a leisure-based attachment to nature requires an appreciation of how the natural world fits within the logic of tourism. Tourism requires a nature that is separate, distant, and exotic—a nature that one goes to see. It also requires a natural world that is marked and collapsed into a manageable canon of sites to which tourists can travel. In this latter sense, nature tourism is not necessarily synonymous with outdoor recreation. Tourism is a site-specific activity; nature tourists are usually out to see particular places, such as the Grand Canyon, and to acquire the experience of having done so. Outdoor recreation, on the other hand, is often activity specific. If one is interested in hiking, camping, or swimming, any number of places might serve the purpose. Although this distinction is perhaps too neat—recreation and tourism are often intermingled activities—it does help in isolating and emphasizing the acquisitive dynamics of tourist behavior.

Nature tourism, like any other type of tourism, is built upon a fluid body of cultural information that marks particular sites as worth seeing. Tourism thus involves what critic Dean MacCannell has called an "empirical relationship between a tourist, a sight, and a marker."[24] Seeing tourism in this way helps to highlight the crucial mediating role of information in shaping tourist desires and itineraries. Tourists are sign readers in an expansive sense; their behavior is largely, though not completely, directed by the various sorts of information available to them.[25]

The interwar period was a crucial one in the development of modern forms of cultural production that undergirded nature tourism. As an integral part of the era's consumer culture, the production and reproduction of tourist information—whether in the form of postcards, magazine articles, photographs, guidebooks, maps, advertisements, souvenirs, titles of distinction (such as "national park"), or literally signs on the side of the road—shaped the American desire to visit nature and inculcated the acquisitive habits of mind central to consumerism. Modern tourism expanded rapidly in the late 1910s and 1920s, when automobiles enabled greater mobility for a larger number of people, and sources of cultural production became more numerous, sophisticated, and broadly cast. Nature became profusely marked and advertised, its representations more widely available than ever before. This contributed to the collective urge of Americans to visit particular natural sites, and to see them preserved. Mass nature tourism, in other words, grew out of the existence of a powerful triad of preserved areas, widely available information about such destinations, and a system of transportation that made these sites available to large numbers of people.

Despite its shaping role, cultural production does not necessarily determine tourist behavior. Tourists are rarely mindless automatons following preordained trails. As important as prepackaged information is to shaping the tourist experience, tourist motivations and reactions constantly spill beyond the prescriptive bounds of guidebooks, maps, and even word of mouth. Indeed, tourist behavior reshapes the cultural production upon which tourism rests, and tourists themselves are important cultural producers, as they relay their itineraries and experiences, in words and images, to other potential tourists. That said, however, it is important to understand that the tourist economy that emerged after

World War I—shaped as it was by multiple forms of advertising, by the proliferation of the automobile, and by the government's developing commitment to road building and recreational development—did create structural constraints that those who sought their leisure in nature had to confront. As tourists headed to and through the premier national parks, for instance, they traveled paths that were becoming well worn. Some tourists may have responded to the urge to chart a new course, but many others stuck to the increasingly well-provisioned tourist routes. As Americans came to know nature through leisure after World War I, they faced the physical and cultural infrastructure of a growing tourist economy.

Working against this confining aspect of tourism, however, was an internal contradiction that defines tourism's modern essence. While tourists by necessity rely on information, many tourists make it their goal to push beyond and find an authenticity unmediated by cultural production. This is a common tourist impulse—to transcend tourism, to leave the other tourists behind, and to seek the "real" place or culture that exists, by definition, at the edges of signification.[26] Having been guided to places described as authentic and exotic, tourists often find themselves surrounded by other tourists and by tourist development that compromises the experience of authenticity. Disappointed, they push the edge of tourism's reach to find places seemingly untouched by tourism yet still meeting tourist expectations. The irony of this quest is that in finding such "virgin" territories, these modern pioneers inevitably pull them into the tourist orbit.

This internal contradiction illustrates two important points. First, it suggests the complexity of tourist behavior and how inextricably caught up in its logic we are. As members of a modern consumer culture, we cannot escape tourism, although we can maneuver within its bounds. Nor can we understand our attraction to nature as a leisure destination, or our preservationist ideals, until we confront the fact that at one time or another we have all been nature tourists, drawn to famous natural sites as pilgrims are to holy shrines. Yet American idealizations of nature have within them strong strains of tourism's contradictory impulse: the search for pureness or authenticity, for a pristine nature devoid of tourists and other signs of modernity. The tourist desire to visit nature is a modern desire, but it also is driven by a deep ambivalence about modernity.[27]

The emergent logic of nature tourism was hugely important in shaping various preservationist ideals during the interwar period, including wilderness. Moreover, the nature tourist's compulsion to push beyond the canon of signified sites had some very real physical impacts. As more Americans visited tourist sites in search of authentic experiences in nature, they began to crowd established destinations. Americans thus became acutely aware that to find real nature, nature unsignified, they would have to go further afield. As they did so, they inevitably brought more use to previously unused areas. Moreover, these use patterns often promoted the sorts of tourist development that many sought to escape in nature. The physical manifestations of this tourist impulse, in other words, contributed to a growing sense that what remained of the American wilderness was vanishing. The early history of motor touring and autocamping reveals this dynamic vividly.

"Roughing It DeLuxe": Autocamping and Motor Touring

Autocamping and motor touring were among the most popular outdoor recreational activities during the interwar period.[28] The *New York Times* estimated that at least five million automobiles a year were being used for autocamping in the early 1920s, a significant figure when one considers that there were only about ten million automobiles in the entire country at the time. Other sources estimated that ten to fifteen million Americans went autocamping each year during the mid-twenties, or about 10 to 15 percent of the population.[29] American motorists engaged in motor touring of various magnitudes, from anonymous evening or weekend drives into the country to well-publicized transcontinental journeys.[30] Before the widespread improvement of roads, touring was something of a sport—a contest of humans (and their machines) against nature. Roads, where they existed, were often of poor quality, frequently impassable, and not well marked. Rural roads were designed to serve the needs of local residents and were not equipped for extensive through traffic. Nor were rural residents initially ready to coexist with urban motorists. For one thing, automobiles spooked horses. This conflict between mechanical and biological forms of transportation had to be resolved before rural roads became fully open to automobiles. Only after an initial period of uneasy truce did horse traffic give way to automobiles, and this only after farmers came to see the advantages (or what they thought were advantages) of owning automobiles.[31]

Motorists venturing into the countryside raised perplexing questions about the "nature" of roads and roadsides as public spaces. This was particularly true before 1920—a period historian Warren Belasco has called the "squatter-anarchist" stage of motor touring—when motorists took to the roads, the rural landscape, and the wilds as if they were a great public commons. Motor tourists subjected rural roads to significant wear and tear, and they were quick to avail themselves of roadside resources. Many rural roads of the time lacked clear demarcations between public and private property, such as fences and landscaped rights-of-way. Motoring parties thus tended to picnic and camp in farmers' fields, often without seeking permission, and their activities took a toll on the landscape. In the first decades of autocamping and motor touring, the roadsides in rural and wild areas received some harsh treatment, as did property owners who had to deal with the consequences. Among these property owners were public land agencies such as the Park Service and the Forest Service.[32]

It is difficult for people today, conditioned by the circumscribed nature of roads as public space, to appreciate how open and liberating roads seemed to the first generation of autocampers and motor tourists. Motorists prized the freedom their automobiles gave them; indeed, "gypsying," as it was often called, was a conscious repudiation of the hotel-railroad complex and the class-based nature tourism of an earlier era. The automobile was praised as a liberator and leveler. It allowed tourists to make their own schedules, to stop wherever they wanted, and to enjoy the landscape less expensively and more in tune with the doctrine of strenuosity. Commentators equated motor touring with roughing it, idealizing it as a self-sufficient and self-contained encounter with nature. It was, in many ways, a rejection of the tourist structures, particularly the canon of sites, that had shaped nature travel in the days before the automobile. Autocampers often celebrated nature as an open space rather than a series of predetermined destinations. They concocted a mythical pastoral countryside that was theirs to occupy, and played at tramping when real tramps were the subject of national scorn. Early recreational motorists challenged the logic of tourism as often as they followed it.[33]

Among the great popularizers of autocamping during this period were a few men whose industrial empires had helped to usher in the modern age. In a series of richly symbolic autocamping trips in the late teens and early twenties, Harvey Firestone, Henry Ford, Thomas Edison, and a number of guest participants motored into the wilds to "rough it" before the eyes of the nation—thanks, in large part, to the press entourage that followed close behind. These trips confirmed for Americans that auto-camping was a leisure activity worthy not only of hoboes, a group whose "leisure" was suspect, but of industrial titans as well.

In August 1918, the famous nature writer John Burroughs accompanied Ford, Firestone, and Edison on a lengthy trip to the Great Smoky Mountains and wrote about it in his essay "A Strenuous Holiday." As Burroughs himself admitted, they were "a luxuriously equipped expedition going forth to seek discomfort."[34] And the discomfort they experienced, though they tended to naturalize it, had as much to do with the encounter between the automobile and poor roads as it did with living close to nature. Burroughs also noted the careful urban-rural diplomacy necessitated by this new machine in the garden. On their second night out, as they picked out a field across from a creek as an ideal camping spot, Burroughs recounted receiving the "reluctant consent" of the widow who owned the land. "She had probably had experiences with gypsy parties," Burroughs guessed, "and was not impressed in our favor even when I gave her the names of two well-known men in our party."[35] As Burroughs recorded his bracing contact with nature, he and his party could not escape the fact that they were traversing and intruding upon an owned and worked landscape.

Three years later, in yet another of their touring trips, Firestone, Ford, and Edison invited the new president, Warren G. Harding, to accompany them.[36] The president's trip, which was covered heavily by the press, further publicized autocamping and provided apt symbolism for burgeoning federal support of outdoor recreation. Indeed, Harding was the first of a string of interwar presidents who led by example and publicly situated their leisure in nature.[37] Harding was not the first president to go camping; Theodore Roosevelt had gone on some widely publicized trips. "What distinguished Harding's outing and excited the public," historian Allan Wallis has posited, "was its equipment and style."[38] The same distinction could be made for outdoor recreation generally during these two eras. Automobiles allowed Americans to go into nature with more gear and gadgetry than had the campers of previous eras. In just a few short years, camping had become a very different experience.

While many motorists, in their enthusiasm for getting back to nature, failed to note that their experiences with the natural world were being mediated by a powerful new technology, others were quite clear that the automobile, as mediator, was a key part of the experience. Compared to the railroad, the automobile gave its users a more intimate experience in the natural world.[39] It also afforded motorists the ability (usually) to return safely to the civilized world—or, increasingly, to carry more of the civilized world with them. Indeed, an entire material culture developed to inform and facilitate the motor touring experience. The Coleman Lamp Company, for instance, augmented its business in gas appliances for the farm home by developing kindred products for autocampers. Proclaiming their products as "The Smooth Way to Rough It," Coleman began marketing camp stoves in 1923.[40] Other companies produced tents, camp chairs, camp clothing, specially packaged foods, and a variety of gadgets to smooth the rough edges of autocamping.[41]

A great place to examine the technology of outdoor recreation—the complex of gadgetry attached to motor touring and autocamping—is in the pages of *Popular Mechanics*. During the interwar period, the magazine was replete with articles on and advertisements for the latest in recreational technology. Particularly popular were articles on how to transform the automobile into an outdoor living space. Although *Popular Mechanics* favored publicizing gadgets developed by individual inventors, the magazine also covered commercially available products for the autocamper. The articles and ads in *Popular Mechanics* during the period suggest that Americans increasingly saw their automobiles as technological capsules to be launched into the natural world. And as outdoor recreation became more technological, so it involved greater consumption and a multiplying array of consumer choices.[42]

The 1920s saw several companies getting into the production of specialized auto bodies and tent trailers, and by the 1930s there were fully furnished house trailers on the market.[43] Wally Byam, founder of the Airstream Company, built his first prototype trailer in the late 1920s, and by 1930 he was producing trailers to meet tremendous demand.[44] He was not alone. By 1937, according to a number of observers, trailer production had become the "fastest growing U.S. industry," with factory produced trailers supplanting homemade versions. By the late 1930s, approximately 400 companies were making almost 100,000 trailers a year.[45] In a 1937 article, *Fortune* announced that "200,000 Trailers" would "swarm on the roads this spring." "Whether they betoken a New Way of Life," the article mused, "or a plague of locusts is something that makers, taxpayers, hotelkeepers, and lawmakers are bitterly disputing. This much is certain: trailer making is becoming a $50,000,000 industry."[46] The house trailer gave Americans a new and curious way to know nature through leisure, one marked by an ideal of high-tech mobility.[47]

Along with these technological developments, the interwar years saw a profusion of literature on motor touring, autocamping, and trailer life. Authors such as Frederick Brimmer and Elon Jessup published extensive treatises on "motor campcraft," wedding the tradition of woodcraft to the modern use of the automobile and accompanying gadgetry.[48] While almost all of these authors conceded that the automobile allowed people to get "off the beaten path into the longed-for wilderness," they were overwhelmingly concerned with explaining autocamping as a technological encounter with nature.[49] A common refrain was that proper preparation and outfitting would keep things from getting too wild. With the automobile, Brimmer pointed out, Americans could "rough it de luxe." They could travel in and through nature without sacrificing the conveniences and comforts of home.[50]

The theme of mobile domesticity was a key characteristic of interwar outdoor recreation. Brimmer began *Motor Campcraft*, for instance, by admitting that he was propelled into the outdoors by his wife's apparently fragile condition. "Nerve specialists," Brimmer wrote, "told me emphatically that I must take my wife into the outdoors and keep her *contented* there or else patronize a sanitarium." Nature was to provide the "tonic," according to Brimmer, while the automobile toted all the trappings of the modern home.[51]

There were, as Brimmer's story attested, a number of gendered implications and ironies to autocamping. It was mainly a family activity, a model of companionate leisure that seemed antithetical to the individualism of the strenuous life.[52] Some men regretted the automobile's capacity to bring along the family and many of the comforts of home, but most proponents lauded these very capacities as autocamping's central appeal. Autocamping was a good compromise. While not quite as strenuous as the all-male wilderness hunting and fishing trips popular among elites during the late 1800s, autocamping was a more masculine (and affordable) alternative to the genteel resorts to which these same elites sent their wives and children.[53] Moreover, much of the work of autocamping mimicked housewifery: autocamping allowed men to make a sport of domestic chores, to make women's work into a form of leisure. As writer and autocamper Mary Roberts Rinehart put it, "the difference between the men I have camped with and myself, generally speaking, has been this: they have called it sport; I have known it as work."[54]

Equipment manufacturers and other providers of amenities often tried to sell their goods by appealing to the importance of keeping wives happy on the road.[55] This fascination with gadgetry also occurred within the context of a new domestic ideal: the modern electrified home. Auto-camping was both an escape from and a recapitulation of the modernized home. Autocampers turned to the open road to escape from the ease of modern living and to recreate older domestic routines lost to modernization. Nature was a place where one cooked over an open fire rather than a stove, where one had to find water from its source rather than turning on a tap. Part of the romance of autocamping, then, was in recreating more primitive relationships with basic resources. Yet the technology that accompanied autocamping often functioned to recreate the modern home in nature. There were numerous affinities between camping technologies and the appliances making their way into American homes during the interwar years; in the house trailer of the 1930s, the most extreme example of this trend, manufacturers strove for modernized mobile living spaces that were themselves a celebration of how completely technology had freed people from the whims of natural provisioning.[56] The motorized outdoor recreation of the interwar years promised an extension of the modern domestic sphere without forfeiting strenuosity. The notion of "roughing it de luxe" summed up these tensions quite well.[57]

All of the authors who wrote in this autocamping genre insisted on strict etiquette. The openness of the autocamping experience was a fragile one, they suggested, and those who ignored rules of courtesy jeopardized the autocamper's freedom to use the landscape and commune with nature. But more than common courtesy was at work here; this attention to etiquette was a response to some disturbing environmental problems. Access to wood and water was essential for good camping sites, yet carelessness impaired their future use. Improper sanitation left areas fouled by human wastes and polluted water. Many campers freely cut and gathered wood for fires, and a number of autocamping manuals explained how to make beds out of pine or balsam boughs, a process that surely left its mark. In heavily used areas, soil compaction became a significant problem. In addition, a proliferation of litter, including the ubiquitous tin can, documented the increasing popularity of motor touring. In their efforts to escape civilization and get back to nature, autocampers often found the roadside sullied by the practices of their fellow enthusiasts.

Along with this refuse, the forces of commerce increasingly intruded on the motorist's ability to commune with nature. As more people went touring, a greater percentage of the roadside was devoted to catering to

Once the delicious reward only of the venturesome and resourceful—

Now anyone may have it for a few gallons of gasoline and an eye full of dust

The Advance of Civilization, 1926. This cartoon illustrates the ways in which the automobile and motor tourism were changing the landscape and the outdoor recreational experience for Americans during the interwar years. Modern wilderness advocacy arose in response in these forces. (*New York Herald Tribune,* August 15, 1926, sec. III, p. 6) their needs. In a 1934 article, *Fortune* announced that "The Great American Roadside" had become a $3 billion industry, built upon "the restlessness of the American people." Hot dog and ice cream stands, autocamping and cabin camps, billboards and gas stations dotted the major tourist routes. These and other manifestations of the new commercial spirit transformed much of the open space of the roadside into commercial space. Other facilities such as restaurants, inns, and roadside attractions competed for the business of motor tourists, as entire regions tied their economic future to tourist development.[58]

One of the more fleeting examples of such development was the municipal autocamp, an institution that appeared overnight in the early 1920s and was particularly evident in the American West. There were between 3,000 and 6,000 of these camps, run by towns and cities, during their heyday in the early 1920s. Local governments, in conjunction with booster and tourist organizations, provided these stopping points as alternatives to unregulated use of the roadside. Initially, almost all of these camps were free of charge, the assumption being that the costs of setting up and maintaining such facilities was returned in the commerce that motor tourists brought. Many camps had developed sites and, in some cases, modern cooking and laundry facilities. The largest and most famous camp was Denver's Overland Park, a 160-acre plot with 800 lots, which opened in 1915. Far more common, however, were smaller and more modest camps in town centers throughout the Midwest and West. After a brief reign, privately owned camps, tourist cottages, and motels replaced these free municipal camps.[59]

The early 1920s also saw the emergence of a state parks movement whose vitality stemmed from the needs of autocampers and motor tourists. State parks emerged as an important alternative to the increasingly commercial nature of roadside provisioning. The National Conference on State Parks, started in 1921 at the behest of Stephen Mather and the National Park Service, was the leading force in this movement. Its motto, "A State Park Every Hundred Miles," was a clear indication that the convenience of the motorist was paramount; indeed, one of Mather's reasons for pushing such a system was to provide motorists with autocamping links to the western national parks.[60] State parks were not a unique creation of the interwar era, but they saw a vast expansion during the period. As sociologist Andrew Truxal wrote in 1927, "the automobile has made possible an extent of tourist travel heretofore unknown and the movement for state parks which is a phenomenon of the present time, is a movement to meet the needs of this touring populace."[61] As of 1933, by Jesse Steiner's estimate, 70 percent of America's state parks had been created since 1920. The New Deal gave the movement an even more dramatic boost.[62]

To aid in and shape this autocamping and motor touring boom, groups such as the AAA (founded in 1902), the National Park Service, and a variety of highway associations and booster groups published maps and guidebooks that laid out itineraries for motorists. By the mid-twenties, motorists had a number of cross-country routes to choose from, many of them marked, partially paved, outfitted with tourist facilities, and promoted by the highway associations that supported them. The Lincoln Highway, supported by the Lincoln Highway Association and funded largely by automobile manufacturers, was perhaps the best known of these routes. These groups, and the maps, guidebooks, and marked routes they produced, were key to reshaping the motor tourist experience in interwar America.[63]

As road conditions improved, the sense of intimacy with the landscape faded. A modernized infrastructure increased average speeds, decreased the number of stops, and eliminated the natural obstacles, such as boulders, mud, and chuckholes, that had been important to the sport of early motor touring. As motorists glided effortlessly across the terrain, they no longer felt the resistance or drag of nature; they enjoyed cinematic scenery while traveling between structured destinations.[64] The escape from tourist structures that had marked early touring gave way to a system of modern roads and facilities that recreated the hotel-rail complex in a new form. The nature one drove through was no longer as important as the nature at which one arrived. "In one ironic sense," Warren Belasco has concluded about these changes, "tourism rapidly returned to a resort-railroad mode: fast, utilitarian travel between stereotyped spectacular sights."[65] As this new regime took hold, the roadside became a scenic rather than a recreational space, as groups such as the Motorists' League for Countryside Protection launched campaigns to discourage littering, flower picking, and other activities destructive of roadside aesthetic values.[66]

As rural roadsides were closed off to autocampers and as the short-lived municipal camp gave way to commercial camps and motels, federal sponsorship of outdoor recreation grew in kind, filling an important niche. So did public expectations that federal lands would provide the inexpensive services and pristine experiences that motor tourists had come to expect from nature. The appeal of autocamping was premised on the existence of a public roadscape. As that roadscape disappeared behind billboards, fences, "no trespassing" signs, commercial facilities, and landscaping, the public lands became an important refuge for those still interested in roughing it. The increased use of the national parks and forests, then, was not only a result of the access provided by improved roads and the automobile; it was also a product of the closing off of the rural roadsides that had briefly served as public recreational landscapes. Increasingly rigid and visible distinctions between the public and private landscape propelled Americans onto the public lands.

Public lands administrators had to make concessions. On the national forests, reluctant foresters developed campgrounds with cursory facilities such as toilets and fire pits to combat problems with water pollution and fire. The National Park Service, though generally enthusiastic about the motoring crowds, quickly found it necessary to confine autocampers to specially developed areas. Indeed, by the early 1930s, plant pathologist E. P. Meinecke had alerted officials to a new set of problems brought about by autocamping:

> The enormous increase in the number of automobiles driven into, and parked in, the camps intensified the effects of heavy use to such a degree that first the ground cover, then the shrubs and young trees, and finally the larger shade-giving trees began to die, and that many camp grounds finally became useless and had to be abandoned because they no longer offered the pleasant surroundings that the visitor seeks in camping. *The driving of automobiles into the camp itself is by far the most damaging element.* The soil surface becomes compacted and hard, and plant roots are injured. The automobile, in turning and moving about among the trees and shrubs, causes a great deal of mechanical injury, and at last a point is reached where the cumulative effects on plant life become so great that suffering and then death ensues.[67]

Meinecke's solution was a general blueprint for campsite development that called for loop roads with numerous "garage spurs" into which campers could pull their vehicles. As today's autocamper will immediately recognize, Meinecke's plan remains the dominant one for developed sites in the national parks and forests. As historian Linda Flint McClelland has concluded, "Meinecke's recommendations revolutionized camping in the national parks and forests in the 1930s," and this was in direct response to the ecological impacts of autocamping.[68]

Some other aspects of the boom in outdoor recreation after World War I deserve brief mention. Hunting and fishing, which had transmogrified into leisure activities over the previous half-century, were themselves affected by technological developments that increased participation. Between 1920 and 1930, the number of licensed hunters in the United States grew from four to seven million, and Jesse Steiner estimated that Americans were spending $750 million a year on hunting, fishing, and related expenses by the early 1930s.[69] The automobile allowed hunters and fishers a greater radius in their search for productive spots, and, as with autocamping, this greater mobility may have worked to close off portions of the agricultural landscape to these activities. After World War I, hunters could get their hands on more efficient firearms, and this, in combination with a more extensive agriculture and a concerted national effort at wetlands drainage, raised serious concerns about game and fish stocks.

In the motorboat—the aquatic version of the automobile—those who fished had a greater ability to move quickly across the nation's waterways. Outboard motors, which first appeared after the turn of the century, experienced a revolution in design after 1920. Early motors were loud, heavy, and not particularly powerful. But the Johnson Company changed things in the 1920s by introducing a significantly lighter outboard and then a heavy but more powerful motor. Spurred on by advertising that claimed for the motorboat the same possibilities for escape promised by the automobile, outboard sales soared in the 1920s. But, as with the automobile, the motorboat gave Americans increased access to the nation's undeveloped lakes and waterways at the cost of undermining the wild qualities sought by those who preferred the solitude of canoeing and other nonmechanized forms of travel.[70]

Summer camps flourished during the interwar period, as did a variety of scouting movements and organizations.[71] Nature study continued as a popular activity, and nature photography grew in importance as developments in photographic technology allowed more Americans to make their own pictures easily and affordably. Hiking gained adherents as a second generation of hiking clubs sprang up across the United States.[72] Summer homes also trickled down across socioeconomic lines, as the automobile put prime vacationlands within reach of many and replaced the genteel resort hotels with the middle-class summer homestead.[73] One could even purchase "canvas cottages" that functioned as inexpensive, portable summer residences.[74] All of these developments were consistent with the transformation of outdoor recreation into a mass activity, and most spoke to the importance of the automobile in achieving that transformation.

As a technology that allowed Americans a new knowledge of nature, the automobile was a profoundly ambiguous force. It brought millions of Americans out of towns and cities and into a variety of natural settings, but it also proved a vector for modernity. The automobile both facilitated the great popularity of outdoor recreation and forced Americans to confront the implications of that popularity.

The National Conference on Outdoor Recreation

During the interwar years, the federal government took a more active role in discussions about outdoor recreation and its promise. Nowhere was this clearer than at the meetings of the National Conference on Outdoor Recreation (NCOR). Organized by President Calvin Coolidge, the two large gatherings in 1924 and 1926 were the era's answer to Theodore Roosevelt's Governor's Conference on Conservation in 1908. The 1908 conference had brought together forty-four governors and numerous experts on conservation in an effort to forge a national policy on natural resources. But, famously, almost no one spoke for those interested in outdoor recreation.[75] Sixteen years later, the tables were turned as hundreds of delegates arrived in Washington to craft a national recreational policy. Such a policy effort was unprecedented, and it spoke to both the cultural centrality of outdoor recreation and the pressing need federal officials felt to define the roles that the public lands and their supervisory agencies would play in providing recreation. The NCOR meetings signaled a new era in American environmental politics in which recreational politics emerged as a central if complex force.[76]

In his speech opening the first NCOR meeting in May 1924, Coolidge laid bare a number of themes that unified the presentations. First, the conference would aim at putting outdoor recreation "within the grasp of the rank and file of our people." Although there were several agendas in this pursuit, most of the participants accepted the democratization of outdoor recreation as an appropriate goal. A second broad theme was the therapeutic promise of outdoor recreation. Coolidge mentioned the "physical vigor, moral strength, and clean simplicity of mind" that came with recreation in nature. Finally, he spoke of the historical juncture represented by this national embrace of, and reckoning with, outdoor recre-

ation. At a time when fewer Americans were making a living doing physical labor, and those who were often found themselves confined in fac-
tories and engaged in drudgery, physical activity in the out-of-doors
seemed a pressing national need.[77] Former Secretary of State Elihu Root
made the same point, suggesting that the NCOR confronted "one of the
most important and necessary readjustments of American life."[78] As one
delegate noted, the extension of leisure was "full of menace."[79] But most
felt that the perils of leisure could be offset by the ability of public nature to mold recreation into a fulfilling experience. This was the synthesis that many hoped a national recreational policy would achieve.

But these shared commitments and an acute sense of the historical
moment aside, NCOR delegates were far from united in their visions of
outdoor recreation's national importance. Indeed, a major lesson of the
NCOR was that outdoor recreation had become a magnet for all sorts of
reform-minded Americans. Much of the reformist sentiment at the
NCOR meetings was conservative and custodial, and there were parallels
between this debate and the discussions about the reformist possibilities
of urban parks that had taken place over the previous half century. But
there was also content distinctive to the NCOR, a product of the national
level of discussion. Correlations between outdoor recreation and the nation were particularly emphasized. In connecting outdoor recreation
with "Americanism," conference participants united a number of important cultural themes. One was the martial vigor that supposedly came
with physical exertion and the challenge of surviving in unmediated
contact with nature. Roosevelt's "strenuous life" was given a new twist
after the nation's experience during the First World War, as a number of
participants made connections between a national outdoor recreation
policy and military preparedness.[80] Firsthand experience with the American landscape was also linked to national identity and pride. This theme
had been prominent throughout the nineteenth century, but it gained
steam in the "See America First" campaigns launched in the mid-teens.[81]
A third theme, less overt than the first two, was the importance many
participants assigned to public order and national loyalty in the wake of
the labor unrest following World War I. Outdoor recreation could serve
the nation, many thought, as a guard against class conflict. Nature could
be a new safety valve to mitigate workplace monotony and strife.[82]

In contrast to the caucus of nationalists and moralists in the crowd,
there was another group of delegates for whom federal sponsorship of
outdoor recreation promised needed leverage in the fight to preserve

scenery and wildlife. Wildlife protection was a particularly hot topic, and the NCOR distinguished itself with a sophisticated debate about the sources of depletion. Some, such as Will Dilg of the Izaak Walton League, bemoaned the toll taken on game populations by pollution and habitat loss. Federal involvement, they hoped, would reverse these trends through habitat preservation. Others, such as the reformed hunter William Temple Hornaday, argued for stricter regulation of hunting and fishing, which he insisted were the primary sources of game depletion. Hornaday expressed particular concern about modern hunting technologies. "As every one knows," he said, "the automobile has become a fearful scourge to the game of our land, by enabling at least 2,000,000 men of the annual army of hunters to cover about four times as much hunting territory as they formerly could comb out with their guns."[83] The continued recreational use of American wildlife, these advocates argued, depended on more aggressive federal efforts to preserve habitat and species.

Another group of preservationists sought, through comprehensive planning, a more thorough and varied system of federal recreation areas. Robert Sterling Yard of the National Parks Association (NPA), for instance, spoke of the great national opportunity presented by the remaining public domain to craft such a system. Yard was mainly interested in preserving national park standards and promoting the national parks as educational arenas, but he also recognized that the proliferation of motor tourists and autocampers required a federal response. Yard's most tangible contribution to the conference was to suggest a survey of the nation's scenic and recreational resources, a survey completed under the combined leadership of the NPA and the American Forestry Association (AFA).[84]

A third identifiable group at the conference was the boosters, those who argued for seeing the developmental possibilities of outdoor recreation or who came to the conference to promote specific projects. Not surprisingly, most were interested in boosting motorized outdoor recreation. Dr. S. M. Johnson of the National Highways Association (whose rather apocalyptic motto was "A Paved United States in our Day") argued that Americans' recreational embrace of the automobile had not been met by an appropriate federal commitment to improve roads. Johnson called for a national program to develop a road system that reflected the recreational value of motoring, and the NCOR later adopted a resolution that recognized and affirmed the importance of federal road-building efforts to outdoor recreation.[85] The NCOR also commissioned the U.S. Chamber of Commerce, the era's preeminent booster group, to survey "the recreational possibilities of our highways," a report that they delivered to the 1926 meeting.[86] Most of the booster groups and their delegates saw clearly that federal sponsorship of outdoor recreation would mean commercial and infrastructural development for particular areas, and they hoped to profit from such development. More broadly, the strong booster presence, in conjunction and ostensible coalition with the other interests at the NCOR, suggested that recreational preservation on the public lands had become good politics.

The first meeting of the NCOR served to air these dominant views, but little was accomplished in terms of concrete policy. There were too many competing agendas. During the second full meeting, convened in 1926, tensions rose to the fore. Chauncey Hamlin, chairman of the NCOR, began the meeting with critical words for states' rights objections to such national planning for outdoor recreation. Some saw the federal government overextending itself in seeking a prominent role in recreational policy and development. But it was a disagreement over what recreational development meant and how the federal government ought to support it that proved most divisive. One group of delegates stressed what Paul McGahan of the American Legion referred to as "the conservation of the human being."[87] This caucus wanted the NCOR to make a stronger statement on the need for a wide variety of active recreational facilities such as playgrounds and athletic facilities that would cater to the largest number of Americans possible. For them, outdoor recreation was an outgrowth of urban recreation, and they were nonplused by the first meeting's strong focus on national parks, wildlife refuges, and other areas of distant nature. The preservationists, on the other hand, argued that the federal government should assume as its primary charge the protection and promotion of the unique recreational opportunities afforded by the remaining public domain—opportunities that were not necessarily of the active, urban kind. This disagreement, which emerged most pointedly in a debate over the NCOR's final resolutions, was overdrawn; the delegates themselves came to recognize that the two agendas could work together. But there were a number of delegates who saw in the final resolutions an emphasis on preserving nature that neglected human needs.[88]

Ethelbert Stewart, the Commissioner of Labor Statistics, made one of the most pointed statements on this split. He warned that the "City

Beautiful" movement, with its emphasis on aesthetic rather than recreational development, was quickly transforming open spaces in urban settings. At the same time, Stewart said, employers were slackening in their efforts to provide for the recreational needs of their employees, shunting their responsibility onto the very municipal governments keenly turning parks into manicured gardens. As active recreational spaces retreated from urban centers and sprang up in the hinterlands, workers were often the ones who lost access.

Stewart saw this process recapitulating itself in the NCOR resolutions. "We sympathize with you in your desire to protect wild life," Stewart insisted:

> but we ask you to sympathize with us in our attempt to furnish some sort of outdoor recreation for the factory girl and the factory man—the vamp matcher in a shoe factory who must handle 720 pairs of shoes an hour, which means 1,440 operations all exactly alike every 60 minutes for eight hours a day; the toe-cutter in the sheep-killing beds in our slaughter houses who must handle 400 sheep an hour, which means cutting off 1,600 toes every 60 minutes for nine hours a day—this is the wild life with which we have to deal and whom we would like to save from the insanity that follows industrial fatigue, by providing them a place where they can get a little fresh air in the evening, where they can eat a picnic dinner Saturday afternoon or on Sunday, with no "Keep off the grass" sign to molest or make them afraid.

Here, Stewart spoke eloquently on a number of important themes. First, he suggested that the degradation of labor was one of the major reasons why working-class Americans needed more and better leisure opportunities in nature. The experience of nature garnered through slaughterhouse labor, Stewart intimated, was not particularly uplifting. Second, he suggested that preservationists had a blind spot. In pushing for the protection of lands and species threatened by the nation's industrialization, they had marginalized the needs of the industrial laborer. Third, Stewart argued that efforts to shape working-class leisure in nature toward certain social ends were manipulative and condescending. Workers needed open spaces; they could decide for themselves how to use them. As democratized as outdoor recreation had become by 1926, Stewart made it clear that the needs of American workers were not being met. He thus asked the NCOR to endorse a parallel effort to secure more space for working-class recreation.[89]

Some who joined Stewart's call may have shared his empathy for the nation's working classes, but most seconded his sentiments for more self-serving reasons. Delegates from playground associations, groups promoting organized athletics, and urban-based reform organizations such as the YMCA were among Stewart's staunchest supporters, as were delegates from groups with martial interests, such as the American Legion. More interested in the physical and moral fitness of the nation than in the numbing routines of the American worker, these delegates nonetheless applauded Stewart's critique of the NCOR's focus because it served their vision of a national recreational policy based in personal and national uplift.[90]

This line of dissent was compelling, but most of these objections missed one major point: the NCOR, because it was sponsored by the federal government and was aimed at achieving a federal recreation policy, was mostly focused on the recreational potential of the public lands. These areas were generally large, wild, and distant from urban centers and populations. The NCOR might have paid more attention to the lack of affordable access to such areas, but the federal government, its sponsor, was not in the best position to build urban parks and playgrounds. The provision of such facilities, though perhaps of national import, was better handled by state and local government. Stewart himself recognized this bind when he implored the federal government to transfer public lands near major urban-industrial centers to states and municipalities for the development of working-class recreational areas. "Of course," he continued, "the trouble with this is that the public has very little or no land to transfer in the case of most of the cities."[91] The outdoor recreational opportunities that the federal government was in the best position to promote and develop were of a different variety.

Like Ethelbert Stewart, Aldo Leopold came to the 1926 NCOR meeting to address a topic—"wilderness conservation"—that had been neglected at the 1924 meeting. Leopold began his speech by challenging a purpose that animated many in the audience. "Recreation," he said, "has in many instances been overpromoted." Americans, Leopold insisted, echoing one of Stewart's points, no longer needed "to be urged to play outdoors," or to have their recreation shaped to particular ends. Rather, "the best way to teach [Americans] how to play outdoors is to provide them some outdoors to play in." A federally sponsored wilderness policy, he suggested, could provide Americans with a "fundamental recreational resource" for a more basic form of outdoor recreation.

Leopold knew that many in the crowd would assume that wilderness recreation was expensive and elitist—precisely the sort of thing Stewart had protested against. Indeed, Leopold figured he would be lumped with those among the delegates alleged to have little interest in human welfare. He sought to address this presumptuousness directly:

> It so happens that some people go into the wilderness accompanied by hired guides, large pack trains, and other expensive trimmings. The ill-informed tend to jump to the conclusion that this is the sort of thing the wilderness idea seeks to perpetuate. They are exactly wrong. The man who can afford these expensive trimmings, and prefers to have them, can also afford to go to the ends of the earth where there is still lots of wilderness left. . . . It is the other kind of man, the man who can not afford to travel farther, and who must seek his wilderness near home or not at all, whose standard of living is endangered by the impending motorization of every last nook and corner of the continental United States.[92]

This was an eloquent plea for wilderness as an egalitarian space, and in making it Leopold intimated that mechanization was the era's most pressing threat to the wilderness qualities of the public lands. In an era of autocamping and motor touring, those who sought wilderness recreation might be few. But they nonetheless deserved the opportunity to find such recreation; it was a public good, Leopold thought, that was worthy of federal protection. More than that, wilderness preservation fit squarely within federal jurisdiction. Most of the nation's remaining de facto wilderness was in federal ownership, and the federal government was in an excellent position to preserve such areas as part of a national recreational policy.

As the comments of both Aldo Leopold and Ethelbert Stewart attested, the NCOR was a complicated venue. Although outdoor recreation had burst upon the American scene as an important component of national life by the 1920s, there was by no means a consensus on how recreational opportunities should be pursued and developed at the federal level. While the NCOR represented a challenge to the dominance of utilitarian control over the federal lands, its major lesson was that recreational advocacy was contested terrain. Nonetheless, federal involvement in recreational development during the 1920s marked a distinct departure from the previous era. As Jesse Steiner concluded, "A new era in the history of recreation began when the government accepted re-

sponsibility for the provision of public recreational facilities."[93] This federal move into recreational development was a trend essential to framing the wilderness movement that emerged during this period.

Outdoor Recreation in the Depression and New Deal

The recreational trends of the interwar era were largely forged during its
prosperous first half. But beginning in 1929, the nation slipped into a
prolonged depression that had a profound effect on conservation policy
generally and recreational development in particular. The national craze
for autocamping subsided somewhat—though less than one might
think, and not for long—as a more troubling version of auto-itinerancy
captured the public imagination. President Hoover reluctantly inched
the government into public works projects that his successor, Franklin
Roosevelt, adopted much more widely and enthusiastically. As the federal government took an even more direct role in recreational development, a new alchemy of work, recreation, and nature emerged that both
built upon and reshuffled 1920s trends.

During the 1920s, the nation's affluence produced increased leisure
time; during the 1930s, increased "leisure" was the result of the Depression itself. The National Recovery Administration, during its brief existence, promoted industry codes that shortened the workweek for most
industrial workers to forty hours, a near universal standard ever since.[94]
But the more conspicuous and troubling source of free time was the lack
of employment opportunities; fully half of the American workforce was
unemployed or underemployed during the Depression's worst years.
Franklin Roosevelt was quick to fix upon the natural world, and the public lands in particular, as the perfect setting to make use of this new
"leisure" and to restore American confidence in the nation and its economic system. Roosevelt built such signature programs as the Civilian
Conservation Corps (CCC) on the foundation of the nation's growing
recreational interest in nature and the supposed salubriousness of outdoor life. Work in nature, he thought, would heal the nation's psyche
while rendering concrete conservation accomplishments. And many of
these accomplishments, not surprisingly, were in the area of public recreational facilities. Others, although not explicitly recreational, helped
open up the public lands to increased recreational use.

The national parks felt the impact of New Deal programs almost immediately. In its first two years of operation, the Public Works Adminis-

The Hetch Hetchy Valley before the dam was built. The battle over Hetch Hetchy, which has long symbolized the peak of Progressive Era conflict between utilitarian conservationists and preservationists, also signaled a turning point in American environmental politics. During the interwar era, questions about recreational development often overshadowed tensions between preservation and use. (National Archives and Records Administration [NARA] 79-G-8-B-1)

Road to Cooke City, Yellowstone National Park, 1922. Early motor tourists celebrated the automobile as a technology that brought them into direct contact with nature, but more often than not the sport of motor touring involved contending with poor roads and mechanical failures. (National Park Service Historic Photograph Collection [NPSHPC])

Motor tourist showing off her park entry stickers, Yellowstone National Park, 1922. Developing consumer habits marked nature tourism during the interwar period, as Americans increasingly saw recreational nature as an experiential commodity and as they used stickers, postcards, and other souvenirs as markers of their consumption of nature. (NPSHPC)

Members of the Western Auto Studebaker Camp Inspection Tour on the Auto Log, Sequoia National Park, early 1920s. The Auto Log was developed as a tourist site after a sequoia fell from natural causes. Since the 1910s, it has been one of the richest symbols of the intertwined American love affairs with automobiles and nature. (NPSHPC)

No Camping sign, San Isabel National Forest, Colorado, 1919. During the early years of autocamping, motorists used the countryside freely, helping themselves to the landscape's resources and often ignoring private property boundaries. As a result, signs such as these began appearing to discourage campers. (NARA 95-G-43377A)

Camping near Camp Curry, Yosemite National Park, 1927. With private landowners increasingly hostile to them, autocampers spread themselves freely throughout the national parks, camping in areas such as Yosemite's meadows. By the late 1920s, however, plant pathologist E. P. Meinecke began noticing the extensive damage to vegetation done by automobiles—through soil compaction in particular. In response, he pioneered the field of modern campground planning. (NARA 79-G-17-C-1)

Did You Put Out Your Fire? Sawtooth National Forest, Idaho, c. 1925. The escalating recreational use of the national forests in the late 1910s and 1920s came with some unfortunate results. One was an increased risk of forest fires caused by careless auto-campers. The Forest Service responded by providing simple fire pits, and by posting such reminders. (NARA 95-G-204244)

Close-up of roadside tin cans and other trash, San Isabel National Forest, Colorado, 1919. This photo, taken by Arthur Carhart, illustrates another unfortunate result of heavy recreational use of the national forests—camping areas littered with tin cans and other trash left behind by autocampers. (NARA 95-G-43375A)

Breakfast hour in campground, Glacier National Park, 1932. As opposed to the back-country hunting and fishing trips of the late nineteenth century, which were primarily a male province, or the elite resort hotels of the same era, which catered to women and children, autocamping offered a middle-class, family-based model of outdoor recreation, though women still found themselves responsible for much of the work. (NPSHPC)

Sagebrusher, location unknown (probably Yellowstone), 1923. Early autocampers used their automobiles for more than just transportation. Often they doubled as sleeping quarters, with tents rigged over and around them. (NPSHPC)

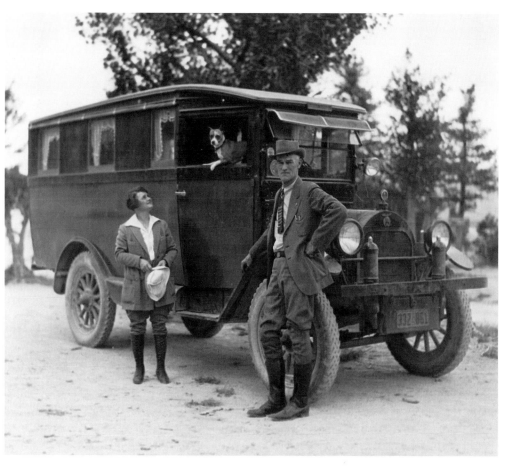

Visitors with mobile home, Yellowstone National Park, 1925. By the mid-1920s, motorists had custom-built the first prototype mobile homes, and industrial production was not far behind. By the mid-1930s, the mobile home industry was one of the fastest growing in the United States. (NPSHPC)

Dedication of the Going-to-the-Sun Highway, Logan Pass, Glacier National Park, July 15, 1933. Going-to-the-Sun Highway, which crossed Glacier from east to west, was an engineering marvel and one of the most important new park roads of the late 1920s and early 1930s. It was a product of the increased funding for park roads that began in the mid-1920s and climaxed during the Depression and the New Deal. (NARA 79-G-5-F-769)

tration pumped $17 million into Park Service coffers for road and trail development.[95] Moreover, almost every national park had a CCC camp and funds to conduct a variety of projects. By the end of 1933, there were approximately 35,000 CCC enrollees and 2,300 supervisors at work in 175 camps in the national and state parks.[96] Two years later, Park Service Director Arno Cammerer announced a vast expansion of emergency conservation work (ECW) in areas supervised by the Park Service. Six hundred camps had been allotted to the Park Service for the CCC's fifth enrollment period, 118 of which were to be in national parks and a staggering 482 in state parks. This CCC labor helped create more than 700 new state parks in the United States.[97] Indeed, the commitment of federal workers and financial resources to the development of state parks, albeit under the leadership of the Park Service, was a departure from the policies of the 1920s, when the federal government encouraged but did not directly aid the recreational work of the states. Federal programs also devoted considerable resources, particularly through the Works Progress Administration, to municipal recreational developments such as parks and playgrounds—an outcome that would have made the dissenters at the NCOR proud. The New Deal thus saw a fuller assumption of federal responsibility for recreation at all levels.[98]

As with other New Deal programs such as rural electrification, the federal government took upon itself the responsibility of extending to all Americans the opportunities—in this case recreational—that many had attained during the prosperous 1920s—a direct echo of Calvin Coolidge's challenge at the first meeting of the NCOR.[99] The Park Service, for instance, embarked on a project to transform areas of marginal land, purchased by the Resettlement Administration, into Recreational Demonstration Areas (RDAs) with group camps and other facilities for low-income Americans. The Park Service also used emergency funding to further promotional efforts. But where Park Service promotion during the teens and twenties had been designed to achieve agency strength through increased visitation to the parks, Park Service promotion after 1933 was directed at local and regional economic reliance on park tourism. "The objective," Donald Swain has said of this New Deal promotion, "was to benefit local economies near the parks, assist the concessioners in the parks, . . . and indirectly stimulate the national economy."[100] Promoting park visitation became a matter of economic recovery.

The New Deal was a flush time for the Park Service. Not only did its budget grow dramatically, but the Service also expanded its jurisdiction over a variety of other federal and state holdings. But there were concerns about all this expansion. In the first years of the New Deal, the Park Service was overwhelmed by ECW recruits but lacked adequate supervisory personnel. In his 1935 report, Park Service Director Arno Cammerer noted that the extensiveness of work relief in the national and state parks had necessitated the addition of more trained supervisors to ensure the "protection of the natural features."[101] By and large, these trained supervisors were landscape architects. Although the Park Service had long seen landscape architecture as a necessary form of expertise, they had hired only a handful before the Depression. By the mid-1930s, the Park Service had hundreds of landscape architects on staff, a testament not only to a new appreciation of their skills but also to the transformative labor being undertaken on the public lands. Most of this design labor went into developing recreational facilities and "naturalizing" intrusive roads. Landscape architecture solidified itself as the dominant professional culture of the Park Service during the New Deal.[102]

The scope of New Deal development was similar on the national forests. By the summer of 1933, there were approximately 200,000 CCC workers in forestry camps supervised by the Forest Service. Although the initial plan was to put most of this force to work performing traditional forestry functions such as planting trees and fighting fires, the Roosevelt administration soon realized it could use this labor to construct infrastructural and recreational improvements too. The Forest Service also used New Deal labor to increase the road mileage within the national forests. In particular, CCC workers built and maintained truck trails designed to improve fire-fighting access—roads that inevitably led to greater recreational access as well.[103] But in contrast to the Park Service, the Forest Service initially kept its recreational developments simple, and it exhibited a noticeable distrust of landscape architects.[104] During the early New Deal, the Forest Service stuck to the development of basic campgrounds and trails, relying on trained foresters for supervision rather than on landscape architects. By 1935, however, these priorities began to change, as the Service hired a landscape architect for its Washington office and created a Division of Recreation and Lands.[105]

New Deal conservation projects and work relief programs thus created a vast public landscape, much of it devoted to recreation.[106] They built scores of urban and regional playgrounds and parks, they helped nationalize and develop several of the nation's beaches, they built scenic parkways, and they expended considerable sums of money in the national parks, state parks, and national forests, outfitting them with roads, trails, and a variety of other facilities. Finally, the Taylor Grazing Act of 1934 began the process of federalizing the remaining unclaimed public domain under the Grazing Service (which became the Bureau of Land Management in 1946), creating yet another recreational outlet for Americans.[107]

New Dealers also incorporated much of the rhetoric of outdoor recreation—its supposed therapeutic and nationalistic qualities—into their justifications for work relief programs. Even those projects that were not explicitly recreational, such as planting trees and combating soil erosion, were invested with the promise of rejuvenation that Americans associated with outdoor recreation. Just as outdoor recreation during the period often mimicked work, so the conservation work relief of the New Deal, as a character building but necessarily temporary escape from a reeling private economy, bore a substantial resemblance to recreation. Moreover, New Deal conservation activities exposed hundreds of thousands of Americans to outdoor living. The New Deal was thus a climax in the nation's growing recreational interest in nature during the interwar period, and New Deal activities on the public lands set the stage for an even greater expansion of recreational interest and development after World War II.

Conclusion

During the interwar years, Americans came to know nature through leisure en masse. The period saw a quantitative increase in outdoor recreation that separated it from the previous era, when only a select group of Americans got their leisure in nature. Equally important, however, was a qualitative change in the ways Americans knew nature through leisure. The automobile was the preeminent agent of change in both cases. It allowed more Americans to escape town and city life, and also to travel to remote areas with greater ease. The automobile facilitated popular activities such as autocamping and motor touring, and it allowed Americans to augment their outdoor recreation with gadgetry that, along with the automobile itself, made leisure in nature a technological experience. While it empowered Americans in their desire to pursue the strenuous life, the automobile also changed the family and gender dynamics of leisure in nature. Finally, the recreational use of the automobile altered large portions of the landscape itself. Taxpayers and booster groups contributed to new and improved roads, many of them marked for recreational use, and entrepreneurs developed the roadside to cater to motor tourists and to profit from the nation's new recreational mobility. State governments created state parks to meet the needs of motor tourists and autocampers, while federal agencies developed the national parks and forests to accommodate motorists and their recreational needs. While many saw the automobile as merely a tool for putting Americans back in touch with nature, it was much more than that. Automobiles—and the Americans behind the wheels—changed the very nature of outdoor recreation.

These changes brought about by the automobile were rooted in a series of deeper economic and cultural developments whose temporal parameters extended beyond the interwar period. As a nation, America had become more affluent, and this affluence, though by no means universal, did expand leisure opportunities. As their country became urban and industrial, Americans sought leisure in nature. And as labor and leisure became segregated, so did the physical spaces in which Americans worked and played. These physical and temporal separations mirrored the increasing separation of production and consumption in American life, and the shift from a producer to a consumer ethos. One of the ways this change in values manifested itself was in a diminution of the cultural importance of work, and an obsession with what many called the "new leisure." Leisure in nature assumed some of the cultural workload that had been borne by labor in nature during the nation's early history.

Americans became consumers on an unprecedented scale in the early twentieth century, and their interest in nature was allied with this shift. But their feelings for nature were much more complex than this simple association suggests. Many Americans did become consumers of recreational nature. They spent their leisure time and their hard-earned money going back to nature. Nature tourism became an important industry, as outfitters provided consumer goods that Americans used in their return to nature and as recreational nature became an advertised destination. Yet, despite this intrusion of consumer habits and structures, Americans continued to find in nature a source of resistance to the increasingly commercial world that they inhabited. Seeing wild nature as a haven from the economic imperatives that drove the nation was nothing new. From Jefferson to Thoreau to Muir, American intellectuals had used nature to critique economic trends that threatened the nation's civic moorings. But by the interwar period, Americans on all points of the political spectrum lauded nature as a source of civic virtue. Indeed, as gatherings such as the National Conference on Outdoor Recreation suggest, Americans saw recreation in nature as serving all sorts of important public functions. There were diverse ideological agendas tied to the popularity of outdoor recreation during these years, as various factions vied to define and shape the impulse.

The wilderness advocacy that arose in this period was but one line of reasoning in a cacophonous discussion over what the nation's growing recreational interest in nature, and particularly public nature, meant and how it would impact the natural world. In an era marked by an increasing confusion and conflation of democratic politics and consumer choice, seeking leisure in nature was almost necessarily a process that mixed the private interests of entrepreneurs and consumers with the public interests of government and citizens. Outdoor recreation was both democratized and commercialized during the interwar years. For the wilderness advocates who are the subjects of the following chapters, the trick was in distinguishing between the two.

CHAPTER 3

A Blank Spot on the Map:
Aldo Leopold

Aldo Leopold was, to borrow from the frontier vocabulary of which he was so fond, a pioneer in the cause of wilderness preservation. Through his writings and initiatives within the Forest Service, he pushed forward the cause of wilderness preservation like no other person of his time, at least until Bob Marshall began his crusade for wilderness areas in the 1930s. James Gilligan, in his influential dissertation on the development of Forest Service wilderness policy, canonized Leopold as the "Father of the National Forest Wilderness System." Although this conclusion has elicited some challenges, most notably Donald Baldwin's defense of Arthur Carhart's claim to the title, the idea of wilderness preservation generally is considered among Leopold's unique contributions to American environmental thought and policy.[1]

Leopold's early wilderness idea, however, was a highly contingent one, rooted in his Forest Service milieu and tied to the technological and cultural developments of the 1910s and 1920s. This context was crucial in shaping Leopold's perception that by the early 1920s a particular type of preservation should be undertaken in the national forests of the American West. His wilderness idea evolved as he broke his ties with the forests of the Southwest, and, in a different landscape, began to think and write about wilderness in more abstract terms. In fact, after a flurry of writings on wilderness in the mid-1920s, he drifted away briefly as he grappled with the complexities of game management and agricultural land use. Wildness, rather than wilderness, became his major concern during these years. But his wilderness advocacy was reawakened by the New Deal's emergency conservation initiatives, of which he became an incisive critic. Indeed, it was largely this governmental assault on America's remaining wild lands that prompted him to join in founding the Wilderness Society.

Leopold was undoubtedly one of the originators of the idea that led to the Wilderness Act of 1964. But rather than seeing his thought through the lens of that legislative achievement, I examine his wilderness thinking on its own terms. I think we have missed much of what Leopold was saying about wilderness because we have been so interested in tracing the lineage of current thinking. What exactly did Aldo Leopold mean by "wilderness" when he first started using the term, and what was new about the idea? How did wilderness preservation differ from the national park ideal, and why did Leopold feel the need to promote a new ideal so soon after the Park Service's formation? Was his wilderness idea an upshot of the ecological sensibility that he helped to bring to American conservation, or did other factors and concerns shape his advocacy?

Many scholars have dismissed Leopold's early wilderness advocacy as immature because it was not imbued with the ecological outlook that made his later thought so influential. Craig Allin, for instance, concluded that Leopold's "initial motivation towards wilderness preservation was more a matter of practicality than principle," and Susan Flader has stated that "Leopold's initial vision of wilderness was not exactly 'pure' . . . and there is little evidence in his early writing, even in the battery of articles he wrote in 1925 promoting a national wilderness system, of ecological thinking or any rationale beyond recreation." Roderick Nash saw Leopold as "acting with only a vague rationale" in the early 1920s, when he first proposed that the Forest Service preserve wilderness areas.[2] These conclusions are built on the assumption that the intellectual significance of Leopold's wilderness advocacy was in its ecological insights. Later in his career, he did embrace an ecological argument for wilderness, and it is important to appreciate that development. But in focusing on his later thought, and assuming that it was somehow purer, scholars have overlooked many of the reasons why he felt so passionately about wilderness in the first place.

Leopold's early rationale for preserving wilderness was far from vague and unprincipled. Aware of ecological arguments being offered by others, he chose not to employ them. Instead, he justified wilderness preservation with a strong critique of trends in outdoor recreation. His wilderness idea took form in a mold fashioned by the era's recreational boom, and it would retain that shape throughout his life. It was to stem the growth of road building, to control the automobile, and to temper recreational development of America's public lands that Leopold first suggested the need for wilderness preservation. Where John Muir had railed against the producers who sought to transform nature into raw materials, Leopold criticized the consumer trends that distinguished his era.

Forestry and Outdoor Recreation in the Southwest

Aldo Leopold's sensibility was distinctively midwestern, but eastern training and western field experience had a profound impact on his early environmental thought. He was born in 1887 in Burlington, Iowa, a bustling town along the bluffs of the upper Mississippi River. Although agriculture had transformed much of the region, the bottomlands and surrounding hills were rich in wildlife, particularly the waterfowl that migrated through the area each year. With this abundance as a backdrop, and under his father's tutelage, Leopold became an avid hunter and naturalist. In 1904 he headed east for his education, first to the Lawrenceville School in New Jersey for a year, and then to Yale, where he earned a bachelor's degree in 1908 and a master's degree from its new forestry school in 1909.[3] Having absorbed the scientific and civic principles of Progressive conservation, he then moved to the Southwest to join the Forest Service.

Between 1909 and 1913, Leopold and his colleagues worked to cruise, map, and otherwise rationalize the region's remote forests. Although there was little market for public timber, foresters, trained on the assumption of timber scarcity, were eager to begin the practice of sustained-yield silviculture.[4] But foiled agronomic designs marked the first decade of the Forest Service's existence. The national forests contained approximately 20 percent of the nation's standing timber but contributed only about one percent of the nation's yearly cut.[5] The Forest Service had failed to achieve Gifford Pinchot's dream of aggressively regulating private forestry practices, and its own timber resources languished. Visions of waste haunted Forest Service officials. Timber that was "ripe for the ax," to use the language of their annual reports, stood uncut, maturing and dying before it could be put to use. Fires, particularly in inaccessible areas, insects, and diseases stole precious timber, mocking foresters' goals of efficient management.[6]

The trigger itch that foresters felt emanated from the fiscal requirements of national forestry and the vagaries of the market. The national forests were supposed to be self-supporting, with timber revenues paying for their administration. But with high start-up costs and little demand for national forest timber, supervisors found it difficult to pay their bills. Accusations of market tampering further undermined the ability of the Forest Service to support itself. When public foresters did sell their timber, the private sector accused them of lowering prices by glutting an already crowded market. Adequate private supplies and decreasing market demand thwarted efforts to get the timber out of the national forests. The only thing that foresters could do was plan for the day when their timber would be needed. In general, this meant surveying and taking inventory of forest resources, and building an administrative infrastructure of roads, trails, and telephone lines. Lacking the desired fighting orders, Leopold and his colleagues bided their time with activities that effectively opened up the forests of the West. They also turned to a different, more localized notion of their public mission, with neighborly relations as a stand-in for national reform.[7]

The permanent public reservation of lands in the American West has rarely occurred without some expression of local concern. Visions of an activist federal government providing effective stewardship have clashed with the more venerable homesteading tradition, in which public ownership was but a whistle stop on the way to privatization. In the early 1900s, the formation of western national forests also presented more concrete concerns to residents living near or within them. National forests kept a great deal of land off the tax rolls, limiting the revenues of local governments. New national forests also had to absorb numerous existing uses and perceived rights—particularly grazing claims—that predated national forest status. Finally, many westerners saw these public reservations as physical obstacles to the West's economic growth.[8]

Foresters were generally responsive to these concerns. National forests shared revenues with localities in lieu of lost property taxes, and they gave liberal (often free) access to resources such as timber for local consumption. The Forest Service also set up a system of permits to accommodate, with some necessary reforms in practice, preexisting and other requested uses. Without a national timber market to serve, and under pressure to be neighborly, foresters devoted considerable time to local issues during these early years. Leopold's experiences on the Apache National Forest in Arizona and the Carson National Forest in New Mexico mirrored these trends. Before 1915, the West's national forests were little more than regulated local commons.[9]

Recreational trends, spurred by growing use of the automobile, soon challenged this localism by introducing a new and perplexing national mission. In June 1913, while Leopold was back in Burlington recovering from a case of nephritis, John Muir lost his beloved Hetch Hetchy Valley to San Francisco water interests in one of the most scrutinized episodes in conservation history.[10] The Hetch Hetchy decision, which defined the split between preservationists and conservationists that has framed conservation history ever since, was a Pyrrhic victory for the forces that saw nature as only a store of resources. National debate over the proposed reservoir within Yosemite National Park stirred strong preservationist passions in the nation. One result was the creation of the National Park Service in 1916, an important watershed in the official recognition of scenic preservation as a legitimate form of land use.[11] But with a growing national zeal for outdoor recreation, and with Stephen Mather at the helm of the Park Service, park preservation quickly came to mean development and use of a different sort.

In 1914, Leopold returned to a Forest Service on the brink of a confrontation with the outdoor recreation age. After a nine-month stint in the Office of Grazing, he was assigned to oversee the district's new work on fish and game, recreation, and publicity. The position itself heralded a new (though often halting) direction in Forest Service policy. Leopold saw the position as an opportunity to develop a game policy for the national forests of the Southwest, but other recreational duties effectively divided his attention.[12]

In the mid-1920s, the national parks were still few and far from the centers of population. The national forests, on the other hand, were extensive and closer to population centers (particularly with the advent of the Weeks Act of 1911, which appropriated funds for the purchase of eastern national forests). Moreover, national forests allowed hunting, fishing, and other activities that the Park Service either prohibited or strictly regulated. Increased automobile ownership conspired with a growing grid of roads and trails to produce a rapid increase in recreational visitation to national forests. These visitors represented a new user constituency, raising questions about the role of national forests that many foresters were unprepared to answer.

Leopold's first major assignment in his new position was to visit the Grand Canyon, then a National Monument administered by the Forest Service, and report on recreational conditions there. The result was a 1916 report co-authored with Tusayan National Forest Director Don P. Johnston, and with illustrations by noted landscape architect Frank Waugh. The major concern expressed by Leopold and Johnston was that unplanned commercial development threatened to undermine the recreational and scenic values of the area. They cited various abuses: "discourteous treatment by business permittees"; "offensive sights and sounds such as electric advertising signs, megaphones, soliciting, etc."; unsanitary conditions; unfair business practices; views blocked by establishments such as the photographic studio of Emory and Ellsworth Kolb (who had recently begun showing moving pictures of the canyon to augment their trade in prints, postcards, and other simulacra); and dangerous conditions, including the unfortunate rage for rolling stones down into the canyon.[13]

Although the Grand Canyon had been a popular tourist destination for decades, visitation increased rapidly in the mid-teens.[14] The Leopold-Johnston report noted that of the 380,000 visitors since 1900, a total of 100,000 had come in 1915 alone. This influx was largely due to the "See America First" campaigns and other promotional efforts launched to coincide with travel to the Panama-Pacific and Panama-California Expositions in San Francisco and San Diego. One of the highlights of the San Francisco Exposition was a scale model of the Grand Canyon, displayed by the Santa Fe Railroad, which advertised both the scenic values of the Canyon and Santa Fe's role in getting people there.[15] The Santa Fe was not the only party with a commercial interest in the area. Livery services, souvenir shops, campgrounds, and hotels all took advantage of lax Forest Service permit rules to set up shop. More disturbing, however, was the use of mining claims to gain a stranglehold over the South Rim.[16] Forest Service policies that facilitated private commercial development of public lands contributed to the chaotic scene described in the report.

Leopold and Johnston sought to resolve the conflict between scenic appreciation and the provision of services—between the democratic and nationalistic interest in the canyon and local efforts to profit from that interest—through strict regulation, zoning, and the establishment of business standards. They insisted that a spirit of public service rather than private gain should guide commerce on the canyon rim. Although unwilling to move beyond regulation to outright Forest Service control of visitor facilities, both men expressed concern over the commercial development that accompanied increased tourist traffic to the West's federally owned treasures. Commercial tourism had come to the canyon rim as a filial extension of the broader economic culture, providing novel challenges to the Grand Canyon's maintenance as a public space.

The Grand Canyon was an exceptional case for the Forest Service, which was unaccustomed to dealing with such sites or such extensive tourist development. In 1919, the area became a national park, its ad-

ministration transferred to an agency better prepared to handle these issues. But the phenomena that Leopold observed on the canyon rim
manifest themselves in less boisterous ways in the region's other forests.
The holding pattern in which foresters found themselves made the facilitation of recreation seem a sensible interim activity, particularly in
terms of revenue. Recreational developments also fit cleanly within the
Forest Service scheme of allowing permitted access for local use. Few
imagined that national forest recreation would grow to rival timber extraction as a highest use.

Forest Service Recreational Policy and Planning

Two pieces of federal legislation in the mid-1910s raised just such a
specter. The first was the Term Permit Act of March 1915. From its inception in 1905, the Forest Service had allowed the construction of summer cabins and resorts on its lands, though the short-term nature of the
permits and the tacit understanding that resource needs could thwart
their renewal kept development sparse and simple.[17] The Term Permit
Act increased the duration of recreational permits to up to thirty years
with options for renewal, and it thereby encouraged greater capital in-
vestment in private recreational structures. In a sense, the act provided
for recreational homesteading, though without the prospect of gaining
title to the land.

The Term Permit Act opened the way for a rapid increase in the construction of summer homes, municipal and private camps, resorts, and
hotels in the national forests. By the mid-1920s, less than a decade after
the act took effect, the number of permits for summer homes had
quadrupled. National forests near large cities saw the greatest increase.
The Angeles National Forest, north and west of Los Angeles, had 1,329
permits for summer residences and resorts in effect as of June 1920. These
included a 23-acre municipal camp, run by the city of Los Angeles, with
61 cabins for use by the city's residents. Although the Angeles was an exceptional case, term permit developments soon dotted many national
forests. They were made accessible, of course, by the automobile.[18]

Among Leopold's duties as head of recreational policy was the oversight of term permit development. Prior to and immediately following
the Term Permit Act, most cabin and resort development had been hap-
hazard. People located on a site of their choice, applied for a permit, and
built. As the pace of development quickened, however, many of the

most desirable locations became crowded. This was particularly true of streamside and lakefront areas, where vacationers vied for prime spots and where concessionaires were quick to supply services. The need for order drew foresters into the process.

In May 1916, Leopold jotted down some "Notes on the Lake Mary Public Use Area" that offer an idea of what was fairly characteristic of term permit oversight. Lake Mary, ten miles east of Flagstaff, Arizona, had become a popular destination for the city's population. In response to this growing use, Leopold drew up a plan featuring various sorts of access to the lake. He suggested limiting development of residences to the south shore and reserving the north shore for campers who might otherwise get crowded out. There was already a roadhouse at one end of the lake that rented cottages and rowboats, and sold horse feed, gasoline, bait, and foodstuffs. The proprietors also operated a dance hall. Aside from planning cabin areas and campsites, Leopold mentioned the need for a road to make the entire lake accessible, and suggested the wisdom of advertising the lake's charms to the residents of Flagstaff. He seemed to have few concerns about such developments, though he did give campers a nod by reserving areas for them. Moreover, he was careful to mention the continued availability of the area's timber. A passing comment, however, may have alerted foresters to potential incompatibilities between recreation and silviculture. "It goes without saying," Leopold noted, "that no further cutting within sight of the lake should be allowed."[19]

Although several historians have portrayed Forest Service recreational development as a response to an organized and active National Park Service—and in the minds of some top administrators, such an institutional rivalry was a powerful rationale for action—it seems more appropriate to see foresters' new interest in recreation as a reaction to user demand.[20] Most Forest Service officials, particularly those in the field, had little choice but to provide for visitors who were coming of their own volition. The public had learned to take advantage of the many openings that Forest Service policy allowed for recreational use and development. Moreover, recreation provided needed revenue and reaffirmed foresters' roles as guardians of a useful public resource at a time when timber remained underutilized. Existing use patterns, largely brought about by the automobile, encouraged foresters to institute guiding policies for development even though many remained skeptical about embracing a recreational mission.

The beauty of the permit system was that it allowed foresters to treat recreational development as they would any other special use—as a secondary and incidental matter that did not force them to reconsider the priority of silviculture. But the rapid growth of recreation did in fact divert attention from silviculture. Leopold's plea for the trees ringing Lake Mary was but one sign that recreational development and silviculture would clash. Moreover, the term permit system eventually undermined itself by allowing for the monopolization and overdevelopment of some of the best recreational areas in the national forests. To those who saw such access to the national forests as a public good, this trend toward private control was troubling.

A series of Federal Aid Highway Acts also had a pronounced effect. The first one, in 1916, provided $75 million over ten years for road development throughout the United States. Of this $75 million the Forest Service received the disproportionate sum of $10 million, or $1 million a year, and additional acts in 1919 and 1921 increased the funding. The 1921 act distinguished between two types of roads within the national forests: "forest development roads," designed specifically for administrative purposes like fire control, received an additional $5.5 million; "forest highways," designed to augment state and county road systems and connect local communities, received an additional $9.5 million. The Forest Service and the Department of Agriculture viewed these road improvements as a way of increasing their administrative control over forest lands, of reducing the isolation of communities in and around national forests, of linking these communities to an emergent postal system, and of helping localities with small tax bases pay for road improvements. But Good Roads boosters quickly latched on to these initiatives, in part to create tourist infrastructures. This new round of road improvement was one reason that national forest visitation quadrupled between 1917 and 1924.[21]

The interwar era was a time of aggressive road improvement throughout the nation, but in most cases road funds went to improving and coordinating existing systems.[22] An exception was in the national forests, where road building had a transformative impact after 1916. Forest Service roads increased from a few thousand miles in 1916 to almost 90,000 by 1935, with about 20 percent of this mileage in the form of forest highways.[23] One result was a revolution in access for motorists seeking outdoor recreation and prime sites for term permit development. These road improvements, in conjunction with the Term Permit Act, made the mid-1910s a watershed in the recreational development of the national forests.

Although foresters continued to cling to sustained-yield timber production as their guiding ethos, increased recreational activity prompted
official efforts to survey the recreational potential of the national forests.
In 1917 the Forest Service took an unprecedented step by hiring landscape architect Frank Waugh to complete a recreational survey of the national forests. It was the first move in a short-lived affair between
foresters and landscape architects.[24]

Waugh's report, *Recreation Use on the National Forests* (1918), painted a
vivid picture of the ubiquity of recreational activity in the national
forests. He mentioned "the large number of camps" created for automobile tourists, the popular picnic grounds near urban areas, and the "several hundred" communities of summer vacationers. This heavy use led
Waugh to argue that the Forest Service should see recreation as a primary
use, the equal of timber production and watershed protection. This did
not mean a radical reorientation of the Forest Service mission. In only a
few cases, Waugh maintained, would recreational use be the exclusive
use. But this conclusion did imply that the Forest Service ought to assume a direct role in developing its recreational resources, rather than
simply relying on the term permit system. Waugh's other major suggestion was that the Forest Service hire "landscape engineers," professionals
trained in recreational and aesthetic design.[25]

The Forest Service was hesitant to take Waugh's recommendations to
heart. His conclusions meant looking at the forests as a different sort of
resource. Moreover, the Park Service was making noises about recreation
being its responsibility, and many foresters were unwilling either to get
into such a conflict or take on new responsibilities. To some, term permit development and road and trail building only complicated their jobs
by populating the forests with aesthetes who grew queasy at the thought
of logging or other extractive uses. Why, then, should foresters get directly involved in recreational development? Waugh's two major contentions—that foresters treat recreation as a primary use, and that they
hire landscape architects to plan for this use—were thus tough ones for
many foresters to stomach. And as the experience of Arthur Carhart illustrated, the Forest Service, though willing to flirt, resisted committing
to recreational planning and development in a substantial way.

In 1918, Assistant Forester E. A. Sherman decided to accept Waugh's
logic and hire landscape architects, largely because a couple of districts
had expressed an interest in having them on staff. Arthur Carhart, who
had received a landscape architecture degree from Iowa State College in

1916 and had just been discharged from the armed services, was hired and assigned to District Two, which at the time comprised the Rocky Mountain and northern plains states (including Minnesota's Superior National Forest). Carhart reported to the district office in Denver to begin work on March 1, 1919.[26]

Recreational use of Colorado's national forests had increased rapidly as towns and cities on the Front Range grew, and as more plains farmers got automobiles and good roads to drive them on. Carhart's first big project, an extensive recreation plan for the San Isabel National Forest in south-central Colorado, came in response to these developments. In two written plans for the area, Carhart wove a vision of a vast recreational space, with roads and trails, and public facilities such as campgrounds, cabins, and resorts. He even suggested building a trolley system to move people through the forest. His plan was for a national forest thoroughly dedicated to recreation, one developed to provide uniform public access without the commercial intrusions. Like the Leopold and Johnston report on the Grand Canyon, Carhart's plan was strongly rooted in a particular conception of the public good. It provided few opportunities for private development, advertising, or profit, and those commercial facilities that might spring up would be pressured to adhere to strict business standards. Another goal of the plan was to ensure that the "promiscuous spotting of summer homes" did not monopolize the most scenic areas. Carhart was the first person to provide such a comprehensive recreational plan for a national forest, and by relying on a private organization for financial support he was able to complete some of the development.[27]

Like Leopold, Carhart was responsible for supervising developments under the Term Permit Act. In 1919, he visited Trapper's Lake in Colorado's White River National Forest to plat sites for several hundred summer homes. The lake's relative isolation, even today, is a testament to the pervasiveness of the demand for summer residences at the time. While completing the survey, Carhart concluded that littering the lakeshore with summer homes would not only compromise its natural beauty but would restrict access to an important public recreational resource. As a result, he proposed that no summer homes or hotels be developed along the lake's shore. Instead, they would be kept at a distance of at least half a mile, with trails providing access to an unmarred lake. Carl Stahl, Carhart's supervisor, accepted the suggestion. It was this act of preservation that historian Donald Baldwin claimed as the first application of the wilderness idea.[28]

While a half-mile strip around a lake may not conform to more grandiose notions of wilderness, Carhart recognized that the multiplying private goals of recreational users threatened the public nature of Forest Service lands. By pitching his proposal in opposition to monopolistic forms of recreational development, Carhart shared with Leopold what would be a cornerstone of the latter's argument for wilderness preservation. Yet in his various memoranda on Trapper's Lake, Carhart never used the term "wilderness." He was concerned with protecting "natural scenic beauty" from "the presence of man-made structures" and "monopolization."[29] Leopold would propose something far more sweeping than restraining development in the immediate vicinity of scenic areas. Yet he and Carhart did share common concerns about the threats posed to the national forests' public recreational resources by private development.

Despite his accomplishments on the San Isabel National Forest and at Trapper's Lake—accomplishments that were somewhat antithetical given the aggressiveness of the former plan and the restraint that marked the latter—Arthur Carhart grew frustrated with the lack of support for his efforts.[30] As recreational activity swelled in the region's forests, he faced two specific limitations to his effectiveness: a lack of funding and a clash of professional cultures.

The Forest Service had a hard time convincing Congress to provide funding for even the most cursory recreational developments, despite the desperate need for them. In October 1920, Assistant Forester Leon F. Kniepp wrote a circular letter to the Forest Service's District Foresters mentioning the potential inclusion of $50,000 for recreation in Congress's 1922 budget. Kniepp wanted to know if the foresters could use the money, and how they might use it. The responses overwhelmed him. District foresters returned lengthy reports documenting the recreational use of their forests, and all said they were in dire need of money. A. S. Peck's 48-page report on District Two—Carhart's district—was the most impressive. Peck went to considerable trouble to explain why the national forests had come under such heavy recreational use in the years after World War I. He cited six factors in particular: (1) the "War Factor" and the military's idealization of outdoor life, (2) "Disappearing Local Camping Areas," (3) the "Increase of Wealth among Large Bodies of Citizens," (4) "the advertising of railroads, tourist bureaus, and other agencies, including the National Government," (5) "Automobiles," and (6) "Good Roads." Peck said District Two needed $35,000 just to meet basic needs. "We are actually in the position," he wrote, "of having to catch up with the recreation use of the Forests and take care of a demand established and proceeding without direction in spite of us."[31]

The responses from other District Foresters were not as detailed as Peck's, but each expressed an immediate need for funds to improve heavily used autocamping areas, to prevent problems with sanitation and fires, and to plan more term permit developments. Most asked for funding in the vicinity of $15,000 to $20,000 a year, making it clear that the proposed appropriation of $50,000 would be entirely inadequate. Indeed, a number of the District Foresters were embarrassed that, unable to keep up with recreational demands, they had been forced to rely on assistance from municipalities and private groups to provide basic services. But this embarrassment should not be confused with full-fledged enthusiasm for turning the national forests into recreational areas. Most District Foresters made it clear that they simply hoped to control the impacts of existing recreational use, and to make the forests useful for recreation only insofar as it did not interfere with their other priorities.[32]

As Kniepp and District Foresters exchanged letters about potential appropriations, they also broached the subject of, in Kniepp's words, "specifically dedicating parts of Forests to recreational use."[33] In this effort, they were caught between a number of constituencies. On the one hand, there was pressure from citizen groups to preserve scenic spots in the national forests from resource extraction. Oregon's Mount Hood was the most prominent example. On the other hand, the discussion of such designations irked Park Service officials who thought that the protection of scenery was their job. In January 1921, in the midst of this intra-agency discussion, Arthur Carhart and Stephen Mather squared off at a National Conference on State Parks meeting in Des Moines, Iowa, over whether the Forest Service had a right to enter the recreation field.[34] Most foresters, though protective of Forest Service lands, were eager to avoid the appearance of duplicating Park Service efforts, and Kniepp was concerned that specific recreational designations within the national forests would further confuse the public as to the missions of these two federal bureaus. The Forest Service recognized the recreational value of its lands, and they were willing to facilitate recreational use, but they ultimately balked at designating lands specifically for recreation. Moreover, almost all the District Foresters responded negatively to Kniepp's query about hiring "recreational engineers." Most insisted they had sufficient staff and that hiring landscape architects would only drain money away from urgently needed developments. But beneath this pragmatic logic lay a deeper conflict of professional cultures, one between foresters and landscape architects, that festered throughout the interwar period.

This whole recreational discussion became moot in the spring of 1921 when Congress struck the $50,000 recreational appropriation from the Forest Service budget.[35] Without funding, and wary of overstepping their institutional bounds, most foresters were content to continue to issue permits, and, with a little benign guidance, to let recreational users fend for themselves.[36] In the absence of funding and with foresters icy to landscape architecture, Arthur Carhart resigned his position with the Forest Service effective December 31, 1922. A brief but important chapter in the history of recreational planning on the national forests had ended, and a new one began. In a subtle but important way, the refusal of Congress to provide funds for recreational development on the national forests set the stage for Aldo Leopold's case for wilderness.[37]

The Wilderness Idea Emerges

Aldo Leopold briefly left the Forest Service in 1918 to become secretary of the Albuquerque Chamber of Commerce. He saw the position as an opportunity to further game conservation in the region, but he quickly grew disillusioned with the Chamber's boosterism. Indeed, he later identified boosters as accomplices in the destruction of what he came to call wilderness.

Leopold left the Chamber of Commerce in mid-1919 and returned to the Forest Service in the position of Chief of Operations for District Three.[38] Between 1919 and 1923, as Carhart pushed the Forest Service to embrace recreation, Leopold's major task was to oversee the Southwest's national forests. During the course of his routine inspections, he began to dwell on two distinct themes or concerns. The first was grazing and its relation to erosion in the Southwest. Hispanic and Anglo ranchers had settled the area in the 1890s and early 1900s, bringing with them hundreds of thousands of cattle and sheep that transformed the region's grasslands, crowded out game species, and caused serious erosion.[39] By the early 1920s, as the range deteriorated, Leopold was making sophisticated ecological observations about grazing and its environmental impacts in the Southwest.[40] The other prominent theme raised in his inspection reports was recreational use of the region's forests. In certain cases, Leopold made connections between the two, suggesting that the ecological damage from grazing adversely affected recreational values.[41] The case Leopold soon made for preserving wilderness, however, drew very little from these ecological observations.

Ironically, Leopold's first published discussion of wilderness had little to do with the need to preserve it. Indeed, it almost worked at cross-purposes with his later efforts. In "The Popular Wilderness Fallacy," published in 1918, Leopold attempted to counter a major misconception about game conservation: that the decline of wilderness would mean the end of wild game. This notion, he argued, was a destructive fallacy insofar as it painted game conservation as hopeless in the face of civilization's march across the continent. Civilization did not preclude the continued existence of wildlife, and, conversely, the preservation of the nation's remaining wild lands would not alone save wildlife. The article is important in its recognition of the adaptability of wild game to changing environmental conditions and its foreshadowing of Leopold's later writings on game management. But it also suggests, in a roundabout way, that Leopold's interest in wilderness preservation did not arise out of his concern for the preservation of wild game. In fact, despite his avid interest in getting the Forest Service to recognize the value of their lands for game production, Leopold's early wilderness proposals did not focus on game preservation.[42]

Arthur Carhart's efforts were a more important stimulus to Leopold's wilderness thinking. In December 1919, Leopold stopped in at the District Two office to visit with Carhart, and the two spent the day discussing the preservation of Trapper's Lake and how such a policy might be applied elsewhere. Carhart later jotted down some thoughts on preserving areas within the national forests and sent them to Leopold. In his "Memorandum to Mr. Leopold," he again stressed that scenic monopoly through private development threatened public access. One of his major themes was that, by continuing to rely on permits and refusing to get more involved in recreational development, the Forest Service was shirking its public duty. There was a limit, Carhart pointed out, to the number of lakeshores and other scenic areas in the national forests. As recreational use of the forests continued to increase, the Forest Service needed to reserve its "scenic territories" before many of them were similarly developed.[43]

As Leopold read Carhart's memorandum, he must have realized that the two of them, despite their like-mindedness, had some fundamental differences. A forester by training, Leopold was less interested in the landscape architect's scenic notions; though not immune to the charms of scenery, his support for wilderness preservation was rooted in a different aesthetic. As a hunter, he was keenly aware that the dynamic technological duo of the automobile and efficient firearms had made hunting a modern activity and, in combination with road development, had opened even the most remote areas to modern hunters. His love of hunting and his fears about its modernization shaped his percolating wilderness idea. Moreover, while he agreed with Carhart that foresters ought to protect public resources from recreational monopolization, he felt uneasy about the types of public development that Carhart proposed. Such aggressive development delimited the options of those who sought outings in areas free from modern amenities. Finally, Leopold was still dedicated to responsible resource management and use, and he was wary of seeing the Forest Service get too deeply involved in recreational development. For these and other reasons, his early wilderness idea diverged from Carhart's vision of preservation.

Leopold's first explicit statement on the need to preserve wilderness came in a 1921 article entitled "The Wilderness and Its Place in Forest Recreational Policy." He began by making the case, as Waugh and Carhart had before him, that recreational use of the national forests had become, in many places, the highest use to which such lands could be put. This, he continued, raised the question of whether a policy of industrial development should continue to guide all Forest Service actions, or whether the forests should be seen as repositories of other sorts of resources. But Leopold devoted most of the article to making a further distinction. The question was "whether the principle of highest use does not itself demand that representative portions of some forests be preserved as *wilderness*." He continued:

> That such a question actually exists, both in the minds of some foresters and of part of the public, seems to me to be plainly implied in the recent trend of recreational use policies and in the tone of sporting and outdoor magazines. Recreational plans are leaning toward the segregation of certain areas from certain developments, so that having been led into the wilderness, the people may have some wilderness left to enjoy. Sporting magazines are groping toward some logical reconciliation between getting back to nature and preserving a little nature to get back to. Lamentations over this or that favorite vacation ground being "spoiled by tourists" are becoming more and more frequent. Very evi-

dently, we have here the old conflict between preservation and use, long
since an issue with respect to timber, water power, and other purely economic resources, but just now coming to be an issue with respect to
recreation.[44]

Leopold here made a deeper distinction that was crucial to his notion of
wilderness. His first distinction—between recreational and industrial
use—is an easy one to fix on as the main innovation of his argument.
Clearly, he was interested in seeing national forests preserved for recreation, a relatively novel idea at the time. But this was not what Leopold
meant by wilderness. What he meant by wilderness was preservation
from certain forms of recreational development, particularly road building and term permit facilities. The problem that he sought to solve was
that the Forest Service had, in many cases, been too generous to those
seeking modern recreational access. "The argument for wilderness
areas," he concluded, "is premised wholly on highest *recreational* use."[45]
Leopold acknowledged that the seemingly radical aspect of this suggestion lay, in part, in his use of the term "wilderness." "It is quite possible," he warned, "that the serious discussion of this question will seem
a far cry in some unsettled regions, and rank heresy to some minds." To
counter such fears, he provided a definition of wilderness and a sense of
the scope of his suggestion. "By 'wilderness,'" he said, "I mean a continuous stretch of country preserved in its natural state, open to lawful
hunting and fishing, big enough to absorb a two-weeks pack trip, and
kept devoid of roads, artificial trails, cottages, or other works of man." He
admitted, "The majority undoubtedly want all the automobile roads,
summer hotels, graded trails, and other modern conveniences that we
can give them." He was painfully aware of how an agglomeration of consumer preferences could work a new sort of tyranny of the majority: "But
a very substantial minority, I think, want just the opposite." Again, the
minority right Leopold argued for was not simply the right to use the national forests for recreation; indeed, this was already the majority's demand. It was the right to enjoy large stretches of open country that
lacked modern recreational conveniences such as roads, motorized transport, hotels, cottages, and graded trails. The national parks often failed
to meet these criteria, Leopold noted, because they were "being net-
worked with roads and trails as rapidly as possible."[46] To understand
Leopold's wilderness proposal—and Carhart's plan for Trapper's Lake for
that matter—one must move beyond the simple dichotomy of preserva-

tionists versus conservationists and appreciate how thoroughly concerned with recreational development early wilderness advocacy was. Although not all of his contemporaries understood his argument in these terms, Leopold proposed wilderness to save public landscapes from the popularity of the automobile, outdoor recreation, and the improvements that accompanied both.

When it came to resource use in wilderness areas, Leopold made some important distinctions. He ended his 1921 article by nominating an area in New Mexico's Gila National Forest as a prime candidate for wilderness designation. The region was largely undeveloped and unsuitable for agriculture, he noted, and thus the economic impact would be limited. But he stopped short of suggesting that such a policy would prohibit all resource use. A wilderness policy, he noted, meant that the Gila's "timber would remain inaccessible and available only for limited local consumption." The industrial development of timber would have to be sacrificed; indeed, the sanction against roads would guarantee such a result. But Leopold was not necessarily averse to limited timber cutting for local uses. Cattle also grazed the area, but "the cattle ranches would be an asset from the recreational standpoint because of the interest that attached to cattle grazing under frontier conditions."[47] This acceptance of grazing was likely a pragmatic and diplomatic choice given the article's intended audience and the political power ranchers wielded in New Mexico. But it also indicated the extent to which the exclusion of roads and modern developments, not local uses, was paramount in Leopold's vision of wilderness. As his comment about ranching indicated, he hoped to preserve frontier conditions, not ecologically pristine conditions. Indeed, his definition of wilderness seemed closer to the old census definition of the term—an area with less than two people per square mile—than to current notions of untouched areas whose resources are absolutely off limits.

In his formal proposal for giving the Gila area wilderness status, Leopold shifted his emphasis somewhat. In his "Report on the Proposed Wilderness Area," submitted as an appendix to a general inspection report of the Gila National Forest, Leopold suggested two potential names for the preserved area: either the "Gila Wilderness Area" or the "Gila National Hunting Ground." His stated objective was "to preserve at least one place in the Southwest where pack trips shall be the 'dominant play.'" In describing the function of the preserve, he leaned almost exclusively on its importance to wilderness hunters, and he justified the need for such an area by citing how other suitable areas in the region had been opened up to automobile travel.[48] He later recalled that when he arrived in the Southwest there had been six areas in the region with wilderness qualities similar to the Gila's, but five had been divided by roads so rapidly as to pique his interest in seeing at least one area remain roadless.[49]

Leopold's formal wilderness proposal also reflected a policy context in which the recreational ambivalence of foresters and the jealousies of the Park Service were paramount. For instance, although Leopold predicted that the Gila's timber resources would be prohibitively expensive to get at, he sweetened his proposal by suggesting that wilderness preservation might be more akin to traditional Forest Service stewardship than would more aggressive recreational development. "Is it not possible," he asked, "that an untouched reserve of stumpage for a possible national emergency might be a good thing?"[50] Although he prized inviolability more than this comment suggests, his argument resonated with some foresters who saw wilderness designation as a way of stemming road building at a time when silviculture was dormant and when Congress refused to provide adequate funding for meeting existing recreational needs. Moreover, by highlighting hunting, Leopold distinguished his national forest wilderness proposal from national park preservation. His proposed wilderness policy, in other words, provided a way for foresters to simultaneously embrace recreation and avoid many of the pitfalls of doing so.

For Leopold, the essential quality of wilderness was how one traveled and lived within its confines. In a candid memo from the early 1920s in which he expounded on his broader plans for "a wilderness area program" for the entire public domain, he defined wilderness as "an area publicly owned and permanently dedicated to public use for some distinctive form of outdoor recreation or study requiring primitive means of subsistence and travel in a wild environment." Such primitive living required that substantial acreage—in this memo, he suggested at least 250,000 acres—be kept free from roads, automobiles, and other modern conveniences. This was the clearest difference between his brand of preservation and Carhart's. "Timber cutting and grazing under forestry principles," he reiterated, "are allowable (except if within a park) provided no permanent roads accompany them, and provided that precautions to protect recreational values are observed." Leopold's wilderness idea, while it involved sanctions against modern and intensive forms of resource development, was not entirely hostile to the productive use of resources. It was not, in other words, simply a consumer ideal that stood in opposition to producers' activities; it was more subtle than that.[51]

Leopold ended his memo on a broader wilderness program by making a pitch for scientific preservation—for the protection of "relatively small areas sufficient to preserve a sample of virgin conditions" in each state. But at the heart of this pitch was a clear sense of the difference between such preservation and wilderness. In fact, Leopold pointed out that this idea for scientific preservation was not his, but had been "recommended by the American Ecological Society." Although confused about their name, Leopold knew that there was a movement afoot to preserve what ecologists referred to as "natural conditions," a movement fomented by members of the recently founded Ecological Society of America (ESA).[52]

Ecologists and Preservation

If any group had an ecological approach to preservation during the interwar period, it was the Ecological Society of America. The ESA was founded in 1915 at an informal meeting of ecologists in Columbus, Ohio. Most of the era's prominent ecologists—including Victor Shelford, Henry Cowles, Frederick Clements, and W. C. Allee—were involved in putting the organization together. In 1917, the ESA inaugurated a Committee on the Preservation of Natural Conditions and charged it with assembling a list of "typical areas which ought to be preserved in various parts of the country." That effort culminated in the publication of the *Naturalist's Guide to the Americas* in 1926, which, weighing in at over 700 pages, provided a thorough inventory of potential areas for preservation in the United States, Canada, Mexico, and Central America.[53]

Although the ESA initiative arose at the same time as the movement to preserve wilderness areas in the national forests, there were important differences between these two efforts. For the era's ecologists, "natural conditions" existed in those areas where humans had not intervened in or altered the processes of succession and climax.[54] Although ecologists recognized that wholly undisturbed conditions were not easily found, they strove to preserve for scientific study areas of minimal human disturbance. When they spoke of "natural conditions," they meant something akin to laboratory conditions, where natural processes were uncomplicated by human impacts. As such, their notion of preservation was much more ecologically specific than that of most of their nonscientific contemporaries who used similar language. Even Aldo Leopold, an increasingly sophisticated scientific thinker, meant something more general in his 1921 definition of wilderness when he called for preserving areas in their "natural state."[55]

A second important characteristic of the ecologists' model of preservation was their desire to save as many "representative types" of such natural conditions as possible, meaning distinctive plant and animal communities representative of full adaptation to climate, topography, and other determining factors. This was the motivation behind their impressive combing of the North American continent. In a 1920 paper, Victor Shelford, Chair of the Committee on the Preservation of Natural Conditions, listed twenty-five specific "types of conditions" worthy of preservation, including tundra, northern conifer forest, deciduous forest, oak grove savanna, grassland, and a variety of other desert, forest, and wetland forms.[56] The ESA's goal was to preserve examples of each of these representative communities as ecological reference sites.

The final important characteristic of this ecological model of preservation was its scale. While ecologists worked to identify a great variety of worthy areas, the ones they sought to preserve were generally small compared with Leopold's vision of wilderness. While he talked in terms of hundreds of thousands of acres, ecologists usually sought areas in the tens or hundreds of acres. There was some divergence between plant ecologists, who focused on small areas, and animal ecologists, who studied larger habitats. But, by and large, when the ESA pushed for preserves of "natural conditions," they sought small areas.[57]

An article by G. A. Pearson, which appeared in the ESA's journal *Ecology* in 1922, exemplified these differences. In "Preservation of Natural Areas in the National Forests," Pearson, who was also employed by the Forest Service in the Southwest, cited Leopold's published wilderness proposal of a year before, but he clearly distinguished Leopold's advocacy from his desire to preserve "typical areas which would remain as examples of the primeval forest after the mass of virgin stands has disappeared." "To the true lover of the outdoors," Pearson continued, summarizing Leopold's position, "relative inaccessibility by modern modes of travel and the rigid exclusion of all things artificial may seem essential requisites; on the other hand, the scientist who goes into the woods to study rather than to play would welcome a few modern conveniences." Ecologists, Pearson made clear, were less committed to primitive living than they were to pristine natural conditions.[58]

During the interwar years, professional ecologists presented a model for preservation that diverged from Leopold's recreational model in terms of scale, variety, degree of human influence, and accessibility, and that model led to another distinct strain of advocacy. In the 1940s, when the ESA began pulling back on financial support for its Committee for the Preservation of Natural Conditions for fear that advocacy would tarnish its reputation as an organization of disinterested scientists, Shelford and other disgruntled ESA members formed a splinter group, the Ecologists' Union, to carry forward their preservationist agenda. In 1950, the Ecologists' Union changed its name to The Nature Conservancy, a group whose preservationist ideal was recognizably tied to what professional ecologists were doing in the 1920s, and recognizably different from the wilderness ideal eventually embraced by the Wilderness Society.[59] The separate and sometimes conflicting goals of Leopold and the ESA belie a simple correlation between an ecological sensibility and the rise of wilderness advocacy.

Boosterism and Tourism

In proposing wilderness preservation, Aldo Leopold was primarily concerned with the physical impacts of road building and modernized recreational development, but he also raised important questions about the relationship between nature and tourism. Boosters who attempted to tie regional economic development to tourism were particular subjects of his wrath. In a 1923 address to the Ten Dons, an Albuquerque civic society, he reflected on the growth of tourism and its connection to the booster spirit he had known firsthand as a member of the Chamber of Commerce during World War I.

Leopold found several elements of the booster ethos distasteful and destructive. The first was its reliance on advertising as a mode of communication. In the most concrete sense, he saw billboards as an aesthetic blight. Moreover, these signs reflected a relationship between the region's scenic wealth and commercial health. Leopold lambasted the boosters' rejection of "labor, frugality, and natural increase" in favor of "attracting tourists and capital from elsewhere, and extracting appropriations from public treasuries." Such an economic tack was particularly disheartening to Leopold because it occurred at a time when poor land use practices degraded the land. "The typical booster," he lamented, "is entirely out of contact with the most fundamental of his boasted assets, the soil."[60]

"Growing away from the soil," Leopold lectured the Ten Dons, "has spiritual as well as economic consequences which sometimes lead one to doubt whether the booster's hundred per cent Americanism attaches itself to the country, or only to the living which we by hook or crook extract from it."[61] What troubled him most about the booster spirit was that while residents of the Southwest should have been paying attention to the lessons of their land use failures, boosters focused attention on how profitable tourism might be. However poorly they had managed their lands, farmers and ranchers remained dependent on the soil's continuing productivity. But boosters were busy encouraging a relationship with the landscape that had little to do with a sustainable match between people and resources. Boosters urged national parks and scenic highways to attract government and tourist dollars. They advertised the scenic virtues of their surroundings as "tourist bait" rather than catering to the recreational needs of the city's citizens. Although they homed in on a growing national interest in outdoor recreation, they had little interest in the health of the land.

Leopold did not mention wilderness preservation in his speech on the booster spirit, but the connections were clear. Wilderness preservation was, in part, a way of saving the public nature of the region from the designs of its boosters. To see Leopold's wilderness proposal as an effort to save areas for tourism and from the impacts of local users, as some historians have, is to miss both his accommodation of limited local resource use and his strong critique of the forces pushing for tourist development.[62] This discomfort with tourism and recreational boosterism proved a common concern among the founders of the Wilderness Society.

The Public Case for Wilderness

In 1924, with his wilderness proposal for the Gila National Forest inching its way toward fruition, Aldo Leopold accepted a position with the Forest Service's Forest Products Laboratory in Madison, Wisconsin. He left the open spaces of the Southwest to return to a landscape less spectacular, more productive, and much more thoroughly transformed by human labor. Eventually, he would shift his intellectual focus to the environmental problems of his home region. Yet Leopold's four-year tenure at the Forest Products Lab (1924–28) coincided with a period in which the theme of wilderness dominated his writings. The mundane pursuit of wood waste may have bent his mind to such thoughts, but he was also spurred on by the political opportunities of the mid-1920s. Dismayed by the failure of the first meeting of the National Conference on Outdoor Recreation to address the question of wilderness preservation, Leopold wrote a series of articles in 1924 and 1925 (he produced six of them, five of which were published) aimed at garnering public support for a national wilderness policy. His speech at the second NCOR meeting in 1926 was part of this effort.

In these wilderness articles, Leopold continued to emphasize road building, recreational development, and boosterism as threats to wilderness. In response to charges that his ideas were elitist, he refined an argument for wilderness preservation that hinged on both minority rights and the rights of those of lesser means to enjoy the wilderness. Finally, he made a spiritual case for wilderness as an environment of preserved mystery in an otherwise explored and charted world. Together, Leopold's essays of the mid-1920s offer a thorough look at his early wilderness thought.

The most evocative of these essays was not printed. The *Yale Review* rejected his manuscript "The River of the Mother of God," but the piece deserves mention because it explored an important tension at the heart of the wilderness idea. Leopold began the essay by lamenting the incorporation of El Rio Madre de Dios (the River of the Mother of God, which flowed out of the Andes and disappeared into the Amazon basin) into the modern world. For years this particular river had haunted him as a symbol of "the Unknown Places of the Earth" and "the Conquest that has reduced those Unknown Places, one by one, until now there are none left."[63] Like the Gila, it remained a blank spot on the map, and this lack of knowledge about the area was its greatest appeal. Leopold had to admit that his interest in wilderness grew out of the same romantic fascination with mystery that had driven the exploration and conquest of such areas. Wilderness was a construct that necessarily existed in relation to these phenomena; in a culture without a tradition of exploration and conquest, it would have made little sense. Yet his call to preserve wilderness was deeply critical of the destructive results of exploration—results that he portrayed in this essay in epistemic terms. Wilderness preservation involved much more than a celebration of the explorer's urge, offered as the "Second Great Age of Discovery" waned; it was a gesture of self-control, an admission that the explorer's urge ultimately undermined the values that drove it.

Leopold believed that the urge to explore was a basic one, and that

there was a relationship between declining opportunities for true exploration and the rise of a strong national interest in outdoor recreation. "Is
it to be expected that [the unknown] shall be lost from human experience," Leopold asked, "without something likewise being lost from
human character?"[64] Like many of his generation, he thought such a
psychic loss was imminent. His hope was that wilderness preservation
would extend opportunities to exercise the explorer's urge without allowing destructiveness to follow in its wake. In promoting wilderness
preservation, Leopold tried to have it both ways—to preserve the romance of exploration while jettisoning the destructive logic that came
with it. The result, as he understood only too well, was a peculiarly modern amalgam.

Leopold ended "The River of the Mother of God" on more familiar terrain, with a diatribe against "the Great God Motor" and "that Franken-
stein which our boosters have builded, the 'Good Roads Movement.'"
The movement to open up wild areas by roads, Leopold wrote:

> has been boosted until it resembles a gold-rush, with about the same re-
> gard for ethics and good craftsmanship. The spilled treasures of Nature
> and of the Government seem to incite about the same kind of stam-
> pede in the human mind. In this case, the yellow lure is the Motor
> Tourist. . . . [W]e offer our groves and our greenswards for him to camp
> upon, and he litters them with cans and with rubbish. We hand him
> our wild life and our wild flowers, and humbly continue the gesture
> after there are none left to hand. But of all the foolish offerings roads
> are to him the most pleasing sacrifice. . . . And of all the foolish roads,
> the most pleasing is the one that "opens up" some last vestige of virgin
> wilderness. With the unholy zeal of fanatics we hunt them out and pile
> them upon his altar, while from the throats of a thousand luncheon
> clubs and Chambers of Commerce and Greater Gopher Prairie Associa-
> tions rises the solemn chant "There is No God but Gasoline and Motor
> is His Prophet."[65]

The allusion to Gopher Prairie made clear the influence of Sinclair Lewis
and his critical portrait of Middle American values. Indeed, Leopold frequently invoked Lewis's most famous protagonist, George Babbitt, when
lambasting the spiritual hollowness of 1920s boosterism and consumerism.[66] It was precisely this spiritual void, he thought, that drove
the opening up of the nation's remaining wilderness.

In March 1925, the editors of *Sunset Magazine* published Leopold's

"Conserving the Covered Wagon," perhaps his most explicit case for wilderness as a form of historic preservation. He juxtaposed the covered wagon and the automobile as symbols of the traditional past and the technological present, representing a historical disjuncture between the modern and the primitive that was pivotal to his conception of wilderness. He recognized the importance of the automobile and roads in restoring, for "millions of city dwellers," contact "with the land and with nature." He was not against all recreational road building, but like the forces that drove pioneering, "good roads mania" and "unthinking Boosterism" threatened to undermine their own resource base by pushing roads into every remaining roadless area.[67] Preserving these historical environments that he called "covered wagon wildernesses" would offer a more intimate way of knowing nature. Writing for a magazine noted for promoting Western tourism, Leopold urged *Sunset*'s readers to protect the wilderness from their own modern urges.[68]

If Leopold was after a pristine quality in wilderness, it was more historical than ecological. But while his rhetoric romanticized primitive living, it was not a facile celebration of the American frontier experience. Despite his own Turnerian proclivities (for a brief period he lived two doors down from Turner in Madison), he understood the difference between the grim realities of pioneering and the simulated modern adventure of wilderness recreation; he realized that he was invoking a mythic past to support a modern policy. But he was also adamant that primitive forms of travel and living were better ways of knowing nature through leisure than were motorized forms. Americans, he insisted, needed to understand both the positive and negative aspects of pioneering. "If we think we are going to learn by cruising around the mountains in a Ford," he wrote, "we are largely deceiving ourselves."[69]

Leopold also grounded his argument for wilderness preservation in a call for democratic access. Look at the outdoor magazines, he implored his readers in "Conserving the Covered Wagon," and see where wilderness recreation could be had and by whom. The world's remaining wilderness was exotic and distant, and only those who could afford to pay thousands of dollars for travel and guides had access to it. "It is the opportunity, not the desire, on which the well-to-do are coming to have a monopoly," Leopold insisted. Saving some wilderness in each state would make the wilderness experience more accessible to those who desired it.[70]

Leopold worked hard, in "Conserving the Covered Wagon" and later writings, to make a case for wilderness as a democratic form of land use, and there is no reason to doubt his sincerity. Certainly he never intended to use wilderness designation as a way of keeping people out of a particular area. He ardently believed in making wilderness areas as accessible to as many people as possible, largely by preserving more of them nearer centers of population. But he implored his readers not to confuse democratic access with mechanized access. Indeed, he insisted that access itself was a qualitative proposition. Building a road into an undeveloped area changed its very nature. Leopold tried to get his readers to think much more critically about what such access meant. Nonetheless, if he was going to make a truly compelling case for wilderness as a democratic ideal, as he hoped he could, he had to find a constituency for wilderness. His wilderness articles of the mid-1920s were aimed at achieving such support. But ironically, in the absence of vocal popular support for wilderness preservation and given the growing popularity of motorized recreation, Leopold consistently had to fall back on an argument for minority rights. In his thinking, as in the thought of a number of the other founders, there would be a nagging tension between wilderness as a democratic ideal and wilderness as a minority preference, a tension that he thought was the product of the conflation of democracy and consumer choice.

In "A Plea for Wilderness Hunting Grounds," which appeared in *Outdoor Life*, Leopold preached a similar message to hunters. With the rapid proliferation of roads, he wrote, the opportunities for hunting in "a wild roadless area" were diminishing. "I am trying to make it clear," he insisted, "that a wilderness hunting trip is by way of becoming a rich man's privilege, whereas it has always been a poor man's right." Preserving wilderness would help reverse that trend. Leopold concluded this piece, as he did the others, by urging his readers to write to the President's Committee on Outdoor Recreation to encourage the NCOR to adopt wilderness preservation as part of any national recreational policy.[71]

The wilderness pieces Leopold wrote in 1925 varied with their audiences. He waxed poetic for the literary *Yale Review;* he reminisced about the nation's pioneer past for the readers of *Sunset Magazine;* he celebrated wilderness hunting in *Outdoor Life*. In "Wilderness as a Form of Land Use," Leopold continued this trend, describing his broad policy vision for the land managers who read the *Journal of Land and Public Utility Economics*. Wilderness could be part of a "balanced land system." Economic necessity may have doomed much of the continent's wilderness, but intelligent land planning dictated that a few remnants be saved. The remaining de facto wilderness areas, Leopold pointed out, were "disappearing not so much by reason of economic need, as by extension of motor roads." "Generally speaking," he continued, "it is not timber, and certainly not agriculture which is causing the decimation of wilderness areas, but rather the desire to attract tourists." It was the social impulse to return to nature, not the economic drive to transform it, that threatened to obliterate what wilderness was left, and behind this social impulse was a federal government willing to subsidize road building and recreational development without adequate attention to planning. The federal government had a terrific opportunity, Leopold wrote, to preserve portions of the remaining public domain as wilderness, if only the land managers could distinguish between it and other types of recreational development.[72]

In "The Last Stand of the Wilderness," published in *American Forests and Forest Life,* Leopold addressed his forestry colleagues, urging them to expand the effort begun in the Gila. A few decades earlier, he wrote, foresters had responded to the potential exhaustion of the nation's forests. By the 1920s, "the next resource, the exhaustion of which is due for 'discovery,' is the wilderness." He urged his forestry colleagues to treat the disappearance of roadless areas with the same sense of responsibility and public service as they had the specter of timber famine. And, again, he made it clear that wilderness designation would not necessarily preclude silviculture. In response to the rhetorical question of whether wilderness designation allowed for continued cutting, Leopold wrote: "If the conditions are such that cuttings would leave motor roads in their wake, I would say 'no.' But in the Lake States, much logging can be done over the lakes without any trunk roads, so it seems to me possible, by skillful planning, permanently to use much of the remaining wild country for both wilderness recreation and timber production without large sacrifice of either use."[73] As with grazing on the Gila, timber cutting in places such as Minnesota's Superior National Forest could coexist with wilderness because it did not violate the central tenet of excluding roads.

In the wake of this flurry of wilderness writings, critics began to take notice of Leopold's advocacy. "The Last Stand of the Wilderness" elicited a rebuttal from a forester named Howard R. Flint. Flint's reply, "Wasted Wilderness," appeared in *American Forests and Forest Life* in 1926, complete with a brief last word by Leopold. Flint spent the lion's share of his response outlining the case of 30,000 acres of dead lodgepole pine in

Yosemite National Park. The culprit was insect infestation, and the result, as Flint's title indicated, was "waste." Flint charged that, by preserving forests as scenery and ignoring both their life cycles and their resource potential, the Park Service's hands-off policy had wasted resources and degraded scenery. The dead stand of pine, Flint warned, stood as an example to the "sentimentalist" that preservation was not timeless. "Thirty thousand acres of dead trees in one patch is quite a sizable monument to—to what?" Flint asked. "Shall we call it the wilderness idea, sentiment, or thoughtless waste?"[74]

Flint's argument was a telling one. That he chose to focus on a grove of trees in a national park suggested that he saw wilderness preservation and park preservation as synonymous; he was slow to pick up on the recreational concerns that made Leopold's wilderness idea unique. That said, however, Flint understood that wilderness preservation was going to preclude most timber cutting. But his characterization of wilderness advocates as sentimentalists who saw forests as timeless scenery rather than dynamic natural communities was off base, as Leopold made clear in his rebuttal. Flint, however, saved the real bombshell for the end of his article. "Perhaps if one closely analyzes the arguments of the true 'wilderness' advocate," he posited, "it will become apparent that it is not roads but people he objects to. Perhaps he wants the 'wilderness' to himself and the elect few, and objects to roads because they inevitably bring other people."[75] There was a fine line between minority rights and elitism, and Flint forced Leopold to tread it.

Leopold responded with a couple of points. First, he tackled Flint's argument that preservationist sentiment was out of touch with basic forestry principles. He sympathized with Flint's concerns about sentimental views of timeless nature, and he admitted that some might "support the wilderness idea without really understanding it"—without recognizing that wilderness preservation was about primitive recreation, not scenic preservation. But he also felt that wilderness preservation could correct the sentimental and scenic misapprehensions that concerned Flint. Decay itself was a potent lesson for sentimental nature lovers; "they must learn," Leopold wrote, "that even untouched nature always harvests, always destroys, and always kills." But Leopold primarily wanted to emphasize the recreational need for wilderness, given the proliferation of roads: "For every square mile of wild country permanently dedicated to wilderness, there have certainly been hundreds of square miles of wild country permanently motorized by the construction of new roads." Road building and "the ultimate universal motorization of the forests" posed the most notable threat to wilderness, and he wanted to make sure Flint understood that.[76]

Leopold countered Flint's charge of elitism by making a case for the experience of *"facing nature alone,* without any hotels, guides, motors, roads, or other flunkeys or reinforcements." In response to the charge that he sought to exclude the masses, he recounted an incident that occurred in "Canada's blessed remnant of wilderness canoe country." He and his son were on a portage when they encountered a party of five guides laden with gear and goods, followed by five tourists carrying only cameras and fishing rods. Leopold objected to these people not because they invaded his wilderness, but because *"they had bought their way,* instead of *working their way"* in. "Now it doesn't matter," Leopold concluded, "whether people buy their way into the wilderness by building roads, or riding in motors, or hiring guides. The result is the same—it spoils the whole flavor for those who are trying to use the same area for a wilderness trip."[77] As Leopold's equation of "building roads" and "riding in motors" with buying one's way in suggests, wilderness recreation was for him a low-consumption activity. It was certainly true that getting to wilderness areas would involve some consumption, and more than likely an automobile. This meant that the wilderness experience would remain out of the orbit of many. But if Americans of modest means could not get to wilderness areas, they would not be able to get to those places even if roads were built through them. Thus wilderness recreation was no more elitist than motorized recreation. There was, however, a widespread belief that wilderness areas would cater only to those who could afford guides and other expensive trimmings. Unfortunately for Leopold, some of the most ardent supporters of wilderness preservation were the dude ranchers and wilderness guides who relied on such a wealthy clientele.[78] Leopold offered the above vignette in an effort to counter that impression, and to emphasize that he found the resort to wealth as a means of access a debasement of the wilderness experience.

But Flint had suggested a larger dilemma. The wilderness experience, as Leopold described it, was premised on solitude. Although Flint was wrong to conclude that Leopold wanted to keep a certain class of people out, he was right that Leopold wanted to get away from people as well as automobiles and roads. A wilderness area would not accommodate as many people as, say, a developed national park of similar size. Thus, at a time when only a small segment of the population was interested in wilderness recreation, it was logical to assume that there were exclusive intentions behind the wilderness policy. Unfortunately, all Leopold had to fall back on was an argument for minority rights that effectively reinforced the elitist image of wilderness advocacy that he fought so hard to dismantle.

Toward a Wilderness Policy

While Leopold defended wilderness preservation against its critics, his campaign was making some headway in various policy arenas. *Recreation Resources of Federal Lands,* one of the final reports of the NCOR, quoted Leopold extensively in urging a national wilderness preservation program. The Forest Service also moved toward a more definite policy of wilderness designation. Several new wilderness areas were added to the Forest Service system in the 1920s, though all were created at the district level and existed at the discretion of supervising foresters. Nonetheless, an important change had occurred. Discussions that had focused in the early 1920s on whether foresters should be directly involved with recreational development in the national forests had shifted to considering whether foresters ought to exclude such development as a matter of preservationist policy. Although some foresters perceived wilderness preservation as elitist and a locking up of resources, to the extent that it excluded roads, recreational developments, and the administrative headaches that went with them, wilderness may have seemed the lesser of two evils.[79]

In William B. Greeley, head of the Forest Service from 1920 to 1928, and Assistant Forester Leon F. Kniepp, Leopold found two powerful forestry officials concerned about recreation and willing to explore the wilderness idea as a national policy initiative. In the October 18, 1926, edition of the *Service Bulletin,* an in-house Forest Service publication, Greeley made an extensive statement on the possibility of preserving "Wilderness Recreation Areas."[80] Greeley's major concern was that wilderness designations would mean "a perpetual exclusion of economic uses," and, though he appreciated Leopold's recreational goals, he worried that such a policy might compromise "the industrial and community relationships of the National Forests." Foresters had already altered traditional forestry techniques and priorities to provide for recreation by restricting grazing and cutting in areas of scenic import such as Trapper's Lake. But wilderness differed from such cosmetic measures. "A 'wilderness,'" Greeley wrote, "means something big." By and large, he was uncomfortable with setting aside sizable areas whose resources would be off limits to future development.

But Greeley was willing to compromise. He suggested that the Forest Service adopt a wilderness policy that addressed Leopold's immediate concerns and left areas open to future resource use. He outlined a four-stage process. First, district foresters would locate areas that were roadless and offered unique opportunities for wilderness recreation. Second, once located, such areas would be protected from road building and term permit development. Third, wilderness recreation would be recognized as the dominant use in such areas, and "the usual protection will be afforded camp grounds, forage required for saddle and pack stock, spots of special beauty, and the like." In other words, logging and grazing would be prohibited in key locations within wilderness areas but could continue in a limited way in other locations provided they did not require roads. Finally, such areas might in time be opened up to more extensive exploitation if the nation needed their "economic resources." Greeley's policy preserved areas only from recreational and administrative developments.

In putting a halt to recreational developments, Greeley's proposal gave Leopold much of what he wanted. In most areas to be designated as wilderness, Leopold also hoped for sanctions against future intensive resource development; he hoped, in other words, that wilderness designations would be permanent. But there is little evidence that he was concerned over the impermanence of the Greeley policy. Perhaps he thought the possibility of future resource use was moot since most of the eligible areas had few economic resources or were so inaccessible as to make their development prohibitively expensive. Moreover, because the resources in other areas of the national forests were still largely undeveloped, Leopold may have thought that the need to develop such areas in the future was, at best, a distant possibility, particularly if sustainable forest management proved a success. After all, the promise of sustainable forestry was that a limited area of forest land, if properly managed, could provide for the nation's timber needs. What Greeley and other foresters feared was that permanently preserved wilderness areas would become equivalent to national parks, which might work to remove them from Forest Service control. And Leopold, though he wanted to restrict the behavior of foresters and resource users within the national forests, still saw wilderness as a form of preservation uniquely adapted to Forest Service administration.

Unlike the preservationist position that emerged from battles over national park integrity, Greeley's wilderness policy, built as it was upon Leopold's advocacy, targeted a different set of invasions. "The policy," Greeley concluded, "boils down to outlining areas where the Service will build no roads and issue no permits. The methods of administration in other respects will doubtless vary between different areas, with the general conception that the wilderness program does not require exclusion of economic uses and contemplates dealing with them in the most common sense way under each set of local conditions."[81] For those who failed to appreciate the differences between wilderness and national park preservation, it seemed that the Forest Service was offering a cynical proposal. To national parks advocates, for whom preservation meant the prohibition of extractive uses, it seemed that Greeley's wilderness policy proposal was just a stay of execution. Similarly, Leopold's acceptance of limited grazing and logging in wilderness areas suggested to some contemporaries, as it did to later scholars, that he had yet to fully appreciate the value of wild land preservation. But to see Greeley's wilderness policy as cynical, or Leopold's wilderness idea as half-baked, is to miss much of what they were after. To the extent that Forest Service wilderness policy thwarted roads and other tourist-oriented development, Leopold saw it as a success.

In the midst of this Forest Service policy discussion, the elitist criticism reemerged. In 1928, two months after a special "wilderness number" of the *Service Bulletin* had appeared, Manly Thompson, a forester from District Four, responded with "A Call from the Wilds." "What makes the wilderness wild?" Thompson asked. "Exclusion of the hoi polloi," he answered. "How can we exclude said hoi polloi?" he continued. "Keep the wilderness inaccessible."[82] Thompson chastised the Forest Service for taking seriously the demands of a minority that seemed intent on excluding the majority, and he urged the Service to drop the policy from consideration. He countered the cries of wilderness advocates with the utilitarian credo of "the greatest good for the greatest number." De facto wilderness, Thompson said, was fine while it lasted, but once the public demanded other uses of these areas, the wilderness would have to yield to development, whether industrial or recreational.

Leopold was moved to respond. Although not his most eloquent rebuttal, "Mr. Thompson's Wilderness" did effectively reiterate that a wilderness policy was in no way meant to keep a particular class of people out. Rather, it was to preserve a type of experience that would be lost if the remaining public wild lands were motorized and modernized. But Thompson had again raised the dilemma posed by arguing for a supposedly democratic policy from a minority rights position.

Another response to Thompson's article came from a young forester named Bob Marshall. With "The Wilderness as Minority Right," Marshall forcefully entered the debate on the side of preservation.[83] Although Marshall and Leopold had not met, their common case against Thompson represented an important confluence of wilderness thinking and, as it turned out, a temporary passing of the torch of advocacy. In the coming years, Marshall too worked hard to show that protecting the minority rights inherent in the wilderness was a democratic thing to do.[84]

In the summer of 1929, the Forest Service debate over wilderness preservation culminated in Regulation L-20, which formalized the policy outlined in Greeley's 1926 *Service Bulletin* article, though only after incorporating some of the concerns of skeptical foresters. The result was a fuzzy regulation that rejected any "hard and fast rules, principles or standards."[85] Nonetheless, the Forest Service had its first formal system for preserving wilderness areas. Regulation L-20 required that district foresters identify potential areas for preservation and file a report with the chief of the Forest Service, with plans for how each area would be administered. Although roads and recreational permits were strongly discouraged, the ultimate decision on how undeveloped an area would be remained with the district forester. Roads for fire prevention were allowed, though the regulation encouraged foresters to be mindful of "primitive values." Finally, the establishment of an area did "not operate to withdraw timber, forage or water resources from industrial use, since the utilization of such resources, if properly regulated, will not be incompatible with the purposes for which such an area is designated." In deference to ecologists, L-20 also suggested that foresters set aside "research reserves" for scientific study.[86] In this sense, L-20 represented the coming together of Leopold's recreational advocacy and the ecological advocacy of the ESA's Committee on the Preservation of National Conditions.

In writing up Regulation L-20, Leon Kniepp changed the name of the preserves from "wilderness areas" to "primitive areas." In a separate article addressing nomenclature, Kniepp insisted that "wilderness" was a misnomer since many of the areas had been "prospected, grazed, logged, or otherwise occupied or utilized for a half century, threaded with trails and telephone lines, bounded by highways, scrutinized daily during the fire season by lookouts and now transversed frequently by airplanes." Because he equated "wilderness" with terms like "virgin," "pristine," "primeval," or "unexplored," Kniepp wanted to make clear that these areas were not always pristine and would not necessarily be administered as such.[87] "Primitive," on the other hand, implied "conditions characteristic of earlier stages of the Nation's growth," echoing Leopold's historical definition. Such semantic gymnastics were partly an indication of institutional rivalries, as the Park Service also fiddled with their designations and terms, but they also reflected broader confusions about what words like "wilderness" and "primitive" meant. Given the public's proven ability to reshape land use priorities on public lands, clarity was important. Regulation L-20 served as the Forest Service's formal mechanism for wilderness designation for the next decade.

That same year, developments in the Gila National Forest drew Leopold's attention west again. There were reports of a threatened deer irruption in the Gila; a similar irruption had occurred a few years earlier on the Kaibab plateau, north of the Grand Canyon.[88] Leopold's response to the news indicated his growing understanding of the importance of predators in keeping game populations in check. In particular, he argued that the disproportionate ratio of bucks to does might have been related to the elimination of larger predators and, subsequently, the better fortunes of smaller predators. He counseled managers to concentrate their predator control on coyotes and leave the mountain lions alone for a while.[89]

A Forest Service committee, put together to pursue a course of action on the Gila, proposed a number of solutions. One was, as Leopold had recommended, to ease up on mountain lion control. Another was to allow a greater take of deer by hunters. But most disturbing to Leopold was the committee's recommendation that a road be built, cleaving off a considerable portion of the wilderness area, to provide access for hunters. By 1931, the road had been built.[90] Almost two decades later, in a moving if not entirely accurate recollection, Leopold rued his inability to see the value of predators in maintaining healthy wildlife populations. "Here," Leopold lamented, "my sin against the wolves caught up with me":

The Forest Service, in the name of range conservation, ordered the construction of a new road splitting my wilderness in two, so that hunters might have access to the top-heavy herd. I was helpless, and so was the

Wilderness Society. I was hoist on my own petard. . . . Ironically enough, this same sequence of proclaiming a wilderness, erasing the predators to increase the game, and then erasing the wilderness to harvest the game, is still being repeated in state after hapless state.[91]

From the late 1920s on, as Leopold reconsidered the role of predators as game managers, he included their presence as a facet of wilderness. Yet, as the comment above suggests, it was not the absence of predators that had undermined the Gila's wilderness qualities so much as the presence of a road. The elimination of predators had merely precipitated the Gila's violation. Although a greater ecological understanding of the role of predators enriched Leopold's understanding of wilderness, it did not re-define it. As for the Wilderness Society, it was not founded until 1935, well after the Gila crisis.

From Wilderness to Wildness and Back Again

From the late 1920s until his death, Aldo Leopold made his living as the nation's foremost teacher and practitioner of game management. In 1928, he left his job at the Forest Products Lab and a twenty-year career with the Forest Service. After a few tough years working on commis-sioned game surveys and weathering the early years of the Depression, he landed a position as professor of game management at the University of Wisconsin in 1933, the first such position in the country.[92] His writ-ings on wilderness tailed off, as he devoted his intellectual energy to game and the problems of agricultural land use. It was not until the early days of the New Deal that he rejoined the wilderness debate. When he did, he came armed with a more developed conservation aesthetic.

In formulating the bases for the new science of game management (soon to become wildlife management), Leopold reshaped his conserva-tion philosophy in important ways. Although there was considerable thematic continuity between his old and new professions, there were also some important shifts in his thinking that deserve closer atten-tion—not because they were related directly to his wilderness idea but because they illustrate how he saw wilderness fitting into the broader landscape of conservation.

Leopold had been raised professionally within a public lands institu-tion, but his game management duties required him to shift his sights to the privately owned landscape as a potential producer of wild game. This was the biggest difference between his careers in forestry and game management, and it was a contrast that had a profound effect on his thinking. Moreover, his particular take on the nation's game problems led him to emphasize individual responsibility over governmental activism. While many of the era's most outspoken protectionists identified over-hunting as the primary culprit in game depletion, Leopold focused on the importance of habitat conservation and management in maintaining game populations.[93] He stressed factors such as food supply and cover, and lamented the impacts on game of intensified and mechanized American agriculture.[94] The government could restrict hunting, he suggested, but until attention was given to habitat needs the fate of American wildlife would not improve. He also emphasized the unique opportunity Americans had to forge a democratic game policy. By the late 1920s and early 1930s, he had turned his professional attention from preserving wilderness on public lands to propagating wildness on private lands.

Leopold's text, *Game Management* (1933), had a tremendous influence on his newly chosen field. A strongly ecological work, it was also a book about recreational aesthetics. As his career progressed, ecology and recreational aesthetics merged into a singular view of the land. "Game management," he wrote in the very first sentence of the book, "is the art of making land produce sustained annual crops of wild game for recreational use." His sophisticated approach to the subject came packaged in that deceptively concise sentence. Most noticeable was his invocation of an agronomic model, the idea that game was a "sustained annual crop," an important thematic continuity from his forestry days. But equally important was the suggestion that game management was about producing "wild game" for "recreational use."[95]

Although for the most part a technical treatise, *Game Management* was infused with a recreational aesthetic of "wildness." Leopold insisted that the "recreational value of a head of game is inversely proportional to the artificiality of its origin, and hence in a broad way to the intensiveness of the system of game management that produced it."[96] The agronomic model, in other words, had the potential to undermine the recreational value of game. Recognizing that without management game would continue to decline, he advocated an *art* of management—"controlled wild culture"—whereby farmers and other landowners could enhance the land's ability to support game populations without substantially interfering with the autonomy of wildlife.[97] To achieve such a result required a subtle understanding of habitat needs, and of the fine line between the wild and the managed.

In *Game Management,* Leopold identified four types of game: farm game, forest and range game, wilderness game, and migratory game. "Wilderness game," he explained, "consists of species harmful to or harmed by economic land uses, and therefore suitable for preservation only in special game reservations, or in public wilderness areas."[98] But in recognizing a more sophisticated wildlife function for wilderness, he realized that wilderness was a small part of the solution to declining game populations. Wilderness was not enough. Large mammals, including predators, might require wilderness conditions to survive, and as such wilderness preservation was a crucial protective measure, but most wildlife required a different form of restraint on the part of humans—a restraint memorialized in smaller pieces of undeveloped land.

In moving his campaign for wildness out of the wilderness, Leopold continued to stress certain themes. In the same way that his wilderness policy sought to save a few large undeveloped areas from the automobile, improved roads, and booster visions of tourist and government dollars, so Leopold hoped to persuade farmers and other landowners to retain undeveloped portions of their properties for the sake of wildlife. Those who planted crops from fencerow to fencerow, he implied, were no different from those who pushed roads into every undeveloped area. Those small habitats on the margins of fields were microcosms of what he hoped to save in the Gila and other wilderness areas; they represented limits on growth, refuges for noneconomic values, havens for the natural beauty that gave life meaning. Both sorts of preservation stood in opposition to the tendency to maximize profit and seek "salvation by machinery."[99]

In his writings on game management, Leopold also began to stress "the public interest in private land," a concept central to his conservation thought during the 1930s. This public interest, according to Leopold, was particularly clear in the American system of game ownership. Unlike much of Europe, where private landowners had property rights in game, in America wild game was the property of the state. Thus private landowners were guardians of a public resource; to the extent that they protected game and its habitat, they served the public interest. Unfortunately, there were few incentives to provide such protection.

Leopold considered a number of potential solutions to this problem. One was to switch to the European system of individual property rights in game. He rejected this approach because he thought it would restrict access to game and commercialize recreational hunting. Another option was government ownership of game-producing lands. He encouraged this path wherever possible, but stressed its limitations; the government could own only a small portion of the land useful for raising game. The best solution was to develop a system of incentives to encourage farmers, the nation's major landowners, to protect publicly owned game on private lands. Farmers needed to be encouraged, by the state as well as sportsmen and conservationists, to manage and protect wildlife. Rather than physically separating the propagation of public game from the private landscape, Leopold argued that a sound game policy would seek to integrate these categories. He saw in wildness a distinctly public virtue that ought to be protected even on private land. Indeed, he would later make a stirring case that such was the ethical thing to do.[100]

During the early 1930s, there emerged in Leopold's thought a growing dialectic between wilderness and wildness. His commitment to wilderness helped him to see what was missing from the nation's farms and other intensively managed landscapes. In turn, his efforts to get the nation's working landscapes to produce an aesthetic crop as well as an economic one gave him a deeper appreciation of the ways that Americans' recreational connections with nature had become as skewed as their working relationships. In their hurry to maximize production, particularly through mechanization, Americans were draining wetlands, clearing woodlots, plowing through hedgerows, and otherwise eliminating the agricultural landscape's wild margins. To compensate for that aesthetic loss, Americans built roads into wilderness areas to put themselves back in touch with the very forces they had driven from the worked landscape. Both processes were part of the same aesthetic atrophying, Leopold thought. Conservation, he came to believe, was the art of producing and perpetuating wildness, both amid and beyond the edges of the worked landscape. His career in game management led him from wilderness to wildness and back again.[101]

Leopold's return to the Midwest, and his immersion in game management and agricultural conservation, developed in him a set of intellectual tools that proved crucial in making a stronger distinction between the wilderness experience and motorized recreation. As he struggled with questions of land health, and as he delved more deeply into ecology, he also spent more time thinking about aesthetics—about what was beautiful in the natural world, and how that beauty could be part of a landscape otherwise devoted to utility. He came to believe that Americans, in their return to nature during the interwar years, were driven by a shallow notion of natural beauty. Most Americans were after magnificent scenery. They stared through their windshields at the surfaces of nature, unable to go deeper. Such a point of view, Leopold realized, was entirely consonant with a population out of touch with the land as both a place of work and a source of culture. Even farmers, he feared, were losing such connections. For Leopold, preserving wildness in the worked landscape and preserving wilderness on the public lands were both efforts to get Americans to see a more complex beauty in nature.

In 1933, the *Journal of Forestry* published Leopold's "The Conservation Ethic," one of the pieces he later synthesized into his famous essay, "The Land Ethic." The essay marked his return to wilderness politics. He began by suggesting that humans needed to extend ethical consideration to the land, just as they had extended such consideration to all members of their own kind over the previous centuries. This notion has become the centerpiece of Leopold's ethical legacy. But perhaps the more striking aspect of this article was its attention to the historical conditions that made such an extension both necessary and difficult.[102] Machines and engineering, he noted, had come to play powerful intermediary roles in the relations between humans and nature. At the same time that a mechanized relationship with the land had increased the ability to achieve major ecological transformations, that same machinery had blurred our vision of the land's needs and of the dangers inherent in ignoring history's ecological substrate. The more power humans gained over the land, the less they understood it. "We of the machine age," Leopold wrote, "admire ourselves for our mechanical ingenuity; we harness cars to the solar energy impounded in carboniferous forests; we fly in mechanical birds; we make the ether carry our words or even our pictures. But are these not in one sense mere parlor tricks compared with our utter ineptitude in keeping the land fit to live upon?"[103] Leopold's emerging land ethic was not just an extension of ethical consideration to nature; it was also a response to a new set of historical developments that had distanced people, physically and morally, from the land. "We can be ethical," Leopold later wrote, "only in relation to something we can see, feel, understand, love, or otherwise have faith in."[104] By the 1930s, many Americans had lost that relationship with nature, and their efforts to restore it through modern forms of recreation struck Leopold as ethically insufficient. There too, a mechanized relationship reigned.

"The Conservation Ethic" also contained an important critique of the increasing reliance on public ownership as a solution to the nation's environmental problems. "Public ownership is a patch but not a program," he wrote, a limited solution to a problem *coextensive with the map of the United States.*"[105] "Let it be clear," he wrote in an essay published the following year:

> I do not challenge the purchase of public lands for conservation. For the first time in history we are buying on a scale commensurate with the size of the problem. I do challenge the growing assumption that bigger buying is a substitute for private conservation practice. Bigger buying, I fear, is serving as an escape mechanism—it masks our failures to solve the harder problem.[106]

Government conservation, while it offered certain solutions, also fostered a brand of ethical surrogacy: "a kind of sacrificial offering, made for us vicariously by bureaus, on lands nobody wants for other purposes, in propitiation for the atrocities which prevail everywhere else."[107] The most pressing ecological problems were on the millions of acres of private land, and the nation's critical ethical problems were between people and the land on which they lived. Government ownership, whatever good it accomplished, could not rectify this situation. Something more radical had to happen, an ethical reorientation and an aesthetic reevaluation of the human relationship with the land, prodded along by farsighted government incentives and restrictions.

Overreliance on government ownership and initiative had recreational implications as well. It meant "the relegation of esthetic or spiritual functions to parks or parlors," while sacrificing the rest of the landscape to forms of economic production devoid of aesthetic principles.[108] By sequestering natural beauty from economic use, Americans were doing themselves a great disservice, Leopold thought. Too heavy an emphasis on scenic preserves had failed to produce a "revival of land esthetics." Indeed, he felt that the separation of the beautiful from the useful served only to perpetuate poor land use and shallow perception. This is how he summed up the situation in his 1935 essay, "Land Pathology":

> A few parcels of outstanding scenery are immured as parks, but under the onslaughts of mass transportation their possible function as "outdoor universities" is being impaired by the very human need that impelled their creation. Parks are overcrowded hospitals trying to cope

with an epidemic of esthetic rickets; the remedy lies not in hospitals, but in daily dietaries. The vast bulk of land beauty and land life, dispersed as it is over a thousand hills, continues to waste away under the same forces as are undermining land utility.[109]

Leopold was a strong advocate for preserving public recreational grounds, but he recognized that their popularity was partly a sign that Americans were losing the sort of knowledge of nature that came through daily contact. As a result, many saw natural beauty only in spectacular scenery, and they contented themselves with such preservation.

Leopold's criticisms of New Deal conservation were based in experience. During the summer before his first semester of full-time teaching in 1933, he returned to the Southwest to supervise Civilian Conservation Corps (CCC) erosion control projects. Here and elsewhere, he witnessed an unplanned rush to perform conservation work that lacked the sort of vision and integration it deserved. The flurry of federal legislation, he thought, treated "single-track" problems rather than seeking a unified approach to conservation. The Agricultural Adjustment Act, for instance, removed land from agricultural production, but it provided no incentives for conservation of that land, or of new land that might be cleared and put under the plow. The CCC combated erosion by building check-dams at the public expense, yet no legislation required significant alterations in the private practices that produced such erosion. In the field, conservation work often proceeded at cross-purposes. The work of road-building crews contributed to the very erosion the check-dams crews sought to reverse. Tree-planting crews afforested meadows that were vital wildlife habitats. To be fair to New Deal conservation programs, they were often more integrative than earlier federal policies had been. Leopold appreciated the good intentions behind these initiatives, but he still thought that the rush to "conserve" the public lands sometimes did more harm than good.[110] This was particularly the case with wilderness.

While he tried to focus attention on the private landscape and the limitations of public conservation, the New Deal pulled Leopold back into wilderness politics, largely because New Deal work relief projects magnified threats to wilderness as he defined it. If the Forest Service, in the 1910s and 1920s, had constructed or "permitted" the construction of roads, trails, and recreational facilities in ways that seemed irresponsible and poorly planned, then the gung-ho development of the New Deal brought this activity to a crescendo. With money, labor, and the enthusiasm of an auto-bound president, the public domain became a staging ground for the nation's pent-up energies.[111] It was these New Deal projects that awoke in Leopold renewed concern for wilderness. As he later put it, "Wilderness remnants are tempting fodder for those [New Deal] administrators who possess an infinite labor supply but a very finite ability to picture the real needs of his [sic] country."[112]

The year 1935 was an important one for Aldo Leopold. A number of scholars have seen it as a watershed year in his intellectual development.[113] It was the year that he and his family bought a small piece of land in Sauk County, Wisconsin, converted an old chicken coop into a weekend getaway, and began an experiment in coaxing wildness from a piece of degraded bottomland. The family's experiences would form the major content of Leopold's *A Sand County Almanac*. That summer, he accompanied several American foresters on a trip to Germany to study German forest management. He quickly noted that, in their ordered simplicity, the German forests lacked "wildness." He was heartened by the emergence of a "Naturschutz," or nature-protection movement, and hoped the Germans would take the next step. "This impulse to save wild remnants," he observed, in what served as an unintended summary of his own intellectual development, "is always, I think, the forerunner of the more important and complex task of mixing degrees of wildness with utility."[114]

In 1935, Leopold also helped to found the Wilderness Society. A number of scholars have seen Leopold's charter membership in the Wilderness Society as one of the strongest signs that his thinking was becoming more ecological by the mid-1930s.[115] But we need to be careful about causation here. Leopold's thinking did change during the mid-1930s, in part because his efforts to propagate wildness necessitated careful ecological observation. But these ecological insights did as much to divert his focus from wilderness as they did to bolster his case for wilderness preservation. His reticence about federal conservation made this clear. Not long after the founding of the Wilderness Society, Leopold did make some important statements about the ecological importance of wilderness. But to see the founding of the Wilderness Society, and Leopold's participation in it, as the result of ecological concerns is to miss the persistent centrality of roads, automobiles, and recreational development to wilderness advocacy. The impacts of the automobile, not the insights of ecology, spurred the creation of the Wilderness Society and Leopold's willing participation in it.

Although he did speak for the scientific value of wilderness, Leopold continued to stress the difference between wilderness and the preservationist ideal being offered by ecologists. One exchange illustrates this well. In a reply to Marshall's invitation to join the Wilderness Society, Leopold, commenting on a draft statement of principles that stressed wilderness as a recreational ideal, wrote: "I have no criticism of the draft of the invitation, except that I would raise the question of whether we should definitely indicate that the Society includes only those interested in wilderness from the esthetic and social point of view, or whether it should also include those desiring wilderness for ecological studies." On its surface, this sounds like a strong plug for the ecological value of wilderness.[116] Yet the paragraph that followed indicated otherwise. "My hunch," Leopold continued, "is that the ecological group should be included, but with a minor emphasis. Dr. Merriam could tell you who are the most active members of this group. I know it includes Shelford of Illinois." At the time of the founding, Leopold prioritized the "esthetic and social point of view" when it came to making a case for wilderness; he saw ecological advocacy as allied but secondary.[117]

Almost a year later, that stance had changed somewhat. In "Why the Wilderness Society," his contribution to the first issue of *The Living Wilderness,* Leopold stressed more stridently the scientific value of wilderness.[118] The dust bowl and the nation's other Depression era environmental crises had brought into focus, in the year between the founding and that first issue, the important function of wild areas as standards against which environmental manipulations might be measured.[119] That lesson became even more poignant when, in 1936, Leopold took a hunting trip to Mexico's Sierra Madre Occidental. Saved from the extensive grazing to which the Gila had been subjected, the Sierra Madre was a "lovely picture of ecological health, whereas our own states, plastered as they are with National Forests, National Parks and all other trappings of conservation, are so badly damaged that only tourists and others ecologically color-blind, can look upon them without feelings of sadness and regret."[120] Toward the end of his life, Leopold recalled the impact of the trip in stark terms: "It was here that I first realized that the land is an organism, that all my life I had seen only sick land, whereas here was a biota still in perfect aboriginal health. The term 'unspoiled wilderness' took on new meaning."[121] Leopold was indeed coming to see wilderness as an ecological baseline against which to measure human impacts—as what he later called a "land laboratory."[122]

But to see Leopold's wilderness thought as simply "more ecological" obscures as much as it reveals. What was different and important after 1935 was not simply that he made an ecological case for the preservation of nature; ecologists had been doing that for a couple of decades. Leopold's importance lay in the moral and aesthetic implications of bringing ecology to bear on his already well-developed recreational aesthetic. His comment about "ecologically color-blind" tourists made this clear. He hoped to integrate ecology and recreation as aesthetic pursuits, and to combine his thinking on wilderness and wildness into a unified critique of the human-land relationship. To see and experience wilderness, Leopold thought, was to confront the problems of the rest of the landscape rather than to escape from them. And the opposite was true as well. To appreciate and nurture wildness throughout the landscape was to understand the importance of wildness on a larger scale.

Conclusion

Almost two decades after Aldo Leopold first proposed that the Forest Service preserve wilderness areas, he published his most incisive critique of modern outdoor recreation and the type of relationship with the natural world that it encouraged. "Conservation Esthetic," which first appeared in *Bird-Lore* in 1938, was a trenchant statement on the irony that defined conservation during the interwar years—that a growing fascination with motorized outdoor recreation was not only a grave threat to wilderness but also a manifestation of a deeper pathology. Recreation became an issue during the days of "the elder Roosevelt," Leopold wrote, when urban Americans began turning "to the countryside." He continued:

> The automobile has spread this once mild and local predicament to the outermost limits of good roads—it has made scarce in the hinterlands what was once abundant on the back forty. . . . Like ions shot from the sun, the week-enders radiate from every town, generating heat and friction as they go. A tourist industry purveys bed and board to bait more ions, faster, further. Advertisements on rock and rill confide to all and sundry the whereabouts of new retreats, landscapes, hunting-grounds, and fishing-lakes just beyond those recently overrun. Bureaus build roads into new hinterlands, then buy more hinterlands to absorb the exodus accelerated by the roads. A gadget industry pads the bumps against nature-in-the-raw; woodcraft becomes the art of using gadgets. And

now, to cap the pyramid of banalities, the trailer. To him who seeks in the woods and mountains only those things obtainable from travel or golf, the present situation is tolerable. But to him who seeks something more, recreation has become a self-destructive process of seeking but never quite finding, a major frustration of mechanized society.[123]

Here were the origins of the modern wilderness idea in a nutshell.

"It is the expansion of transport," Leopold concluded, "without a corresponding growth of perception that threatens us with qualitative bankruptcy of the recreational process. Recreational development is a job not of building roads into lovely country, but of building receptivity into the still unlovely human mind."[124] It was this "qualitative bankruptcy of the recreational process," particularly as it manifest itself on the national forests and other public lands, that prompted Leopold to propose wilderness preservation, and to join the Wilderness Society as a founding member. Wilderness preservation, as Leopold conceived it, was not meant to be a complete solution to the problems of a nation that was "growing away from the soil." But neither was wilderness a distraction from those problems. It was merely a way to thwart a nation bent on opening up the few remaining undeveloped portions of the federal estate for modern recreational ends. By advocating wilderness preservation, and by insisting that Americans get out of their automobiles to experience wilderness, Leopold hoped that he would help Americans see the needs of the rest of the landscape more clearly. He was not alone among the founders of the Wilderness Society in harboring such a hope.

CHAPTER 4

Advertising the Wild:
Robert Sterling Yard

As the only founder to come from National Park Service circles, Robert Sterling Yard was an anomaly among the foresters and hiking activists who made up the Wilderness Society's original membership. He was the lone envoy—though an alienated one—from an agency that had defined nature preservation in the early twentieth century. Yard brought to the Wilderness Society twenty years of grappling with the national park ideal in an era of substantial change. In particular, he came burdened with questions and disillusionment about the intermixture of advertising, recreation, and tourism—forces he had once thought so promising in a national program of preservation. Yard began his conservation career in the mid-teens as a promoter of the national parks and a defender of their scenic magnificence; by 1935, he embraced wilderness advocacy as an alternative to scenic preservation and the recreational developments his promotional efforts had facilitated. His change of heart helps to explain why, less than two decades after the creation of the National Park Service, preservationists coalesced around wilderness as a new paradigm.

Yard was not one of his generation's great thinkers, and my claims for his importance should not be misinterpreted as such an assertion. He was a quixotic character whose stodginess and obsessive attention to semantics—what many have called his "purism"—made him resist change when the majority of his preservationist colleagues were busy crafting positions more amenable to the growing popularity of outdoor recreation. At times, Yard's greatest distinction was his thorniness, and his career is remembered more for the principled enthusiasm of his one-man crusade for standards than for any formidable intellectual legacy. But his stodginess, his purism, made his journey to the wilderness ideal a revealing one. Although his critics saw him as elitist and blind to political realities—and to a great extent he was both—Yard was able to focus more clearly than most on the ways Americans were revaluing nature in the midst of an emerging consumer culture. His response was a shift away from sublime scenery toward wilderness as a preservationist ideal.

A Publicity Man in an Advertising Age

Robert Sterling Yard was born in Haverstraw, New York, in 1861. After graduating from Princeton University in 1883, his first career, which spanned more than thirty years, was as a journalist and publisher. During the 1880s and 1890s, he wrote for the *New York Sun* and the *New York Herald*. In 1900 he moved into publishing, where he spent more than a decade before becoming editor-in-chief at *Century Magazine* in 1913. After a brief stay at *Century,* Yard returned to the *New York Herald* as Sunday editor. It was at that point, in 1915, that his old friend from his *New York Sun* days, Stephen Mather, summoned him to Washington, D.C.[1]

Stephen Mather left a successful career in the borax business in 1914 to accept a challenge from Interior Secretary Franklin Lane, a fellow graduate of the University of California. Mather had sent Lane a letter critical of national park administration, and Lane responded by urging Mather to come to Washington to see if he could do better. Mather accepted with the understanding that his stint as Assistant Secretary of the Interior in charge of the national parks would last only one year. As it turned out, he stayed for a decade and a half, a tenure marked by his frenetic efforts to build a national park system and punctuated by a series of nervous collapses. He was wildly successful at getting Americans into the national parks, and at getting the national park ideal into the hearts of Americans.[2]

Mather entered government service with two goals. First, he sought to create an independent agency to oversee the national parks. Utilitarian conservationists had agencies in the Forest Service and the Bureau of Reclamation that lent them credibility and a clear policy voice. The parks, Mather thought, deserved similar institutional power. Mather's second goal was to publicize the national parks as proper destinations for the masses, and to build a strong national constituency interested in preserving scenic areas from the reigning conservation doctrine of utilitarianism.

Mather arrived in Washington at a propitious time. The Hetch Hetchy battle had heightened Americans' awareness of the value of national parks and increased political support for park preservation. Mather also had a great opportunity to connect the parks with the current upsurge in travel and publicity, for the outbreak of World War I was limiting travel in Europe and American motorists were flocking across the country to the Panama-Pacific and Panama-California expositions of 1915. "See America First" campaigns launched by railroad companies, automobile associations, and tourist bureaus also encouraged Americans to travel to the national parks.[3]

With Yard's help, Mather was able to ride this wave. He hired Yard to run a national parks publicity campaign, though under slightly unusual auspices—Yard's annual salary of $5,000 came out of Mather's pocket, an arrangement that was legal at the time.[4] While Mather drummed up political support for an independent parks bureau, Yard pumped out scores of articles praising the scenic qualities of the parks and their educational, inspirational, and recreational possibilities. He later referred to his production during the late 1910s as "a tidal wave of newspaper and magazine publicity that in time passed far beyond all control."[5] Robert Shankland, Mather's biographer, estimated that more than a thousand articles on national park subjects appeared in various newspapers and periodicals between 1917 and 1919.[6] Not all of these were written by Yard, but he and Mather played a significant role in getting journalists to take the parks seriously. This unprecedented press coverage proved effective; it got the word out, particularly to influential Americans, about the national parks and put pressure on Congress to create an independent parks agency.

The Yard-Mather campaign focused on journalism as a publicity medium. The two recognized the growing influence of mass circulation magazines and their potential for catechizing their predominantly middle class audiences.[7] The newspaper and magazine world had been Yard's milieu for more than three decades, and his ability to promote the parks within this community was one of his greatest assets. While Yard wrote and solicited articles, Mather conducted journalists such as *National Geographic's* Gilbert Grosvenor and the popular writer Emerson Hough on guided tours of the national parks and got them to feature the parks in their writing. In 1916, for instance, *National Geographic* devoted an entire issue to the national parks, while other prominent magazines, such as George Horace Lorimer's *Saturday Evening Post,* gave considerable coverage to the parks.[8]

To publicize the parks, Yard and Mather strove to make them newsworthy. This reliance on the press reflected a Progressive approach to publicity as rational and educational, an approach that was on the cusp of being undermined by the war and the rise of modern advertising. As historian Jackson Lears has pointed out, World War I "marked the high tide of a progressive faith in the beneficent powers of 'publicity.'"[9] Publicity was an essential part of the federal government's wartime mobilization, but what began as a campaign to educate Americans about the war's aims devolved into a more cynical effort to shape what Walter Lippmann would soon call "public opinion." Publicity became propaganda or, worse, the stifling of dissent. In this context, the Progressive faith in the social power of rational publicity yielded to concerns about manipulative advertising and its power to shape and steer a mass society. "Publicity," historian David Kennedy has concluded, "became in the postwar decade little more than an adjunct to the new economy of consumerism, as the fledgling industry of advertising adopted the propagandists' techniques of mass communication and persuasion."[10]

Yard was a consummate practitioner of the "idiom of publicity."[11] He saw it as essentially educational, and assumed that a rational audience was on the receiving end of his messages. But as modern advertising turned to new techniques that emphasized the psychic and social benefits of certain products, and to new fields like behavioral psychology, Yard found himself on shaky ground. During the 1910s, he insisted that his park publicity aimed to educate as well as popularize. He initially found it difficult to separate the two, but postwar events would make such a separation seem necessary.

Yard's most successful publicity initiative during the 1910s was the *National Parks Portfolio* (1916). The bulk of the work was photographs, with text interspersed that lauded the scenic grandeur of the nation's major parks. Yard juxtaposed the well-known spectacles of Yellowstone's geysers and the Yosemite Valley with such less familiar but equally worthy sights as Crater Lake. He spoke glowingly of the undeveloped environments in which these scenic gems were mounted, suggesting that the wilderness setting was vital to national park status. Although its chief purpose was to argue for the national importance of the parks, the volume also provided an explicit tourist itinerary. Yard's efforts were designed, in part, to connect the parks with a sense of national identity and to make park visitation an imperative of American citizenship.[12]

Yard saw an organic relationship between the parks he discussed in *National Parks Portfolio*. He insisted that each be viewed as part of a coherent system based on common aesthetic standards, and to make that point clear he even included one area, the Grand Canyon, that was not yet a park, though he saw it as eminently qualified for such status. Just as conspicuous was Yard's neglect of a number of areas administered as national parks that he felt did not meet such standards—Platt, Sully's Hill, Hot Springs, and Wind Cave National Parks.[13] But the most important aspect of *National Parks Portfolio* was its circulation. Yard and Mather distributed 275,000 copies free of charge to a carefully selected list of prominent Americans, including every member of Congress. They also produced millions of copies of a more modest pamphlet version, *Glimpses of Our National Parks,* for general consumption. Twenty-one railroad companies picked up the bulk of the tab, with Mather covering the remaining costs.[14] The effort generated promotional reverberations throughout the nation.

The most immediate result of all of this publicity was the creation of a National Park Service. On August 25, 1916, Woodrow Wilson signed into law a bill creating such an agency and charging it with its now-famous dual mandate: "to conserve the scenery and the natural and historic objects and wildlife therein, and to provide for the enjoyment of the same in such manner and by such means as will leave them unimpaired for the enjoyment of future generations." Mather's first goal had been achieved.[15]

In the wake of the passage of the National Park Service Act came administrative reshuffling. Mather made Horace Albright, his young assistant, an Assistant Director, and he appointed Yard to run the Service's Educational Division. Mather, who remained in the position of Assistant Secretary of the Interior, initially appointed Robert B. Marshall (no relation to Wilderness Society founder Bob Marshall) as interim Director of the new National Park Service. But Marshall proved ineffective, and Mather let him go in December 1916. As another in a series of National Parks Conferences convened in Washington in January 1917, the Service was without a director.[16]

This vacancy proved divisive. Mather suffered a breakdown in the midst of the conference and had to take an extended leave from his duties. Yard and Albright jockeyed for position in Mather's absence. Yard, a full quarter century older than Albright, who was then only twenty-seven, thought the position ought to be his. But Mather's wife intervened and made it clear that Albright was Mather's choice to run the Service in his absence. Albright handled the situation diplomatically and Yard yielded graciously, but the incident had its scarring effects.[17] Not long after Mather took his leave, another conflict developed, this time between Yard and Oliver Mitchell, Mather's attorney and business manager. Yard wanted to purchase an expensive lantern slide collection for promotional purposes. Mitchell thought the plan frivolous, refused to grant Yard the funds, and encouraged Albright to fire Yard. Albright seems to have agreed with Mitchell's point of view, but he decided against letting Yard go because he feared it would upset the fragile Mather. This conflict suggested that Yard and Albright were moving in different directions. While Albright focused on institution building and tourist development, Yard sought to undergird the parks system with an educational ideology.[18] In these conflicts between Mather's two most important assistants were the beginnings of a divide that would push Yard away from the Park Service in the years to come.

After more than a year of toiling away within his Educational Division of two (he and his secretary were its only employees), Yard found it necessary to look outside the Service for support. In June 1918, he met with several prominent Washington citizens, including the Secretary of the Smithsonian Charles Walcott, and put together a National Parks Educational Committee. Those assembled voted Walcott chairman, William Kent (former California congressman and donor of Muir Woods) vice chairman, and Yard executive secretary. The Committee set itself the task of shaping the system's use along educational lines, which meant crafting informative publicity to draw Americans to the parks and developing programs to enhance the educational value of the visitor experience. For Yard, the national parks were not just about sightseeing and recreation; they were places of inspirational and educational importance.[19]

By April 1919 the Educational Committee had grown to more than seventy members, many of whom were leading figures in the scientific and educational communities. At a meeting that month, members discussed the creation of a more formal organization—a National Parks Association (NPA)—to promote their objectives. Mather supported the formation of the new group, in part because of extenuating circumstances. Congress passed a law in July 1918 that made his financial arrangement with Yard illegal; the salaries of government employees could no longer be paid by private parties. Mather thus agreed to pay Yard's salary as executive secretary of an extragovernmental organization, at least until that organization achieved a solid financial footing. Yard's duties at the NPA would ostensibly be the same as they had been with the Park Service—to promote the national parks and to educate Americans about their proper uses. The NPA, Yard's home for the next decade and a half, was officially formed at a May 29, 1919, meeting at Washington's Cosmos Club.[20]

Mather's push to increase national park visitation and create a constituency for the parks also met with success. Between 1915 and 1920, annual national park visitation increased from 334,799 to 919,504. Only in 1918, when U.S. involvement in World War I commandeered the attention and resources of the nation, did visitor numbers fail to increase. But this decline of about 40,000 visitors was dwarfed by an increase of over 300,000 visitors in 1919. One-third of that increase came with the addition of the Grand Canyon and Lafayette (Acadia) National Parks to the system, but most of the remainder was recorded in established parks. Yellowstone, for instance, saw visitation triple between 1918 and 1919.[21] Buoyed by victory in Europe and the restoration of peace, and informed by Yard's publicity campaign, Americans headed west to glimpse their nation's great natural wonders.[22] But these were not the only forces propelling Americans westward. The automobile had come into its own, and through its use many Americans found visiting the parks within their physical and financial orbits. Others, seeing the birth of a new mass activity, began the steady work of providing for the motorists who headed for the national parks—"the great lodestones of the West," as Mather liked to call them.[23]

The question of whether to admit automobiles to the national parks was a tricky one. Automobiles were regularly admitted to some parks— Mount Rainier, Crater Lake, and General Grant among them—as early as 1910, but their use was strictly regulated. Motorists also had to purchase expensive permits; the fee to enter Mount Rainier in 1910, for instance, was five dollars, a lot of money at the time. In other parks, most notably Yellowstone and Yosemite, automobiles were prohibited in the early 1910s, despite growing public demands for motorized access.[24]

Debates over automobile use reflected a number of concerns. Some worried about the expense of improving roads for motorized traffic. Others argued that the horse-and-carriage experience was essential to seeing the national parks. The greatest problem, however, was the conflict between horses and machines. As late as the mid-1910s, national park officials clearly favored horses over automobiles. In most parks that allowed automobiles, motorists were required to move to the side of the road whenever they saw teams approaching and to remain stationary until the teams passed.[25] But as automobiles brought more visitors to the national parks, and as greater visitation became a central goal in the fight for park protection, the status of automobiles improved. Both park officials and tourists pointed out that motorists contributed substantially to park revenues, and motorists objected to their second-class treatment. As a result, transportation regimes in the national parks changed rapidly.

Yellowstone's history provides the best example. As late as 1915, automobiles were prohibited in the park. The first car to enter did so on August 1, 1915, and, in one of the better ironies of this story, Yard was among its passengers. For the 1916 season, park officials conducted an experiment in coexistence. Yellowstone's Grand Loop was made a one-way road, and, through the use of newly strung telephone lines and checking stations, park officials tried to keep motoring parties and stages apart. For all its ingenuity, however, the system was too cumbersome. By 1917, concessionaires had sold their horses, and Yellowstone was entirely motorized.[26] By 1920, over 13,000 private automobiles entered the park, and people who came to the park by train saw the sights by motorized jitney.[27] Similar transformations occurred in other parks. The Yosemite Valley, which was closed to automobiles as late as 1912, was soon filled with them. In 1916, more people traveled to Yosemite by car than by rail for the first time; by 1918, the ratio was seven to one.[28]

Unlike railroad tourists, who expected and preferred modern accommodations in the national parks, motorists opted for more rustic arrangements. Many prided themselves on bringing with them equipment and supplies to meet their basic needs. Indeed, motorists saw frugality and self-sufficiency as tourist virtues. The national craze for autocamping also had a significant impact in the national parks. Without a planned landscape to shape their behavior, autocampers distributed themselves liberally throughout the valleys, meadows, and forests of the parks. In the process, they helped themselves to timber resources, polluted the water supplies, left litter, and compacted soil. It was soon clear to park officials that the admission of automobiles required a new generation of facilities and visitor services. In 1916, the first full season in which autos were allowed into Yellowstone, the Park Service built four "large automobile shelter camps." Other parks were quick to follow suit.[29]

As both the national parks and automobiles grew in popularity, a wide range of organizations and facilities sprang up to coordinate touring in the West. From its inception, the Park Service had a policy of cooperating with railroad companies, tourist bureaus, chambers of commerce, automobile associations, Good Roads boosters, and other parties interested in incorporating the national parks into their programs. These alliances, a signature feature of Mather's agency-building efforts, were meant to be mutually beneficial. Western towns, for instance, were early and important allies. After the war, there was a phenomenal rise in the number of municipal autocamps, and competition between municipalities quickly raised the quality of the services provided.[30] In their annual reports, Mather and Albright indicated that these municipal camps were both an important model for dealing with motorists within the national parks and an important support structure in getting them there.[31]

During this period, a variety of automobile and highway associations engaged in efforts to designate and sign automobile routes throughout the West. The most important of these groups to the Park Service was the National Park-to-Park Highway Association, organized at a meeting in Yellowstone in the summer of 1916 and animated by a vision of a 3,500-mile highway connecting all of the major national parks in the West. By 1920, Mather was able to announce the completion of this route. Attached to his annual report for that year was a map showing the path of the giant circular roadway, which began and ended in Denver and connected the following parks and monuments: Rocky Mountain, Yellowstone, Glacier, Mount Rainier, Crater Lake, Lassen, Yosemite, General Grant, Sequoia, Grand Canyon, Petrified Forest, and Mesa Verde. Although extremely rough in spots, the road provided motorists with a challenging itinerary that was well marked and fully supported by the municipalities along its route.[32]

Mather called for other efforts to connect parks with improved roads and wayside facilities. The Yellowstone Transportation Company and the Glacier Park Transportation Company had already joined ranks to provide a park-to-park automobile service.[33] Automobile and highway associations throughout the region worked to designate trunk routes, provide adequate signage, and improve the quality of road networks connecting the parks. Municipal, county, and state governments—the governmental bodies responsible for road development and maintenance—contributed to the effort by devoting a percentage of their road funds and labor to major routes. Various good road movements, most of which had grown out of efforts to promote improved farm-to-market access, turned some of their energies to park-to-park connections. Indeed, in 1919, representatives from good road organizations in twelve western states gathered at Yellowstone and founded their own group, the National Parks Touring Association.[34]

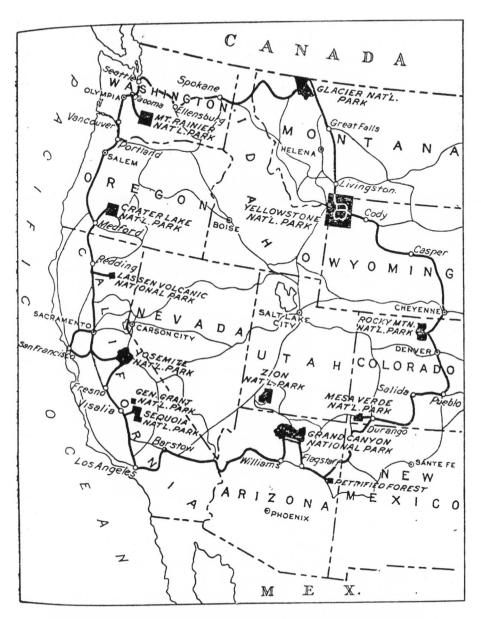

National Park-to-Park Highway. The National Park-to-Park Highway Association, organized in 1916, worked throughout the late 1910s to connect a series of western national parks with a 3,500-mile loop road. The road was essentially complete by
1920. (*Report of the Director of the National Park Service* 1920)

So successful were these local efforts that they soon put the roads and camps within the national parks to shame. As a result, Mather called for greater federal funding for road improvements within the parks, and through those areas outside the national parks that remained in federal ownership. Motorists in the national parks, he pointed out, were asked to pay exorbitant fees to use roads and camps that were markedly inferior to those that were free of charge just outside the parks. Having traveled the Park-to-Park Highway in 1919, Mather was convinced that the stature of the parks required an aggressive federal initiative to develop their roads and other facilities. He also called for the formation of a state parks conference to provide crucial links in the autocamping network.[35] The national parks, those "scenic lodestones" designed to keep resource users at bay, were becoming the centerpieces of a whole new economic regime in the West—tourism.

To deal with current and anticipated developments within the parks, Mather established a Landscape Engineering Department within the Park Service and hired its first employee, Charles Punchard, in 1918. Punchard was responsible for "naturalizing" road, trail, and tourist developments, and for restoring areas damaged by overuse. Creation of this department marked the beginning of a heavy reliance by the Service on landscape architects in mediating between the preservation of nature and increased automobile visitation.[36] It also signaled the beginning of what landscape historian Ethan Carr has called "the most intensive period of such human alterations in the history of the parks."[37] The growing popularity of the national parks in the late teens was inseparable from the growing popularity of automobile touring and the infrastructure that developed around it.

Yard's promotional efforts at that time complemented all of these developments by providing an invaluable body of material for tourists. Yard and the Park Service distributed reams of information and imagery designed to familiarize Americans with the parks, and to get them interested in seeing the real thing. In the Park Service's 1917 annual report, Horace Albright boasted of the Service's "very active campaign of education"; they had distributed 128,000 copies of various bulletins on the parks, 83,000 auto guide maps, 348,000 feet of motion picture film of the national parks, and 117,000 copies of Yard's *Glimpses of the National Parks*. An exhibit of twenty-four enlarged photos of the national parks traveled the public library circuit. Albright even mentioned the Service's inability to keep up with demands for photos and lantern slides, this at the same time that Yard and Oliver Mitchell were having their scuffle over expanding such a program.[38] And the Park Service continued to co-operate with groups such as the Far Western Travelers Association, the See America First League, and the American Automobile Association—groups that linked the civic imperative of visiting the national parks to distinctly commercial and regional goals.[39]

This mixture of public and private motives would increasingly strike Yard as a threat to the mission of the parks. In a reminiscence about these early Park Service years, Yard later explained what he saw as the flaw in his, and Mather's, promotional logic. They had failed to recognize, he wrote, that "the Outdoor Recreation Age, then about to be born unrec-ognized, already was fitting us into the shaping of its own ends." Yard ini-tially had seen the effectiveness of the "hurricane of national park pub-licity" as representative "solely of the people's joy in discovering the existence of so noble a possession as our National Parks System." "It re-quired the perspective of ten years of development," he concluded, "to show us what had actually happened, namely, that we had advertised super-fascinating travel goals to several million potential motor tourists impatient for the long road; for motor touring was then in its very first beginning, awaiting only inspiring objectives and leadership."[40] An audi-ence with desires that transcended his educational motives had greeted the publicity that he had engineered, Yard realized. Moreover, and per-haps more importantly, commercial interests were busy fitting the parks into a new recreational and tourist movement that threatened to sub-merge under a wave of indifferent usage what Yard saw as the high-cultural purpose of the parks. Although he initially had assumed that the tourist landscape developing outside the national parks would serve na-tional park purposes, he became worried that quite the opposite was occurring: that regional boosters, opportunistic politicians, and a variety of commercial interests were plugging the parks into an increasingly sophisticated commercial matrix and distorting the role of the national parks as educational spaces. They had made marketplaces of nature's tem-ples, and they were advertising the wild for their own gain.

The Battle for Standards

"I have no business to be here," Robert Sterling Yard told an audience at a National Parks Conference, "for the reason that I am a tenderfoot." Yard was touring the major western parks, and his lack of experience with outdoor living was apparent to many. In Albuquerque, his traveling companions convinced him that trained dogs were coyotes. Later, when asked to add the name of the park that he represented to a visitor's book, he put "Central Park, New York." "They joshed me all the way across the continent," he told the assembled crowd.[41]

In truth, Yard's park *was* Central Park. He was an urban creature whose attachment to the national parks was abstract and idealized. Although one of his major goals throughout the 1920s would be to distinguish between the functions of urban and national parks—to make it clear that national parks were not the manicured creations that most people thought of when they heard the term "park"—Yard remained committed to a form of cultural custodianship that had characterized debates over urban parks during the late nineteenth and early twentieth centuries. The custodians of Central Park already had been forced to surrender many of their aesthetic and moral goals to popular demands for recreational space.[42] As Yard took the helm of the National Parks Association in 1919, he prepared to fight similar battles in different settings.

Under Yard's guidance, the NPA crafted a broad educational program. In their first *Bulletin,* published in June 1919, Yard explained the organization's objectives. The first was to connect the scenery of the national parks with scientific study in an effort to fathom what made the parks so universally startling. The application of science, Yard and his colleagues thought, could help to counter what critics had widely derided as the sentimentalism at the root of scenic preservation. This goal was to be accomplished by bringing students to the parks, by bringing park scenery (in the form of lantern slides, photographs, films, and scale models) to students, and by seeking the cooperation of a variety of educational institutions. Second, Yard and his colleagues sought to bring the tools of artists and writers to this problem of scenery by encouraging the arts in the national parks. A third goal was to disseminate information on the history and natural history of the parks through a series of publications. A fourth objective was to encourage travel to the parks so as to bring Americans "under the influence of our educational work." Finally, and perhaps most important, Yard and the NPA aimed to develop the national parks into a "complete and rational system" by drawing up a strict set of standards and forcing Congress and the Park Service to adhere to it. Those areas that met such standards would be added to the system, and those that did not would be denied entrance—or, in a few cases, purged from the system.[43]

Civilian Conservation Corps crew with bulldozers, Modoc National Forest, California, 1934. CCC workers unleashed on the public lands during the New Deal did a significant amount of road building, raising concerns among wilderness advocates. (NARA 95-G-290289)

CCC crew, Lassen National Forest, California, 1933. CCC labor was also responsible for building bridges, trails, campgrounds, and other recreational facilities. The founders of the Wilderness Society feared that the federal government was too eager to "barber and manicure wild America as smartly as the modern girl." (NARA 95-G-285484)

Aldo Leopold in the Apache National Forest, Arizona, 1911. Between 1909 and 1924, Leopold worked for the U.S. Forest Service in the Southwest, and his experience dealing with recreational trends in the region's forests led him to suggest national forest wilderness preservation. (NARA 95-G-92605)

Souvenir shop on the South Rim of the Grand Canyon, 1910. In 1915, Leopold, as part of his new recreational duties, developed a plan for controlling commercial development at the Grand Canyon. He and his coauthor Don Johnston were concerned that commercial establishments such as this one were undermining the Grand Canyon as a public space. (NARA 95-G-87379)

Cottage and automobile, San Bernadino National Forest, California, 1922. After the passage of the Term Permit Act of 1915, which allowed for long-term recreational leases on the national forests, Americans eagerly built rustic vacation cottages and quickly monopolized some of the most scenic spots in the national forests. (NARA 95-G-166081)

Canvas cottage, Sierra National Forest, California, 1916. For those who could not afford to construct summer cottages in the national forests, canvas cottages such as this were a good alternative. (NARA 95-G-33501A)

Organizational camp, San Bernadino National Forest, 1916. The Term Permit Act also allowed for the development of organization camps, hotels, and commercial services in the national forests. This particular camp was run by the City of Los Angeles. (NARA 95-G-33365A)

Aldo Leopold during a 1938 trip to Rio Gavilan, Mexico. After visiting the Rio Gavilan in the late 1930s, Leopold noted that Mexico's Sierra Madre was "a lovely picture of ecological health, whereas our own states, plastered as they are with National Forests, National Parks and all other trappings of conservation, are so badly damaged that only tourists and others ecologically color-blind can look upon them without feelings of sadness and regret." By this time, Leopold was developing a stronger ecological rationale for wilderness. (University of Wisconsin–Madison Archives)

Robert Sterling Yard in Yosemite National Park, 1920. In the mid-1910s, Yard was summoned to Washington, D.C., by his friend Stephen Mather to publicize the national parks and to create a national park agency. By the early 1930s, Yard had drifted away from the national parks lobby and toward wilderness preservation, in part because he grew uncomfortable with the results of his publicity work. (NPSHPC)

Automobile passing through the arch at the northern entrance of Yellowstone National Park, n.d. The bill that created the National Park Service in 1916 charged the agency with preserving the scenery and wildlife of the parks and with providing for the enjoyment of the parks "in such a manner and by such means as will leave them unimpaired for the enjoyment of future generations." The coming of large numbers of automobiles to the parks brought the tensions within that dual mandate to the fore. (NARA 79-G-31-42)

Yellowstone Park Transportation Fleet at Mammoth Hotel, n.d. As late as the mid-1910s, most visitors to Yellowstone saw the park by horse and carriage, but, after a brief experiment in coexistence in 1916, concessionaires sold their horses and motorized their fleets. (NARA 79-G-30A-F-30)

Horace Albright, Stephen Mather, and a Park Service automobile, n.d. Mather (right) and Albright, the first two directors of the National Park Service, were eager to develop the parks for motor tourists, and they dismissed Yard's increasingly strident protests against such development as "purist." (NPSHPC)

Young Benton MacKaye, c. late 1920s. MacKaye, a trained forester, devoted his early career to developing a socially informed brand of forestry whose ultimate expression came in a 1919 Department of Labor Report, *Employment and Natural Resources.* Crucial concepts from that report found their way into his 1921 plan for an Appalachian Trail. (Dartmouth College Library)

The Olmstead-Wharton investigating trip, Everglades, 1932. Yard was greatly concerned that the proposed Everglades National Park did not meet scenic standards and was driven by commercial boosterism. In 1932, a National Parks Association delegation, led by Frederick Law Olmsted Jr. (second from left) and William Wharton (middle), got a tour from local park advocate Ernest Coe (second from right). Their published report proved crucial to Yard's eventual approval of the park plan. (NARA 79-G-3-R-1)

Potomac Appalachian Trail Club group portrait, 1928. Independently organized trail clubs such as the PATC were crucial to the construction of MacKaye's Appalachian Trail. From the ranks of the PATC came Harold Anderson (center with pipe), a Washington, D.C., accountant and founding member of the Wilderness Society. (Appalachian Trail Conference)

Benton MacKaye and Myron Avery, n.d. Avery (right) was a leader of the PATC and a tireless blazer of the trail who became the head of the Appalachian Trail Conference, the confederation of hiking clubs responsible for trail construction and maintenance, in 1931. In the mid-1930s, MacKaye and Avery had a bitter falling out over whether the ATC should oppose skyline drives being planned for and built along the Appalachian ridgeline. (Applachian Trail Conference)

Harvey Broome near Siler's Bald, Great Smoky Mountains, 1934. Broome, a Knoxville lawyer and member of the Smoky Mountains Hiking Club, was a friend of MacKaye, a vocal opponent of road building in the Smokies, and a founding member of the Wilderness Society. (Denver Public Library)

Blue Ridge Parkway at Ice Rock, 1938. The construction of the Blue Ridge Parkway, which began in 1935, was one of the major concerns of the Wilderness Society's founders. Among other things, the road forced the relocation of approximately 120 miles of the Appalachian Trail in Virginia. (NARA 79-G-1-L-2)

Bob Marshall, Northern Rocky Mountain Forest Experiment Station, Montana, 1928. Between 1925 and 1928, Marshall lived in Montana and conducted forestry research for the Forest Service. He also fought fires, observed the conditions of the timber industry and timber workers, and explored the region's remaining roadless areas— including one that would later be preserved as the Bob Marshall Wilderness. (NARA 95-G-229349)

Ernest Oberholtzer, c. 1935. While serving as the Department of Interior's representative on the Quetico-Superior Committee, Marshall got to know Ernest Oberholtzer, one of the most important advocates for preserving the Quetico-Superior canoe country (later the Boundary Waters Canoe Area) of Northern Minnesota. Oberholtzer was a founding member of the Wilderness Society. (Minnesota Historical Society)

Harold Ickes at the Wawona Tree, Yosemite National Park, n.d. Among Bob Marshall's many allies in positions of governmental power was Harold Ickes (far left), the Secretary of the Interior under Franklin Roosevelt. Marshall used his influence with Ickes to thwart road building in the national parks. (NARA 79-G-17-D)

Bob Marshall on North Doonerak in the Brooks Range, Alaska, 1939. Bob Marshall first visited Alaska during the summer of 1929, and he spent fourteen months in the town of Wiseman in 1930–31, studying social conditions there and exploring the surrounding wilderness. This photograph was taken during Marshall's final trip to Alaska. A few months later, in November 1939, he died of an apparent heart attack. (Bancroft Library, University of California, Berkeley)

Automobile on a CCC-built road in Olympic National Forest, Washington, 1933. Roads built by the CCC and other New Deal agencies were crucial to further opening the nation's remaining wildlands to motorized recreation. Portions of the Olympic National Forest were made into the Olympic National Park in 1938, which was preserved, thanks to the advocacy of the Wilderness Society founders, on a wilderness model with only minimal road development. (NARA 95-G-280041)

Bernard Frank, Harvey Broome, Bob Marshall, and Benton MacKaye, four of the eight founding members of the Wilderness Society, n.d. (The Wilderness Society)

Echo Park, Dinosaur National Monument, Colorado. The battle to save Echo Park in the early 1950s galvanized the postwar wilderness movement and presaged the passage of the Wilderness Act of 1964. But it was also a battle reminiscent of the Hetch Hetchy controversy four decades earlier. After World War II, the politics of preservation again revolved around debates between preservation and resource development, a political shift that has obscured the recreational politics that defined interwar wilderness advocacy. (NARA 79-G-33-B1-VIII)

Throughout the 1920s, Yard and the NPA defined and redefined national park standards in their efforts to confront a series of threats to the parks. (To a great extent, Yard was the NPA during the 1920s; he was the sole full-time employee, and he set the organization's agenda.) Although Yard's criteria were constantly evolving, there were two central and consistent tenets to his notion of park standards. First, he insisted that national parks be scenically magnificent. Eligible areas had to be so stunning as to constitute sites of *national* importance, thus justifying the federal government's role in protecting them. Yard did not oppose setting aside scenery of lesser caliber, but he thought such areas should not be national parks. This was a stance that most of those associated with the national parks shared, although there were disagreements on how to apply this scenic standard.

Yard shared with many of his generation conventional aesthetic notions like the sublime and the picturesque, but he grounded his search for scenic magnificence in a more specific ideal that emphasized grand geological and biological forces.[44] Only by understanding these processes, Yard thought, could one truly appreciate the nation's supreme scenery. When he looked at the national park system and its holdings, he noted in his 1919 volume *The Book of the National Parks,* one major characteristic struck him. Almost all of the national parks, and certainly all the worthy parks, were areas whose scenery had been forged by powerful geological, or in some cases biological, processes. The glacial sculpting of Yosemite and Glacier; the geothermal activity in Yellowstone; the eroded and scoured Grand Canyon; the volcanic wonders of Lassen, Mount Rainier, and Crater Lake; the gigantic and ancient trees of Sequoia and General Grant: the common denominator was a geological or evolutionary sublime in which scenic magnificence bore a direct relationship to the magnitude of the powers of creation. "[W]e shall not really enjoy our possession of the grandest scenery in the world," Yard wrote, "until we realize that scenery is the written page of the History of Creation, and until we learn to read that page." For Yard, national parks were more than areas of preserved nature; they were preserves that aimed to protect nature's highest achievements. Yard sought to preserve the results of processes rooted in deep time, not dynamic ecological processes. By this logic, only a few areas qualified as national park material.[45]

One of the trickiest issues Yard faced was the subjectivity involved in making scenic judgments. Which areas would fall within the canon of nationally important sites? Who would make such decisions and under what guidelines? The NPA's emphasis on the scientific and artistic study of scenery was largely aimed at unearthing objective scenic criteria, and Yard hoped the NPA would be a crucial resource in making such decisions. Indeed, he clung to a belief that scenic quality was objectively knowable. To admit otherwise would deny the first cornerstone of his attachment to standards.

The second cornerstone of Yard's battle for standards was "complete conservation," an insistence that commercialism be kept out of the parks. In the late teens and early twenties, this meant fighting attempts to cut timber, extract minerals, and develop water resources; commercialism was synonymous with the forces of production. Such uses, Yard thought, were incongruous with park preservation not only because they marred park scenery, but because they undermined the integrity and publicness of these places. The national parks were arenas of scenic display—Yard often referred to them as "American masterpieces"—not to be intruded upon by the workings of the nation's commercial economy. He often admitted, sometimes painfully, that areas of significant commercial value ought to be excluded from national parks so as to avoid conflicts over resources. He sought to save the scenic gems and to protect them from economic activities. Improved access and increased visitation were essential in spreading the cultural influence of the parks. It was not immediately apparent to him that they, too, could be commercial intrusions.

In explaining his doctrine of complete conservation, Yard often advocated the preservation of "wilderness" conditions in the parks. He mentioned the importance of maintaining "museums of the primitive American wilderness" and "reservations of original America, to be preserved untouched forever by the hand of commerce."[46] In making these proclamations, Yard was giving voice to what many would recognize as the wilderness idea. But a few distinctions are necessary here. First, wilderness was, by Yard's logic, a necessary but not a sufficient condition for park preservation. Wilderness lent authenticity to scenic displays, but without the presence of magnificent scenery wilderness conditions did not qualify an area for park status. In 1920, wilderness was, for Yard and most other preservationists, still a concept subsumed by the larger issue of scenic preservation. Second, Yard defined wilderness solely in opposition to resource use. Because he saw the parks as cultural institutions, providing access was a necessary and presumed part of their development. "Except to make way for roads, trails, hotels and camps sufficient to permit the people to live there awhile and contemplate the unaltered works of nature," Yard wrote, "no tree, shrub or wild flower is cut, no stream or lake is disturbed, no bird or animal is destroyed."[47] His use of wilderness as a rough synonym for complete conservation suggested his opposition to all commercial uses of park resources, but it did not preclude the development of a tourist infrastructure. In the late 1910s and early 1920s, Yard was attached to a national park ideal in which wilderness played but a supporting role. And in that supporting role, wilderness could coexist with tourist developments that, a decade or so later, Yard would define as antithetical to wilderness preservation.

During the early 1920s, Yard spent most of his energy opposing a variety of plans to open the parks to water development. In 1920, Congress passed the Water Power Act, which created a three-person Federal Power Commission authorized to grant licenses to develop waterpower on federal lands. To Yard's horror, the national parks were not exempted. As a result, various schemes to get at water resources within established parks gained new life. Yard and the NPA joined with Mather, Albright (who had become Yellowstone's supervisor), and the Park Service to oppose, quite effectively, this intrusion on Park Service control and the threat it posed to the integrity of national parks. In 1921, Congress passed the Jones-Esch Bill, amending the Water Power Act to exempt existing national parks from the powers of the commission, but future park additions still had to be secured against such intrusions one by one.[48]

The Water Power Act and the Jones-Esch amendment set the stage for a crucial debate, involving the proposed addition of the Kings and Tehipite River canyons to Sequoia and General Grant National Parks, that drove a wedge between Mather and Yard. Mather made these additions a priority. The city of Los Angeles, however, had already applied for the right to develop the water in these areas under the provisions of the Water Power Act. The question was whether such areas should be included within the borders of a national park if there were preexisting claims to their water resources. Yard and the NPA parted ways with Mather and the Park Service over a bill that proposed these additions but provided no protection against waterpower development. Mather hoped to get the bill passed and then have waterpower excluded; he was no supporter of such development, but for the sake of expediency he hoped to secure the areas first and fight the water interests later. Yard and the NPA opposed the bill, fearing that it would create a powerful precedent by sanctioning water development within the national parks. Ultimately, Yard and the NPA succeeded in getting the bill amended to exclude waterpower, but as constituted the bill failed. After this show-down, Mather refused to provide the NPA with his annual contribution, forcing Yard and the NPA to chart their own course. Yard and Mather, de-spite similar goals, disagreed over the efficacy of compromise, and their relationship suffered as a result.[49]

As the Park Service and the NPA fought off water interests in the early 1920s, national park visitation continued to increase.[50] With these in-creases, and in response to the new wave of threats to the goal of com-plete conservation, Yard echoed Mather and other Park Service officials in defining the national parks as economic assets. Park advocates had always faced the difficult task of having to defend their "sentimental" attach-ment to scenery against the "hard" economic logic of those seeking to de-velop park resources. By the early 1920s, however, scenery itself had be-come an important economic resource. In a series of articles, Yard argued that increased travel to the parks had done more than lend support to the policy of preserving those areas from commercial intrusions; tourism it-self provided an economic rationale for preservation. Increased travel brought to those areas around parks, and the nation as a whole, a tangi-ble economic dividend, a point that Yard hoped would not be lost on the nation's politicians.[51]

But just as Yard and Park Service officials were enjoying this potent new rationale for park preservation, they faced a new challenge to park standards—proposed additions to the national park system—that further strained relations between the Park Service and the NPA. In January 1923, Yard testified before the House Public Lands Committee on a bill to create an Appalachia National Park. The NPA had been a vocal skep-tic of this and other park proposals, and many in Congress sought a fuller understanding of why the NPA, a strong supporter of the parks, would oppose additions to the system. Yard seized the opportunity to ex-plain his case for standards. He told the committee that he sought to de-fend the policy that parks "shall be scenically magnificent, or remarkable for extraordinary natural manifestations, and preserved in a state of primitive nature." Here, in a nutshell, were his twin tenets of scenic mag-nificence and complete conservation.[52] Yard insisted that the NPA sup-ported the creation of new national parks, but only if they met these standards. New parks must be chosen carefully, he argued, perhaps under the guidance of experts who could make the scenic judgments he thought vital to the process. The Park Service was in the best position to make such decisions; the NPA, according to Yard, sought no such authority. But neither was he willing to sit idly by while Congress made park creation into a spoils system.[53]

During this period, a variety of proposals for national parks originated from congressional delegations whose districts would have profited from the plans.[54] Invariably, Yard thought the areas were substandard and that their inclusion would undermine the integrity of the system as a whole. Allowing such substandard areas opened the way for two troubling outcomes. First, every locale might insist on having a national park so as to profit from the tourist trade. Scenic standards, in other words, were important in keeping national and local interests separate and distinct. Second, substandard areas that lessened the system's prestige would compromise the overall quality of the system, and its ability to draw tourists from at home and abroad. In consumer parlance, Yard worried that poor quality products would tarnish the brand-name reputation of the parks. He was concerned about truth in advertising, about preserving the integrity of his park publicity. Though still visibly enthusiastic about national park promotion, he objected to the inclusion of substandard parks and to pork barrel park politics.[55]

In his testimony, Yard also addressed a perception within Congress that the NPA was trying to redirect park policy and "dictate" to Congress. The NPA, he insisted, was interested only in protecting park standards and ensuring the ability of the responsible government authorities (including the House Public Lands Committee) to make objective decisions about park additions without factional or local political pressure. What rankled Yard was the charge that he and the NPA were "propagandists," for it went to the heart of his conviction about the public role of the parks and his own educational and promotional efforts. "We are not [propagandists]," Yard testified, "in any possible meaning which that abused word has come to have since the great war."[56] To Yard, he and the NPA were purveyors of facts who sought to guard the parks against those who would distort standards and protocol. He saw the matter as partly a battle between his objective publicity and the propaganda of others. But on this point Yard was in a bind. Despite his insistence to Congress that standards had always been an operative part of park designation, he was pushing something new. Never before had there been so many members of that legislative body eager to create parks, and for the first time scenic standards were being used to argue *against* the creation of parks. Standards had previously been the basis for deciding to keep areas free from resource exploitation, but by the early 1920s Yard was deploying them to counter the popularity of outdoor recreation and its reconstitution of park politics. However much Yard wanted to hold to a notion that he and the NPA were defending the parks against misinformed and disingenuous schemes, he was actually in a contest to define the very meaning of national parks in an era of skyrocketing public interest in outdoor recreation. As Yard would later admit, he was engaged in a partisan battle in which there was little difference between publicity and propaganda.[57]

Yard concluded his testimony before the Public Lands Committee with a crucial suggestion that said much about his fight for standards. "The area of travel, of outdoor living, of recreation in the open," he indicated, "has dawned these several years, and is in the beginning of a wonderful explosion." Municipalities, states, and counties had already begun to provide for the needs of recreational users, he noted, but the federal government was lagging behind. Yard suggested that the federal government, through the Public Lands Committee, build a national "recreational system," separate from the national parks, to provide for the burgeoning demands of Americans. "So far," he said, "the National Parks System has had to bear the brunt of this immense demand," which promised only to crowd out the aesthetic, educational, and inspirational qualities of the parks. A separate recreational system, Yard hoped, would siphon off those visitors who sought only recreational escape, and for whom magnificent scenery and complete conservation were of secondary importance.[58]

In 1924, at the first meeting of the National Conference on Outdoor Recreation, Yard made a similar plea. He suggested a federal survey of scenic resources with the goal of putting together a system of recreational reserves. Toward this end, the NPA joined with the American Forestry Association (AFA) to undertake such a survey of the nation's public lands— a survey whose results would appear in 1928. As this effort suggests, Yard was enthusiastic about efforts in the mid-1920s to create a federal recreational policy and hopeful that a fuller accounting of this nation's recreational needs would save the national park system from dilution.[59]

In repeating his call for a separate system of national recreational areas, Yard increasingly praised and cooperated with the Forest Service and its recreational initiatives. This irked Mather and Albright, both of whom suspected the Forest Service of trying to capitalize on the popularity of outdoor recreation by moving in on the Park Service's domain. To Yard's credit, he appreciated the recreational pressures faced by foresters. Although some in the Forest Service leadership did support

recreational development for fear that the Park Service was bent on expanding its domain at their expense, most recreational developments on
the national forests came in response to existing demands that foresters
had little choice but to meet. While Mather hoped to contain most recreational use of the federal lands within his bureaucratic domain, Yard saw
the Forest Service as a necessary ally in his efforts to cleave ordinary
recreational provisioning from the national park program. He was willing to cede responsibility for recreation to the Forest Service, and to reserve for the national parks an educational and inspirational function.
To Mather, on the other hand, such a policy would have meant the loss
of his main political base. Understandably, he rejected Yard's narrow definition of park purposes.

Mather and Albright were not unconcerned about standards; they
often shared Yard's fears about the inclusion of substandard areas. When,
for instance, Secretary of the Interior Albert Fall tried to push through
Congress a plan for an "All Year National Park" in New Mexico, both
joined Yard and the NPA in opposing the addition. Fall's proposal had a
number of suspect elements: the area bordered on Fall's ranch and thus
smacked of a conflict of interest; it was composed of a series of disconnected areas, none of which was particularly scenic; and it sanctioned resource development and included areas where such development had already occurred. This was an easy proposal to oppose, and revelations
about Fall's involvement in the Teapot Dome scandal eventually put an
end to the plan.[60] But Mather and Albright were not so inclined to agree
with many of the other objections Yard made. Both were interested in
building up the park system, and embracing recreation was a sensible
way to achieve this goal. Yard's goal of perfecting the system by adher-
ing to stringent standards did not always sit well with them.

Yard was not necessarily hostile to the wave of outdoor recreational interest sweeping the country in the early 1920s, but he hoped to thwart
efforts to use national park status to lure people to sites of only recreational value. Thanks in large part to his publicity campaign, national
park status had become a powerful advertising tool. As long as it was
being used to signify truly important and worthy areas, Yard was content. But employing the National Park "trademark," as he called it, to advertise substandard areas and to manipulate tourists into visiting them
introduced a form of dishonesty into the relationship between publicity,
the parks, and the people that Yard held sacred. With the advertisement
of national park status, he realized, any area could be made into a tourist

site. The automobile and improved roads only heightened this problem. Motorists seemed bent on visiting as many parks as they could in a season, while devoting little time to contemplating or studying the qualities Yard thought distinguished park scenery.[61] As tourists became more acquisitive, they opened themselves up to promoters who tried to pass off as national parks areas unworthy of such designation. Without attention to education, visitors might not notice the difference. Thus did Yard connect advertising with mass commercial tourism, a form distinct from the cultural tourism he hoped to foster through his publicity efforts and educational programs.[62] The job of park publicity was to distinguish areas of national import from mere roadside attractions. Advertisers, Yard worried, increasingly blurred this distinction.

Elevating the Primitive

During the late 1920s, Yard began objecting more vociferously to the Park Service's recreational priorities. And as he came to recognize recreational development as a threat to the mission of the national parks, he saw such development—particularly roads—as a potential abrogation of complete conservation, a change from the teens and early twenties when such development seemed innocuous. In the process, he became more committed to the "primitive" as the essential park standard, and with the help of others he gave more time to formulating an argument for its importance. He was also exposed during this period to two other preservationist ideals: the wilderness policy being tentatively implemented by the Forest Service, and the "natural conditions" ideal formulated by the Ecological Society of America. Yard borrowed from each in formulating an argument for the primitive.

The Park Service continued to push for tourist development. Yet, despite a growing and more demanding constituency, it received only paltry appropriations for improvements in the first decade of its existence. Mather was impatient to modernize the parks and gain the financial clout to counter Forest Service recreational initiatives. Throughout his reports during the early 1920s, he repeated the need for funding to improve and expand the system of park roads. While the Forest Service had received federal funding to improve roads within its holdings, park officials struggled with appropriations that were inadequate even for basic maintenance.[63]

In 1922, Mather called for a three-year $7.5 million effort to extend and modernize park roads. In April 1924, Congress granted Mather his wish, and beginning in 1925 the Park Service began to redesign park road systems with motorists in mind. In his annual report for 1924, Mather outlined a plan to modernize the existing road network in the parks—a little more than 1,000 miles at the time—and to add 360 miles of new roads. Sensitive to potential objections that the parks would be "gridironed" with roads, Mather insisted that large sections of each park would "be kept in a natural wilderness state without piercing feeder roads" and "be accessible only by trails by the horseback rider and the hiker."[64] He seemed conscious of criticisms that roads were destroying the few remaining wild areas in the West. Others in Park Service leadership positions, however, promoted a balance between provision and preservation in terms that revealed a less favorable assessment of wilderness. Without improved facilities to accommodate public use, park superintendents wrote in a 1922 policy statement supporting park development, a national park would be *merely a wilderness,* not serving the purpose for which it was set aside, not benefitting the general public."[65]

While Yard was not yet ready in the mid-1920s to see the parks preserved as "mere" wilderness areas, he was increasingly inclined to limit road development. He also began showing signs of unease with park promotion. In a 1925 letter to new Interior Secretary Hubert Work, Yard shared his concerns about how the outdoor recreation boom was transforming the parks. "The National Parks System has the stuff already in it for a unique and wonderful national institution," he noted, "but it is so loaded up with indifferent exhibits, and its shining purpose is so hidden behind its official recreational promotion that its great fundamental purpose doesn't get over." He thought he had found in Work someone who appreciated that the national parks should have a "higher status." Yard told Work that the NPA "would gladly give up the job of promoting" the national parks in exchange for a more explicitly educational role. Education and promotion were, for Yard, no longer one and the same.[66]

By the mid-1920s, as he turned more adamantly to education and interpretation as his and the NPA's mission, Yard was in the market for another mentor, someone upon whom he could place the mantle of "leader in the educational field." He found his man in John Campbell Merriam, a paleontologist who was head of the Carnegie Institution in Washington, D.C. Like Yard, Merriam saw the parks as "Super-Universities of Nature," places for the study of nature at its most magnificent. As a paleontologist, Merriam had developed an almost deistic approach to nature that equated the most beautiful and sublime natural spectacles with the powers of a creator whose works were best studied with scientific tools. For Merriam, untouched nature was, to borrow from the title of one of his later books, "the Garment of God." Merriam shared Yard's attachment to geological and evolutionary standards; he too saw the most scenic areas as illustrative of great natural forces. But Merriam expressed these standards with more eloquence and authority, and he provided an intellectual model for steering visitation in an educational direction. As recreational trends muddled park purposes, Yard drew inspiration from Merriam's conviction that scientific study could heighten one's appreciation of scenery and perhaps determine which scenery was worthy of higher appreciation.[67]

As Merriam's thoughts became a prominent feature in the NPA's *Bulletins,* Yard increasingly identified him with advocacy for the "primitive," for saving in the parks "records, nowhere else existing, of uninterrupted evolution."[68] Merriam's emphasis on the primitive had a profound impact on Yard. Whereas in the late 1910s and early 1920s Yard had seen wilderness as backdrop, Merriam's study of the primitive prompted in Yard a deeper commitment to the aesthetic importance of natural areas that remained relatively unaltered by humans. Yard's increased use of terms like "primitive" and "primeval" reflected a growing reverence for environments having an aged and undisturbed quality, even if they were not sublime in the conventional sense. Under Merriam's influence, Yard came to see that the preservation of primitive or primeval nature could be as aesthetically important as preserving spectacular scenery.

This new emphasis on the primitive also gave Yard a tool in his fight against the Park Service's recreational drift. In a preface to an article by Merriam in the 1926 *National Parks Bulletin,* Yard reiterated his attachment to standards in a way indicating that times, and threats, were changing. "Danger to the National Parks comes from three sources," Yard warned:

(1) From industrial companies that want to use the parks for profit; (2) from communities which want to attract profitable motor crowds by offering local national parks developed and maintained at the expense of the national government; and (3) from one-idea enthusiasts for unlimited recreational expansion who call for new and enormous national parks, irrespective of established standards, and necessarily in competition with State parks and National Forests.[69]

Yard noted that industrial invasions were a diminishing concern, and that it remained for the NPA to counter the other threats, which were "just in their beginnings and far more difficult to meet."[70] Merriam's leadership, and his educational agenda, promised some help in these areas.

Yard and the NPA continued to push their comparisons between the national parks and other institutions of high culture, but Yard's references to the parks as "galleries" of "scenic masterpieces" diminished as he shifted his rhetoric to laud the parks as museums and universities. Indeed, he advocated the development of a system of national park museums, modeled on the museum created in Yosemite in the early 1900s, and he and Merriam were instrumental in the creation of the Yavapai Point Interpretive Center at the Grand Canyon, a model interpretive exhibit. Yard also promoted the nature guide idea, which had been initiated in Yosemite.[71] Finally, he encouraged the use of the parks for university and other school courses, and he supported initiatives, many of which bore a striking resemblance to his earlier promotional efforts, to educate the public through the use of lecturers, lantern slides, and other visual aids.[72]

In these educational efforts, Yard and the NPA received some support from the Park Service. In his annual reports, Mather made frequent mention of initiatives to promote education within the parks. But Yard saw much of this as lip service, or as an effort to provide for interpretation without making the more difficult distinctions between recreational and educational uses of the parks. For their part, Mather, Albright, and Arno Cammerer (who became Assistant Director when Albright took the Superintendent post in Yellowstone) were wary of overdoing the museum idea; they saw it as "stuffy" and potentially unpopular. They also had limited funding and understood keenly that larger appropriations would be tied to increased visitation. For Yard, on the other hand, the museum idea was more than an interpretive addition; it was a guiding metaphor.[73] He read the Park Service's skepticism about museums as a clear sign that they prioritized recreation, not education. "They advertise the parks for travel," Yard complained in 1926.[74] As a criticism of the Park Service, this statement reveals how far Yard had strayed from his earlier commitment to park promotion and increased travel. Indeed, in 1925 the NPA officially dropped the promotion of travel, one of its original goals, from its mission statement.[75]

Mather and the Park Service did embrace educational efforts more completely by the late 1920s, in part because of lobbying by Chauncey

Hamlin and the American Association of Museums. In 1928, with the aid of a $10,000 grant from the Laura Spelman Rockefeller Fund, Mather created a committee to survey "the educational possibilities of the National Parks." The committee's report echoed the NPA position on most issues, probably because the committee included a number of NPA members. A year later, with the prodding of the educational committee's report, Mather created an Educational Advisory Board headed by Merriam. Another Rockefeller grant gave the Park Service $118,000 to set up a series of museums in Yellowstone National Park and to complete the Yavapai Interpretive Center.[76] In 1930, the Educational Division, which had been located in Berkeley under the guidance of Chief Naturalist Ansel Hall, moved to Washington, D.C., and became a major part of the Park Service's administration.[77] The park naturalist program also took off in the late 1920s. Perhaps the Park Service was not as indifferent to education as Yard thought.[78]

How did Yard make sense of the fairly rapid embrace of education by the Park Service? Were his suspicions about the Service's commitment to education unwarranted? His strident calls for an educational mission did diminish somewhat in the late 1920s, but there is little evidence that he read the adoption of education by Mather and Albright as a victory. He continued to express suspicion and dismay about their recreational motives. On their part, Mather and Albright gave Yard almost no credit for pushing the Service in an educational direction, though they did acknowledge the NPA and its other members for their assistance. They noted that their adoption of education was a response to the popularity of the limited experiments in the field. But, perhaps more importantly, they moved to where the money was. Mather and Albright were intent on building a constituency for the national parks, and they gladly accepted material assistance that helped in achieving that goal. Both had strong relationships with John D. Rockefeller, the most important park benefactor of the era, and Rockefeller was willing to fund many of these initiatives.[79] This is not to say that Mather and Albright were not interested in educational initiatives; both were. But neither was willing to turn the parks into the didactic spaces Yard envisioned. Nor were they eager to devote scarce resources to the sorts of delimiting projects Yard had in mind. For Albright and Mather, the educational facilities and programs were there to augment, not to shape, visitation. Education would not assume the central role that Yard hoped it would. Never would it be the guiding metaphor.[80]

Merriam's was not the only influence in Yard's growing attachment to the primitive. Yard was aware of the Ecological Society of America's program for preserving "representative" areas of "natural conditions," a program that was somewhat akin to the Yard-Merriam ideal of superlative examples of primitive nature. But where Yard and Merriam emphasized the aesthetics of geological and evolutionary processes, the main goal of the ecologists was to save areas that were particularly instructive of the processes of succession and climax. Even as his attention to the primitive forced him to contend with ecological ideas, Yard still held to scenic park standards that transcended the preservation of "natural conditions."

Yard also knew that the Forest Service, under Aldo Leopold's prompting, was working on a new "wilderness area" policy in which roadlessness, not complete conservation, was the defining component.[81] In fact, Yard corresponded with Leopold as early as 1925, probably in relation to the NPA-AFA survey of scenic resources on federal lands. Leopold, who had just published his flurry of wilderness articles, almost certainly encouraged Yard to emphasize the need for wilderness preservation in the survey results.[82] Soon, Yard was making unusually assertive statements about the need for a wilderness policy. "Because the automobile is rapidly possessing all outdoors, is it necessary to destroy the standards and safeguards of our only educational reservation system," he asked in a 1926 article. "Is it necessary, at irreparable loss, to dilute its primitive museum values with areas of the kinds which state parks are made of?" He obviously thought not. "Five years ago," Yard continued, noting how the times had changed:

> we who defend the National Parks System were charged with opposing the onward march of prosperity. It wasn't true. Now the new enthusiasts charge us with opposing the forward rush of out-door recreation. It isn't true. What is true is this—that, before the National Parks System can be completed, and turned to its highest usefulness, it must be saved, not only from those who would dump it into the channels of business, but from those, who, out of mistaken conceptions of plans and purpose, would reduce it to the general level of the country's playgrounds.

To achieve such salvation, Yard stressed the importance of education. But, because the Park Service had not given education the emphasis he thought it deserved, he also mentioned a distinctly new commitment. "By all means let us foster a national wilderness policy," he wrote, "if only, in the event of the National Parks System's submersion in the still

uncontrolled flood of recreational super-promotion, to discover a classification for the preservation of some samples of this beautiful and very
wonderful earth as God made it." Here, perhaps for the first time, Yard
suggested wilderness designation as a way to preserve areas *from* the Park
Service's "recreational super-promotion."[83] This was a crucial moment in
his transformation from park promoter to wilderness advocate.

The term wilderness had several meanings for Yard, as he mixed his
growing and multifaceted commitment to the primitive with Leopold's
emphasis on roadlessness. "There remain," he wrote, "a few exquisite
wilderness bits outside our National Forests and Parks—here and there
some acres of primitive forest in the immense mountain regions of
the East from which, elsewhere, the primitive has long passed; here and
there a few greater regions in the West, eddies among the rushing current of travel." Here Yard intimated that wilderness might mean different things in different regions. In the East, the preservation of "some
acres of primitive forest" was the primary job of wilderness preservation, while in the West he was more concerned about roadlessness. Yard
also mentioned that the work of scientists had begun to suggest the importance "of the interrelation of living things in the primitive wilderness," a theme that would recur in his later writings. Although he often
used the terms wilderness and primitive interchangeably, he most often
employed the former to indicate a commitment to roadlessness and the
latter to signify pristine nature.[84]

Yard's support for a national wilderness policy was a none-too-subtle
jab at the National Park Service. He essentially suggested that the Park
Service, in its eagerness to promote motorized visitation, could no longer
be trusted to protect wilderness. Increasingly he saw the improvement of
tourist infrastructures in the parks as a violation of complete conservation. Developments in Yosemite National Park were of particular concern.
During the summer of 1926, he traveled to Yosemite and became convinced that the valley was "lost." In a letter to George Bird Grinnell, NPA
president, Yard described how Yosemite had become a weekend resort for
San Franciscans, complete with a jazz band, a bear show, and other attractions.[85] A year later, he referred to Yosemite as "an Asbury Park of the
altitudes," deriding the effects of the Park Service's modernization efforts
and showing a distaste for commercial amusements in such a setting.
Yosemite, he feared, was being "sacrificed on the altar of Gasoline."[86]

Yard believed that Yosemite's newly constructed and newly paved
roads were to blame. These roads brought with them all of the trappings

of 1920s urban culture; they were making Yosemite into a modern place. The only consolation, he noted on a number of occasions, was that paved roads saved "the rest of Yosemite, still an unspoiled wilderness, for the use which Nature intended."[87] In a two-part article on motor tourism in the national parks, he wrote that the price of this first round of modernization was less the disappearance of substantial wilderness than the loss of "the natural quality of the areas of concentration." He hoped that the settling of the boom in outdoor recreation would allow Americans to see that the "ultimate usefulness of our National Parks System is not showing people 'sights,' still less furnishing them recreation, but offering inspiration, horizons and perspectives, elevation of the spirit, and education."[88] But to appreciate the parks as Yard hoped they would, visitors would have to get out of their cars and away from the artificial roadscape. In his second article on motor tourism, Yard suggested that the combination of automobiles and improved roads had made the parks into scenic drives. Much of the increased visitation came in the form of "passers-by" who were eager to see the advertised sights in each park. The impacts of these "habitual motorists" could be contained, Yard thought. "The vital questions," he posited, "are whether it is possible to hold the growing motor invasion to points of concentration and their connecting ribbon of road, and to subordinate road programs to preservation of the System's irreplaceable primitive."[89]

In 1928, Horace Albright wrote an article for the *Saturday Evening Post,* "The Everlasting Wilderness," that was a pointed reply to those such as Yard who argued that the modernization of park infrastructures had destroyed the wilderness quality of the parks. In the wake of the millions of dollars appropriated by the federal government for road and trail improvement, Albright had received, as he put it, "frequent letters from anxious inquirers who want to know if we propose to checkerboard the last wildernesses with highways." His reply was a resounding "no." The great majority of the funds, which that year had been increased from $2.5 to $5 million a year, were being used to improve existing roads. Albright insisted that many areas within the parks would remain untouched by roads. In the lesser-developed parks there were new roads being built, but in places such as Yellowstone and Yosemite, most new funding was going toward paving and landscaping existing facilities. The Park Service, Albright suggested, sought a middle ground between "those who want no roads in the parks and would keep them as unbroken wildernesses reached only by trails, and . . . those who are spokesmen for

automobile clubs, chambers of commerce, and other development organizations, whose appetites for road building are never appeased." The
Service sought only to make accessible the most important sites to the
growing number of Americans who were showing up in the parks,
putting the existing infrastructures under considerable stress. The expense of building park roads, Albright noted, precluded any large-scale
assault on park wilderness.[90]

Why, then, all the concern about wilderness if modern roads had not
threatened such areas? Albright may have underplayed the amount of
new road building that was going on, and certainly there were those
such as Yard who were disturbed by even modest efforts to placate boosters with new roads. But a more subtle process was also going on. That
wilderness seemed to be disappearing within the parks might have had
less to do with the physical expansion of roads into undeveloped areas
than with the ways in which improved roads altered the visual experiences of those who stuck to them. Certainly, part of what Yard found so
jarring and objectionable was how modern the parks were becoming. As
more cars streamed into the parks, traveling more quickly on roads of
urban quality, the modern aura of the roads and associated developments became a dominant part of the experience, challenging the ability of visitors to imagine themselves in a wilderness.[91] While ninety percent of most parks may have remained undeveloped, it was increasingly
difficult for visitors to see beyond the ten percent that did not. Albright
was perhaps right—for the moment—that road-building initiatives in
the parks did not threaten most park wilderness, but he failed to recognize, or was reluctant to admit, how profoundly modernization was
changing the experiences that most people had in the parks. Yard's opposition to recreational development was partly a response to such
changes—changes that helped to define modern wilderness.[92]

Albright and the Park Service faced a tricky task in accommodating
more visitors while also trying to minimize the intrusiveness of development. Yard was not particularly sympathetic to this conundrum. In
the abstract, he was not averse to increased visitation; he wanted everyone to be inspired by the national parks as he was. But his insistence that
the parks serve a specific, high-cultural purpose was linked to a tacit admission that only a certain percentage of the populace was interested in
this purpose. If the "true" message of the parks could be communicated
to Americans, he repeatedly intimated, the infrastructural problems
would take care of themselves—largely by weeding out those who were

merely after recreation from those truly interested in scenic appreciation. In this way, the logic of Yard's battle for standards allowed him to
avoid the complexities of increasing visitation. And when such increases
did occur, and were facilitated, he was too ready to see in them a sinis-
ter omen of eroding standards. This suspicious impulse grew from disillusionment with his own promotional efforts and the realization that
controlling the message of the parks would be even more difficult in an
age of mass recreation. When it came to the politics of visitation, Park
Service officials such as Albright were thus understandably frustrated by
Yard's advocacy.

But Yard's central insight was still an important one: the automobile
was not simply a neutral instrument of visitation but a technology that
had wrought, in a very brief period, a qualitative transformation in the
national parks and how Americans experienced them. It was incumbent
on the Park Service, Yard thought, to think more critically about that.

"It would need a book of its own," Yard wrote in his 1928 book *Our
Federal Lands*, "to present the visible changes the automobile has made
on the face of the country, to say nothing of its effects upon the human
view-point and character."[93] While Yard had not set out to write such a
book, he admitted that his entire discussion of the federal lands and
their various purposes had been shaped by the automobile revolution.
Indeed, it was telling that, less than a decade after he published *The Book
of the National Parks,* he found it necessary in his second book to examine the entire public domain. The motorized invasion of the federal
lands had made recreational planning a necessity. From Yard's perspective, saving the national parks involved taking a broader view of the public purposes of all the federal lands.

The writing of *Our Federal Lands* was motivated by Yard's work on the
joint NPA-AFA survey of the nation's recreational resources. Indeed, 1928
also saw the appearance of the survey results, a report titled *Recreation
Resources of Federal Lands* that Yard had a hand in preparing. The report's
section on the national parks was a compendium of Yard's concerns, and
it included extensive quotations from previous articles he had written.
But perhaps the most striking aspect of *Recreation Resources* was its strong
and detailed case for a national wilderness policy.[94] Both *Our Federal
Lands* and the NPA-AFA report suggest that, while Yard remained concerned with park standards, his preservationist goals had spilled beyond
Park Service bounds. Yard began to push for the preservation of wilderness wherever it occurred.

Eastern Parks and Shifting Standards

In the late 1920s, Yard devoted most of his energy to scrutinizing a series of "Southern" park proposals, as he derisively called them. (Yard employed the "Southern" label to suggest that regional boosters, jealous of
the West's preponderance of national parks, were behind these proposals, but the true regional import of these proposed additions was that
they were in the East.) It was during these fights—over Ouachita National Park in Arkansas, Shenandoah National Park in Virginia, Great
Smoky Mountains National Park in Tennessee and North Carolina,
Mammoth Cave National Park in Kentucky, and the Everglades National
Park in Florida—that Yard came to see wilderness as a solution not only
to road building but also to commercially motivated park making. In
each case, he questioned the scenic standards of the areas under consideration, and behind all of these efforts he saw attempts by local interests
to use the drawing power of the term "National Park" to attract tourist
business.[95]

Despite several attempts, the Ouachita National Park proposal—known
in another incarnation as Mena National Park—never saw the light of
day. Yard charged the congressional sponsors of the bills, Otis Wingo and
Joseph Robinson of Arkansas, with seeking local profits by having the
area "advertised by the title 'National Park' and opened up by motor
roads," and the National Park Service and Congress generally agreed with
Yard's assessment.[96]

The other proposals eventually met with success. In 1926, Congress
passed legislation enabling the creation of Shenandoah, Great Smoky
Mountains, and Mammoth Cave National Parks, pending state purchase
and consolidation of the areas to be included. Yard had particular problems with each of these parks. Shenandoah struck him as the stuff of a
recreation area or state park but not the caliber of a national park.[97] The
mountains were rolling and not particularly grand, many of the views
were of agricultural landscapes, and most of the area had been extensively logged and farmed. The park seemed designed to cater to the recreational needs of Washington, D.C., residents without concern for standards. As far as Mammoth Cave was concerned, Yard had a hard time
incorporating caves into his pantheon of the scenic, and the land to be
preserved above ground struck him as ordinary.[98] He was also dismayed
at how speedily the proposal had passed through Congress, with no objective assessment of the area's scenic attributes. The clamor for "the

tourist business which the fame of the title is supposed to guarantee" seemed to be driving Mammoth's creation.[99]

In Yard's evolving thought on wilderness, the most revealing and influential debates were the ones over the Great Smoky Mountains and Everglades National Parks. The problem was that these proposed parks would mostly be carved from private holdings. In the case of the Great Smoky Mountains, the federal government had agreed in 1926 to create a national park if and when the states purchased a required minimum acreage. Yard thought that portions of the Great Smokies were national park material. The problem was that there were two distinct areas under consideration. The first was, as he put it, an area of more than 100,000 acres of "impressive towering ridges mantled with original unmodified forest" held by lumber companies, and it would cost $30 to $60 an acre. "The balance," Yard reported, was "ordinary Appalachian hill, mountain and valley, lumbered and covered with second growth, worth about $5 an acre." Clearly, the quickest and cheapest route for the states of North Carolina and Tennessee to pursue was to ignore the primitive and purchase the cutover land. "Local sentiment," Yard lamented, "is for buying a whole lot of land cheap, so as to camp enormous motor crowds which are expected to rush into these States from every corner of the country, leaving trails of gold."[100] This situation was resolved when John D. Rockefeller stepped in and provided the funds to purchase the primitive areas along the spine of the range. Yard was relieved, but he remained vigilant about future development in the park.[101]

The Great Smoky Mountains episode of the late 1920s was especially significant for Yard because it prompted him to make primitive values rather than scenic magnificence central to the standards of a park addition. He saw park standards as leaning more heavily toward primitive conditions in the East, where they were rare. But he also began applying this eastern lesson to western parks, recognizing that although the primitive nature of places like Yellowstone and Glacier was incidental to their creation, it had become a central part of the identity of those parks.

The Everglades presented different problems. In 1928, Ernest Coe helped found the Tropic Everglades National Park Association to promote the idea of a national park in south Florida. Yard was initially skeptical. In a letter to Ray Lyman Wilbur, Hoover's Secretary of the Interior, he expressed specific concerns. The first was the coddling Albright had received by Coe's organization on his inspection tour. Yard thought the quality of the area should be determined more impartially. He also cited two letters from university scientists who had worked in the Everglades. Both advocated protecting the area for the sake of the birds and other tropical flora and fauna, but each was explicit that the location was scenically uninspiring. Merriam had apparently made a similar suggestion. Although Yard recognized the need for such preservation, he was not ready to see it happen in the form of a national park unless the area met established scenic standards.[102]

Yard nonetheless tried to remain neutral and await a balanced assessment. Meanwhile, visions of how the area might be developed were leaking out. In a letter to Merriam in October 1930, Yard laid out what he understood to be the three principal features of the development plan: a highway around the cape to connect the east and west coasts and increase tourist traffic; a seaside resort at the cape with hotels and motor camps; and motorboat and houseboat concessions to provide access to the more distant areas of the park. "The entire promotion," he noted, "is commercial in nature." "This scheme," he continued, "merely hastens [development] at the national expense and gives it the advertisement of government ownership and management. Whereas, if the Everglades are worth preserving, this is the kind of thing that a proper reservation might reasonably seek to prevent."[103] What sets this letter apart, aside from Yard's rejection of tourist development in the Everglades, is his recognition that park status could do harm to the area's biological resources. In a sense, he argued that national park status, as an advertisement for tourism in the Everglades, might spoil what was most important about the area. He thus opposed park development in favor of preservation, a departure from the days when he equated the two.

The input of scientists continued to influence Yard's thinking about the Everglades. In another letter to Merriam a few months later, he noted that Dr. L. H. Pammel, an eminent botanist from Iowa State University, favored a large "preservation of the primitive" in the Everglades but ruled out a national park on scenic grounds. Victor Shelford, the University of Illinois ecologist who was the driving force behind the ESA's Committee for the Preservation of Natural Conditions, also apparently favored "a national monument or wild life reservation of considerable size" instead of a national park. Moreover, he raised the disturbing specter that "drainage operations north and east have affected all of the Everglades so that primitive conditions everywhere are passing."[104] In doing so, Shelford challenged Yard's notion of damaged goods. For a long time Yard had fought against industrial invasions that marred a park's scenery and violated the sanctity of its borders. Now, with his attention shifted to the primitive and with the input of ecologists such as Shelford, he recognized that development outside a proposed park could affect the integrity of the natural processes within. The move from scenic preservation to the preservation of the primitive was not without its complexity.

Despite his many concerns, Yard was slowly warming to the idea of an Everglades National Park, but only under conditions that protected the primitive and assured limited tourist development. As these demands suggest, he was in the process of revising his standards, with the primitive displacing scenic magnificence and the specter of tourist development overshadowing the threat of resource exploitation.[105] The Everglades episode also brought into relief Yard's growing unease with the reconfiguration of preservationist politics during this era—with the way politicians and commercial interests had added their support to park preservation in the hope that it would bring government and tourist dollars. "Purists" such as Yard found themselves acting in coalition with parties whose motives seemed less pure. While some preservationists embraced the clout such coalitions lent their cause, Yard rejected these associations. Indeed, his growing attachment to wilderness came in response to this political unease. What he essentially argued, in the Everglades case and elsewhere, was that if proponents really wanted a national park and not just the tourist and government dollars it would bring, they would be willing to preserve it as a wilderness.

By early 1931, Yard was willing to accept a national park in the Everglades under the above conditions.[106] Part of what convinced him may have been an idea of Shelford's. Shelford was now advocating that the Everglades be preserved as a "biological national park," a preserve based solely on its biological importance. As NPA board member Henry Baldwin Ward wrote to Yard in relating Shelford's idea, "it [the Everglades] is absolutely unique biologically; and if we are to use that word at all as part of the definition of national parks or in distinguishing them from less important areas, then it applies here with the fullest force."[107] Yard's response was telling.

The problem, as Yard saw it, was not "an Everglades problem . . . but the problem of whether or not the historic conception and definition of national parks should be changed." "From the beginning," he continued, "the national park standard has been primarily scenic. . . . As Shelford puts it, we now propose to extend the range of national park inclusion so as to cover biology without conforming to scenic standards." The creation of such a new national park category would, he worried, open up a whole new barrel of pork; everyone would find an area of biological importance and want it included in the system. Yet Yard was sympathetic to the cause of biological preservation; he was willing to see scenic standards replaced or added to, so long as there remained standards that distinguished the national parks as unique areas of national importance. He proposed to let scientists decide. "I want to have this area thoroughly seen and studied by experienced men of scientific mind and training," he concluded, "who, from my knowledge of them, I know have the health and standard and uses of the national parks system soundly and principally at heart."[108]

Yard soon completely surrendered to scientific expertise as an arbiter. More importantly, he came to recognize that the primitive, as a scientific standard, was potentially more definitive than scenic quality. "Primitiveness," he felt, "was a quality which the national parks necessarily shared with no other land system; and it could be scientifically determined, which was true of no others. In primitiveness, we have something individual and provable."[109] This apparent objectivity and certainty appealed to Yard as an alternative to his subjective battle over scenic beauty.

An article by Frederick Law Olmsted Jr. and William P. Wharton also did Yard the service of shifting the notion of scenic beauty in a more scientific direction. The NPA, still concerned about the objectivity of the Albright group's assessment, had sent Olmsted and Wharton to the Everglades to report on its scenic attributes. While they found much of the scenery "confusing and monotonous," they also noted "a sense of beauty linked with a sense of power and vastness in nature." Beyond that, they listed two specific qualities that made the Everglades unique. The first was the mangrove swamps, which they characterized as being of a "climax type," the result of visible processes of succession. "The processes of change," they stated, "are so obviously before one's very eyes and occur in so unfamiliar a setting that they do seem to arrest the layman's attention and help him realize that the world about him is not the result of dimly comprehended geologic processes in the past, but is a shifting scene in a continual eternal drama." Again, the authors echoed Yard's conviction that superlative scenery was particularly revealing of the processes of creation, but they replaced the traditional geological and evolutionary aesthetic with an ecological one. The second point of scenic import was the abundance of bird life, a sublime phenomenon the authors described as "no less memorable than the impressions derived from the great mountain and canyon parks of the West." Olmsted and Wharton thus defined the Everglades as an aesthetically important area in terms that Yard could appreciate, even though they diverged from his traditional standards. With the primitive as a new model for park preservation, and with the scenic judgments of Olmsted and Wharton to fall back on, Yard could in good conscience understand the primitive in scenic terms while relying on its impartiality as a scientific standard.[110] With these assurances, Yard threw his support behind the creation of an Everglades National Park, which was authorized by Congress in 1934.[111]

Yard was an inveterate revisionist. As his standards shifted, he tended to read them back into the past, insisting that his convictions were distillates from a long tradition of park making. In the absence of a clear policy statement on park standards, he attempted to enforce what he saw as the weight of history. One of the most revealing of his revisionist documents was "Notes on the History of the Primitive in National Parks," a lengthy unpublished treatise, written in 1932, that amounted to an autobiographical account of his shifting standards. "In the beginning," he wrote, "Congress created national parks wholly of primitive lands, not because they were primitive but wonderful and beautiful." The "primitive," Yard wrote, was "an accident in the national park policy growing out of the fact that the first national parks happened to be unaltered by the hand of man. Sublime scenery was then the recognized motive." Theoretically, he continued, the national parks were the natural repositories of the nation's remaining primitive lands. Though the Forest Service had considerably more primitive land left, that land, according to the Forest Service's utilitarian mission, would eventually yield to economic imperatives. This was the theory, at least. But in practice, the Forest Service was making considerable efforts to save primitive areas while the parks were "forced, from their popularity, to yield increasingly important areas to the housing and pleasure of visiting tourists."[112]

By 1916, Yard insisted, the primitive had become an accepted part of the new National Park Service's mission, though "it was not publicly acknowledged for fear of arousing opposition needlessly in Congress." The accident of their primitive condition was subtly incorporated into the definition of the system, he thought, when Congress directed that the parks be left "unimpaired for the enjoyment of future generations."[113]

This was, in his words, "an intimation of the primitive without commitment thereto." Despite this intimation, early Park Service policy said little about the primitive, emphasizing instead the scenic character of the parks.

What served to distinguish the primitive as an essential part of the national park program, according to Yard, was its rapid disappearance beyond park boundaries. The automobile and auto touring were central to this process. "Civilization," Yard intoned, "has often been called the arch enemy of the primitive, but today we see preservation and use of historical, educational, and inspirational reservations of unmodified nature very definitely part of its program. It is the automobile, agent of material progress, destroyer of deserts, leveler of mountains and annihilator of time and distance that must bear that charge." Yard saw the automobile as uniquely culpable in the destruction of much of the nation's remaining primitive land. But he also intimated that, in destroying the primitive, the automobile worked against the imperatives of civilization itself. Magnificent and sublime scenery had become too intertwined with tourism, recreation, and the developments that facilitated both. By the early 1930s, the primitive was a much more functional standard for shaping the sort of park system Yard envisioned. That was how Yard told the story of his conversion in 1932. But there was one more important chapter to come.

A New Deal for the Parks: A Threat to Wilderness

Despite some signs that the Park Service was taking wilderness sentiment more seriously—a "roadless area" designation for a chunk of Mount Rainier National Park, for instance—the Depression presented opportunities for further development.[114] Although park visitation continued to increase in the first years of the Depression, rail travel and patronage at park concessions dropped off. As visitors opted for a more frugal approach, park hotels and other top-end services struggled. By 1932 a number of facilities were temporarily closed, as for the first time since World War I visitation declined.

In response, the Park Service returned to its promotional past and launched an advertising campaign designed to make the parks into places for renewing national confidence, focal points in Roosevelt's Depression era mobilization. The Park Service declared 1934 "A National Park Year," in large part to bolster the sagging receipts of concessionaires.

The Depression had incited a return to both the publicity and the nationalist rhetoric of the late teens. In an August 1934 radio address from Glacier National Park, Franklin Roosevelt suggested that the slogan "1934—A National Park Year" be changed to "Every Year a National Park Year."[115] Mixing his celebration of the national parks with the prominent mention of other public works projects throughout the nation, Roosevelt made a pitch for the restorative powers of nature and encouraged Americans to visit the parks as a patriotic gesture. By using the national parks as a symbol of the public-spiritedness of his Depression era programs, Roosevelt gave impetus to a new wave of recreation built upon his work-relief projects. Indeed, he ended on a "constructive" note: "We are definitely in an era of building—the building of great public projects for the benefit of the public and with the definite objective of building human happiness."[116] Roosevelt's commitment to development would be the final straw for Yard.

Depression era work-relief projects had a transformative effect on the national parks, as unprecedented levels of funding and labor power were directed toward infrastructural improvements that Yard thought were detrimental to the protection of park standards. By 1931, annual Park Service appropriations for road building and improvement were more than $7 million, with considerable sums going to the completion of new roads in Yosemite (Wawona Road) and Glacier (Going-to-the-Sun Road), and to federally funded parkways such as the Skyline Drive in Shenandoah National Park. In 1933, the Park Service's $7.5 million-per-year road-building budget was augmented by an additional two-year, $17 million Public Works Administration project to improve park roads and trails. The Park Service was also in charge of Civilian Conservation Corps efforts that would, in the course of its first three years, employ over 150,000 workers and 6,000 supervisors in hundreds of camps in the state and national parks. To Yard, all of this spelled a new war on the national parks.

Yard's concerns about the dilution of the system through substandard additions also gained new poignancy as the Depression deepened. As head of the Park Service, Albright made expansion of the system a top priority. In particular, he worked toward his long-held goal of obtaining and developing a variety of historic and military parks, a direction Yard bemoaned. During the early stages of Roosevelt's New Deal, the national battlefields and other sites administered by the Department of War were transferred to the Park Service, along with national cemeteries, national monuments from the Forest Service, a number of historic sites, and the entire capital parks complex. This expansion of park properties, a process that added to the system areas that diverged from his notion of standards, disillusioned Yard even further.[117]

The Depression also severely constrained the NPA's activities. With funds reduced to a trickle and membership plummeting, only four issues of the *National Parks Bulletin* appeared between the end of 1929 and 1935. For a variety of reasons, Yard's role in the NPA also diminished. He took an initial salary cut in March 1931, and another in November, when some of his duties as general secretary were shifted to the new position of "director." Loren Barclay, a New York businessman, filled this position. By the end of 1933, Yard had resigned from his position as executive secretary. He stayed on as a board member and Editor of Publications, while the position of executive secretary remained vacant for three years.

It is not entirely clear why Yard's role in the NPA diminished so markedly. In his history of the NPA, John Miles suggests that the reason was primarily the NPA's inability to meet Yard's salary, but that does not explain why Barclay was hired at the same moment Yard's salary was reduced.[118] Miles also suggests that Yard, who was seventy at the time, was looking to ease out of his central role in the NPA.

Yard may have sought a retreat from organizational politics to devote more energy to publications, but there is also evidence that he was being shown the door, perhaps as early as 1931. By that point, he had become a vocal critic of parkway proposals that threatened the wilderness qualities of the new eastern parks. The Skyline Drive in Shenandoah was one source of concern, but word of a proposed road along the ridgeline in the newly created Great Smoky Mountains National Park really raised Yard's hackles. It was an idea he first got wind of in 1932, and he placed much of the blame for the scheme on Albright's shoulders. "Less than a quarter of this park is primitive," he wrote in August 1932 to Merriam, "and it is the only great parcel of eastern primitive left. To bisect this on its towering crest by a motor road is a crime against the United States. It is an emergency duty to stop it." Yard went on to implicate Albright in a larger plot. "I am persuaded," he fumed, "that it was to get me out of the way for the Shenandoah crest road, and now for this road, that Albright worked his scheme. My successor naturally wouldn't get under the surface; and anyway he came charged to work with Albright."[119] In another letter, Yard suggested that when Cammerer took over the Park Service from Albright in 1933, Albright had assumed a greater role within the NPA and subsequently pushed for Yard's removal.[120] According to Yard, he was being marginalized in an attempt to mold a more cordial relationship between the Park Service and the NPA.

The various skyline road plans had mobilized a number of proposals for an organization to protect the wilderness. In 1930, Bob Marshall had called for such an organization, and in a 1932 letter to Merriam, Yard reported having dined with Marshall. Foreshadowing alliances to come, Yard wrote that "Marshall is a most valuable friend whom I wish we could associate with us."[121] In March 1934, Yard proposed to Merriam that they create an "organization to preserve the primitive." Such an organization, he suggested, "would have no rivals." "With you at its head, a small active board, I its executive secretary, a stenographer-clerk, and a room loaned by one of the scientific societies, it would cost little." He even attached the beginnings of a draft newspaper article announcing the formation of the new organization. "As a public cause," the article stated, "preserving the primitive by battling for national park standards has become insufficient because new systems of National Parks without primitive are appearing. A new nation-wide movement which shall promote study of the primitive, and its protection by name wherever found, inside or outside the National Parks, is due."[122] Merriam supported the idea but balked at a leadership role. Nine months later, with the founding of the Wilderness Society, Yard found himself at the center of just such an organization. For the next ten years, until his death in 1945 at the age of eighty-four, he was the Wilderness Society's most tireless lobbyist. Despite his age, he was still enthusiastic about preservationist politics.

Albright and Cammerer continued through the mid-1930s to view Yard's efforts as obstructionist. In battles over the eastern parks, in debates over including the artificial Jackson Lake within Grand Teton National Park, and in discussions over Olympic National Park, Yard consistently opposed the Park Service's pragmatic efforts to add to the system. For these efforts, Albright referred to Yard and his allies as "purists." In a pointed response, Yard accused Albright of relaxing his vigilance on standards so as to include substandard parks and to develop them in inappropriate ways. "I have not changed my views since I entered the National Park Service way back in the days when Steve Mather first came to Washington," Albright retorted. "You are the one who has altered yours." Albright was right. Over the course of two decades, Yard had altered his views. Although his desire to protect the parks and park standards remained in force, his views had shifted in response to changes that altered the parks and use patterns within them. Yard had wanted to see his commitments to the primitive and wilderness—terms that had separate meanings for Yard, though he often conflated them—as old ones, but Albright pointed out that they were, in fact, quite new.[123]

Conclusion

In questioning the various consequences of advertising the wild, Robert Sterling Yard helped to craft the modern wilderness idea. Yard began his career in the Park Service as a promoter of the national parks. He brought to the position a Progressive's faith in the power of publicity to shape and encourage use. But the rise of the automobile and accompanying cultural developments during the interwar era significantly undermined that faith. As modern advertising lost the educational function of publicity, Yard picked up education and made it his mission. After fighting water developments in the parks and giving expression to the doctrine of complete conservation, he grew disillusioned with the ways in which automobiles, improved roads, commercial tourism, local boosterism, and the nation's sudden affection for outdoor recreation warped his vision of the educational function of the parks. He tried to cling to educational initiatives as correctives, but with little satisfaction. Substandard park proposals popped up everywhere, playing upon the "advertisement" of national park status to lure tourists and their dollars.

Meanwhile, within the parks, development increased, particularly with growing road-building appropriations in the late 1920s and with the advent of the New Deal in the 1930s. Yard searched for the conceptual tools needed to respond to these new threats. While the park ideal— which to Yard meant scenic magnificence and complete conservation— had functioned effectively to defend against efforts to get at natural resources within the parks, it failed to provide guidance in controlling tourist development. Indeed, testaments to the sublime and majestic beauty of the parks, which had been potent ammunition in opposing resource development, had the opposite effect on tourist development. As such developments replaced industrial invasions as the major threats to Yard's standards, his traditional recourse to scenic magnificence and complete conservation was less effective. In response, he turned to wilderness to thwart the inroads of tourist development. For Yard, wilderness was tourist-proof and booster-proof nature—nature that resisted the pernicious effects of advertising. He also elevated the primitive and fell back on the authority of science to help craft a revised set of park standards. Indeed, Yard's advocacy enshrined both the primitive and wilderness (i.e., roadlessness) in equal measure.

In resisting the recreational tenor and developmental imperatives of the interwar years, Yard was accused of clinging to traditional ideas in a modern era, and by and large he was. He continued to believe that the primary purpose of the national parks was spiritual and cultural uplift, and he argued that those who were interested in "mere" recreation ought to find it in other preserved areas such as national forests and state parks. In this sense, he was an elitist. But he was also one of his generation's most perceptive observers of the boom in outdoor recreation, the role of nature in the logic of tourism, and the impact of the automobile on the natural world, in part because he felt that his publicity efforts made him culpable. To dismiss Yard as an old-fashioned elitist would be to miss one of the ironies of this story. Yard—a purist, a cultural custodian, and a cranky obstructionist—found himself allied, as the next two chapters will show, with a bunch of radicals.

CHAPTER 5

Wilderness as Regional Plan:
Benton MacKaye

Benton MacKaye was a visionary. His innovative synthesis of conservation and regional planning made him one of the most important and imaginative environmental thinkers of the early twentieth century. He was also a perceptive social critic whose environmental thought was informed by concerns about labor and community stability. Not only did MacKaye help to craft the modern wilderness idea, but the Appalachian Trail, which he first proposed in 1921, remains an important model for recreational preservation. Few people have had as great an influence on the landscape of preserved open space in America.

Yet MacKaye has been one of the most danced around characters in the American preservationist tradition, a figure who resists categorization precisely because his commitment to wilderness grew amid an intellectual polyculture of forestry, labor activism, democratic socialism, and regional planning. These are commitments that scholars are not used to seeing allied. MacKaye is also frustrating to assess because so many of his ideas either failed to materialize or did so in divergent ways. To some extent, these mixed results can be chalked up to the Achilles' heel of planning: strong visions usually require centralized power for their realization. To MacKaye's credit, he scrupulously avoided the use of such power in the planning process. Whenever possible, he incorporated democratic and decentralized models of implementation, hoping to foster a structured pluralism. The Appalachian Trail (AT) provides a case in point. Its completion in less than two decades was a remarkable example of what can be achieved when a vision is diffused among a group of loosely organized workers. But the transformation of MacKaye's vision into a completed trail came at the expense of certain prominent aspects of his broader regional plan, and his disillusionment with those developments fed directly into the formation of the Wilderness Society. As this chapter will show, wilderness was an idea that MacKaye came to while defending his AT vision against the perils of partial completion.

There are other reasons why environmental historians have neglected MacKaye's thought, some of them quite understandable. His writing was dense and, as his career progressed, increasingly abstract. He possessed neither the literary ability of Aldo Leopold nor the charisma of Bob Marshall. He often worked in isolation, and he never saw his ideas achieve the vogue of those of his colleague, Lewis Mumford. He lacked his friend's skill for self-promotion and the ego that drove Mumford to be so prolific. MacKaye's environmental thought also was imbued with a technological optimism that appears odd in retrospect. He, like many of his contemporaries, invested considerable hope in technologies such as hydroelectricity and the automobile to solve problems of urban concentration, congestion, and the culture's drift from nature.[1] He also had a deep faith in national planning, and in the power of planners and other technical experts to achieve the public good. These commitments lost their political luster in the postwar era, as Americans grew suspicious of government planning and as the very technologies and planning agencies MacKaye had believed in emerged as the bane of the environmental movement. Finally, his environmental thought had a social substrate that suspended it between conservation politics and the politics of social reform.[2] For all of these reasons, MacKaye has not fit easily within the pantheon of preservationist heroes.

Fortunately for MacKaye, environmental historians have turned a critical eye in recent years on the criteria that have defined that pantheon. This process has cleared ground for the emergence of figures such as MacKaye, who mixed social and environmental agendas. Indeed, one cannot fully understand the origins and content of Benton MacKaye's wilderness thinking without grasping the planning context and social critique in which it was embedded. One goal of this chapter, then, is to describe his unique brand of social conservation and to see how it evolved over time. That said, MacKaye's career also suggests that there were serious obstacles to integrating environmental and social reform. Those obstacles did more than litter his intellectual path to the wilderness idea. They defined it.

The Making of a Social Conservationist

The roots of Benton MacKaye's wilderness advocacy reach back, ironically, to a career and body of thought that had little to do with preservation. He was born in 1879, the son of well-known actor and playwright Steele MacKaye. Although they lived mostly in New York City, the MacKaye family took to visiting the town of Shirley Center, Massachusetts, in the late 1880s, and annual summer visits soon yielded to lengthier stays. Thus began a relationship of reverence with this archetypal New England community that lasted until MacKaye's death in 1975 at the age of ninety-six. From Shirley Center, wild country was near. As a teenager, MacKaye tramped through the forests and mountains of northern New England, using his youthful imagination to incubate visions of the region that took fuller form later in his life. From an early age, he came to appreciate the three environments he would later refer to as "indigenous"—the urban, the rural, and the primeval. Existing together, even symbiotically, they would define his ideal of a balanced landscape.[3]

MacKaye was among the founding generation of American foresters. He was the first graduate of Harvard's forestry school in 1905—the same year that Gifford Pinchot founded the U.S. Forest Service. Between 1905 and 1910, MacKaye alternated between teaching at Harvard's new experimental forest near Petersham, Massachusetts, and working as a Forest Assistant with the Forest Service.[4] He also showed signs of socialist leanings during this period. His brother James, the author in 1906 of a socialist tract titled *The Economy of Happiness*, was a prominent influence in this regard, as were Harvard acquaintances such as Walter Lippmann.[5] MacKaye was cut from Progressive cloth; like Aldo Leopold, he was a product of America's experiment in national forestry. But he was also something more; he was among an important cadre of progressives who hoped to push conservation in a more radical direction.

MacKaye made some important early contributions to national forestry. As a Forest Examiner in the early teens, he performed groundbreaking research on the impacts of forest cover on runoff in New Hampshire's White Mountains, research completed amid an intense debate over the connection between deforestation and irregular stream flow. Questions had emerged about the constitutional propriety of the federal purchase of eastern forests then taking place under the auspices of the Weeks Act of 1911; all the previous national forests had come out of the unclaimed public domain in the West. By marshaling scientific evidence that forest cover controlled stream flow, the Forest Service used the constitutional power to regulate navigation as justification for purchase. MacKaye's research provided the evidence necessary for the creation of the White Mountain National Forest, and it confirmed for portions of New England what Gifford Pinchot and others had been arguing on a national level: that there was a direct relation between rapid deforestation and increased flooding and drought.[6]

MacKaye's interest in radical politics grew in tandem with his forestry career. In the early 1910s, after a move to Washington, D.C., he became involved with a group of radical reformers, organized by William Leavitt Stoddard and Judson King, called the "Hell Raisers." Among the other members were Lincoln Steffens, the muckraking author of *The Shame of the Cities* (1906), and the economist Stuart Chase. From the hothouse atmosphere provided by this group, MacKaye nurtured various regional development schemes that combined progressive conservation principles with socialist planning. In 1913 he worked on a plan for the public development of Alaska, and the following year he embarked on a series of plans for the regional development of the Great Lakes cutover and the timber frontier of the Pacific Northwest.[7] In his Great Lakes plan, MacKaye first worked out his concept of "colonization," a model of agricultural settlement rooted in a critique of traditional federal policy for settling and disposing of the public domain. The cutover was an area that had been logged and then abandoned to ferocious slash fires. Subsequent agricultural settlement of the region under individualistic homestead principles had yielded high rates of failure.[8] MacKaye proposed a colonization scheme that stressed government-directed settlement, public ownership of land and resources, planning on the community level, and cooperative marketing. His goal was "to attain an average prosperity, not a segregated one—to develop the country not fast but well."[9] It was a vision he returned to and refined in subsequent years.

By the mid-1910s, MacKaye realized that his interest in colonization was pushing the conceptual bounds of the Forest Service, an agency whose directive was to conserve and develop resources, not human settlements. He increasingly found his efforts more akin to plans being hatched in the Department of Labor, whose officials sought to use the remaining public domain as a safety valve for the nation's chronic unemployment and underemployment.[10] MacKaye was intrigued by their plans, but he was concerned with more than alleviating unemployment;

he sought to align federally sponsored sustainable resource development with fairer and more stable labor regimes. He hoped, in other words, to unite the goals of the Forest Service and the Labor Department in a single integrated plan.

In Assistant Secretary of Labor Louis Post, MacKaye found a potent ally. Post was an economist and single-tax advocate whose journal, *The Public*, was a major outlet for reformist thinking. Post had been an associate of Henry George's, and he remained an avid follower of the Georgist economic philosophy, which targeted for taxation the "unearned increment" that came to landowners through speculation.[11] That MacKaye made such a criticism central to his nascent colonization plans appealed to Post.

In December 1915, MacKaye outlined his colonization ideas in a lengthy letter to Post. The ostensible purpose of the letter was to propose a cooperative effort between the Departments of Agriculture and Labor. Building on Labor's existing initiative, he advocated "a plan for the community settlement of wild lands" that proposed extending the homestead law beyond its "individual pioneer" stage. Public lands were to be withdrawn and made into "farm colony reserves," settled on a community rather than an individual basis, with government retaining title to the land. The government would give settlers equipment, access to start-up capital, and assistance in developing cooperative marketing systems. MacKaye's hope was that this "new homestead principle" would avoid the problems of speculation and failure, which had been particularly acute during the previous quarter century as settlers moved into increasingly marginal environments.[12]

By combining sustainable resource development with the planned settlement of land that would remain in public ownership, MacKaye sought to forge a federal conservation policy that integrated social and environmental concerns. There were good reasons for seeking such a synthesis at the time. In November 1916, MacKaye visited the town of Everett, Washington, in the wake of the famous Everett Massacre, the bloody culmination of organizing efforts by the Industrial Workers of the World (IWW) in the Pacific Northwest. When a group of about two hundred Wobblies attempted to enter the town by sea on November 5, 1916, a mob of deputized vigilantes confronted them on the docks and prompted a melee that left five Wobblies and two deputies dead and about fifty others wounded. MacKaye visited the town just weeks after the episode and took extensive notes on events that led to the confrontation. His experience at Everett, and the reverberations he felt from labor violence in extractive industries throughout the nation, got him thinking about the relation between labor unrest and unsustainable patterns of resource extraction.[13]

There is only scattered evidence that MacKaye was thinking about preservation during this period. In 1916 he did publish an article, "Recreational Possibilities of Public Forests," which provided the first public glimpse at his through-trail idea. In the course of making a general argument for the recreational usefulness of public forests—itself a novel idea at the time—he suggested that a system of existing trails in New England "could be made to connect through the New Jersey and Pennsylvania highlands with the Blue Ridge of Virginia and thence readily throughout the Southern Appalachian Range." Here, in embryonic form, was his vision of an Appalachian Trail.[14] But despite this brief foray into recreational advocacy, he remained focused on colonization, which he believed was the next necessary stage in the Forest Service's philosophical and institutional development.

MacKaye spent the late 1910s drafting colonization bills and trying to get the Forest Service and the Labor Department to work together on the issue. But Forest Service officials continued to insist that colonization did not fall under their jurisdiction. Louis Post requested that MacKaye be transferred to the Labor Department, where he could do his colonization work unfettered by such jurisdictional questions, but MacKaye clung to his Forest Service position, hoping that Post's request would push the Service to support his work. The Forest Service held firmly to its boundaries, however, and reluctantly let MacKaye go to Labor in January 1918.[15]

During 1918 and 1919, while with the Labor Department, MacKaye published several articles and a major government report elaborating his colonization ideas and his innovative conjunction between labor and resource policies. Increasingly, the salient context was the return of soldiers from World War I and the resumption of peacetime production, as policymakers feared that the demobilization of American troops would glut labor markets. Such concerns had prompted the Interior Department, under Franklin Lane, to produce their own proposals for the colonization of public lands—proposals that differed from MacKaye's primarily in their allowance for fee simple landownership.[16] But concerns beyond troop demobilization motivated MacKaye.

Labor violence in places such as Everett pushed MacKaye to critique the limitations of Progressive conservation. He began a February 1918 article in the *Journal of Forestry*, "Some Social Aspects of Forest Management," by pointing to what he called "the problem of the lumberjack." "Our forest schools," he observed, "in their processes of turning out foresters have courses in silviculture, mensuration, dendrology, protection, influences, management, utilization, lumbering, etc.; but the lumberjack himself and the very human problems that go with him do not occur in the curriculum." MacKaye attributed the labor unrest in the nation's timber camps to the "condition of unstable employment in the lumber industry itself," a condition that he thought would continue "so long as the forest industry is conducted as one of harvesting or 'mining' timber and not of reproducing and cropping timber." Transient, male-dominated work camps were a direct result, he thought, of the timber industry's wasteful practice of gutting vast areas and moving on. He lauded the achievements foresters had made in preaching the gospel of timber cropping, but that gospel had not gone far enough. "Compared, however, with the efforts made to advance forestry for the benefit of the consumer and business interests of the country," MacKaye asked, "how much have we done as to the social results of an industry of homeless men?"[17] Progressive forestry, he noted, promised producers and consumers a sustainable wood supply, but it had failed to address the issue of labor.

In response to this blind spot, MacKaye offered a new twist on sustainability—a vision of sustainable communities, planned and built by the federal government on the public lands, which would cater to the needs of forest workers. MacKaye's plan called for the Forest Service to structure working circles that fostered community as well as continuous timber production. At the center of these circles would be towns organized around logging and sawmill operations. Continuous tenure in one spot would enable the family and community stability that existing industry patterns failed to deliver. Central to MacKaye's picture of sustainable communities were democratic processes, public educational facilities, and collective bargaining rights. "Timber cropping," he also suggested, "requires a dependable and long-time form of ownership," which he thought best handled by the government. He thought that "the opportunities in this country for this sort of control exist most extensively in the 150 million acres of National Forests."[18] But he was wary of too heavy a government hand. "Community organization means self-organization," he added in a later article. "Let there be plenty of 'state aid' in getting started, but no paternalistic babying."[19]

MacKaye formalized these thoughts in a lengthy Labor Department report, *Employment and Natural Resources* (1919), one of the most imaginative documents to come out of the Progressive conservation movement.
He proposed a two-part strategy that combined immediate public works
jobs for returning soldiers and other unemployed persons with the development of agricultural, mining, and forestry settlements. He envisioned temporary work relief leading to permanent community development, and criticized projects that had no tangible end beyond
temporary employment. In terms of agricultural settlement, he suggested that the federal government sponsor road-building projects that
would yield to farm building, "conducted on the basis of the community
unit, not the isolated farm-unit." The roads would then provide access
to markets. He reiterated his plan for building sustainable logging and
milling communities, and presented a similar suggestion for developing
mineral resources on the public domain. "The mining community," he
suggested, "organized perhaps in connection with an agricultural unit,
could replace the typical 'mining camp.'" He speculated that these communities would create "a standard of profitable employment and suitable living conditions for the industry in all regions" at a time when
mining was also an area of great labor unrest.[20]

Employment and Natural Resources also gave fuller voice to MacKaye's
assessment of the nation's homesteading policy, a critique reminiscent
of John Wesley Powell's.[21] That policy, he maintained, had three major
shortcomings. First, it assumed that all one needed to succeed was raw
land. This may have been true in parts of the eastern United States where
fertile land and adequate rainfall were more common companions, but,
given the remaining public domain as of 1919, a more comprehensive
and cooperative plan seemed necessary. Doling out marginal land without government assistance or community support was a recipe for disaster. A second and related problem was the individualistic nature of
homesteading. MacKaye thought it more sound, environmentally and
socially, to make the community the basic unit of homesteading. The
third problem with traditional homesteading was that it gave away land
"in absolute fee simple title, minus any restriction."[22] This provision,
which opened up homesteaded lands to speculative pressures and allowed profiteers to amass large landholdings, surrendered the policy's republican intentions to the simple expediency of disposal. The result was
the frequent separation of ownership from use, which led to widespread
tenancy, tax reversion, and environmental abuse. In other words, Mac-

Kaye saw the alienated market in land as a primary contributor to the era's environmental and social abuses. His new homestead idea involved the settlement of lands by cooperative communities, with individual ownership tied directly to use and subordinated to community and resource sustainability.

Employment and Natural Resources was a testament to MacKaye's faith in efficiency, centralized planning, and technical expertise. In this sense, he shared with other conservationists of the era a commitment to what Samuel Hays has called the "gospel of efficiency." But while MacKaye embraced such goals and methods, he was adamant that Progressive conservation not become merely an exercise in efficiency. He intimated that a narrow focus on resource efficiency without a broader examination of social equity allowed conservation to play into the hands of the very business interests whose behavior it sought to reform. He was after a brand of social efficiency tied to equity, not business efficiency and profit maximization, and local control was always his goal.[23] The government's job, he thought, was to structure development in a way that facilitated democratic communities and guarded against the accumulation of private power over labor and resources. Yet he finessed the tensions between democracy and planning in his report, as he would throughout his career. He had an abiding faith in the didactic potential of planning, and he assumed that local democratic control, absent corrupting influences, would produce a community that shared his ideological vision. Thus, while he diverged from many progressive conservationists—or at least our current portrait of them—in avoiding a love affair with methodology, he consistently overstated the ease with which appropriate social ends could be identified and agreed upon in democratic communities.

Nonetheless, MacKaye's diagnosis in *Employment and Natural Resources* was compelling. By insisting that the nation's employment and resource problems were intertwined, he suggested that there was an environmental history to the era's labor unrest. In the nation's transient lumber and mining camps, and in the restless mobility of its pioneer farmers, he saw the social analogues of clear-cut forests, barren fields, and hillsides stripped of their mineral content. Although the legislative measures that came out of his report did not make much headway, *Employment and Natural Resources* offered a radical vision for the future of American conservation, a glimpse of a direction in which federal conservation policy and the development of the public domain could have headed. Unfortunately, the window of opportunity for such a scheme, if there ever was one, did not remain open for long.[24]

In all of his thinking on colonization and sustainable resource communities, MacKaye said very little about preserving nature. His 1916 article on forest recreation had an undercurrent of preservationist sentiment, but there were few other hints of such advocacy at the time. He was interested in work and nature, not outdoor recreation. For him, wilderness was still a blank slate upon which might be fashioned new communities in harmony with the resources that sustained them. In the late 1910s, there was little evidence of MacKaye's future attachment to wilderness. The forces that would spark such an attachment were still in their infancy.

A Radical Interlude

In June 1919, with his Labor Department report gone to press, Benton MacKaye entered a period of professional drifting and radical political activity as he continued to search for a home for his colonization ideas. During the summer of 1919 he was self-employed, writing articles promoting colonization and related issues for various newspapers and magazines.[25] In September, Benton and his wife of four years—Jesse Hardy Stubbs MacKaye, known to most as Betty—moved into a cooperative house in Washington, D.C., which they shared with, among others, Stuart Chase and his family. The living arrangement provided MacKaye with another salon-like setting in which radical politics almost always propelled the discussions. In a play on the name Hull House, Jane Addams's Chicago settlement, they called their living arrangement "Hell House." It was a partial reconstitution of the "Hell Raisers" of the early 1910s. Among those who frequented Hell House during this volatile time were Sinclair Lewis, John Reed, Frederic Howe, and Scott Nearing.[26]

As the Red Scare deepened, and as house members such as Chase became targets of federal investigations, MacKaye and the Hell House cell turned to increasingly radical visions. In March 1920, MacKaye, journalist Herbert Brougham, and architect Charles Harris Whitaker sent a letter to Ludwig C.A.K. Martens, a representative of the Russian Soviet Federated Socialist Republic and the closest thing to an ambassador the unrecognized Bolshevik government had in the United States. The letter offered the services of a group of "competent technical workers from America in industrial and social fields." "In this resolve we act voluntarily," the letter noted, "and without seeking the official support of the American government, well knowing that its disposition is not at present favorable to undertakings of this sort." MacKaye and his colleagues clearly saw a future in socialism, and MacKaye hoped that the Russian Revolution would give him an opportunity to put some of his colonization ideas to work. But nothing came of the entreaty. Two months later, Martens's Russian Soviet Bureau office in New York was a target of the Palmer Raids, which aimed to shut down operations that smacked of a Bolshevik conspiracy in the United States.[27]

In August 1920, Benton and Betty followed their friend Herbert Brougham to Milwaukee to write for Victor Berger's *Milwaukee Leader*, one of the leading socialist newspapers of its time. Berger, a Socialist Party leader, had been convicted under the Espionage Act of 1917, which criminalized the use of the postal system to disseminate ideas and information that undermined the U.S. war effort. The act was a thinly veiled initiative to censor the press, and Postmaster Albert S. Burleson used its force to carry out an assault on organized socialism and labor activism. Berger gained fame when, after being elected to the House of Representatives from the state of Wisconsin in 1919, the House refused to seat him because of his prior conviction.[28] MacKaye had landed in another hotbed. During his brief tenure at the *Leader*, he continued to write about labor and resource issues. But in December 1920, after an apparent rift with the editor, he left *The Leader* and headed back east to search for yet another venue for his ideas.[29]

MacKaye moved to New York City and got work with the Technical Alliance, a group founded by Howard Scott, later a major figure in the Technocratic Movement of the 1930s, and inspired by Thorstein Veblen and his treatise *The Engineers and the Price System*. Veblen had envisioned, somewhat quixotically, the emergence of a revolutionary vanguard of engineers that would take the reins of industry from business interests and rationalize production for the public good. When that vanguard failed to coalesce on its own, Scott sought to organize it. The result was the Technical Alliance, a sort of radical consulting firm. They hoped to offer its services to industries making the shift away from the price system, but, after only a few contracts with the IWW, the group disbanded in May 1921. MacKaye was intellectually homeless again.[30]

About the same time, in April 1921, MacKaye's life was changed drastically by his wife's suicide. Betty MacKaye had been an activist on a number of fronts: she was a suffragist and a devotee of Alice Paul; she

shared her husband's dedication to socialism; and she was a vocal paci-
fist and advocate for international disarmament. She also had a history
of mental instability. A number of friends indicated that she bore her political commitments in ways that were physically and mentally exhaust-
ing. Betty had a breakdown in early April 1921 (not her first) and was
being taken to recuperate in Croton-on-Hudson when she slipped away
through the crowd at Grand Central Station. Her body was later found
in the East River.[31]

As a result of his wife's tragic death, and because of the repressive
political climate that followed the Red Scare and the Palmer Raids,
MacKaye moved in a different intellectual direction after April 1921. The
Harding administration had come to Washington and displaced the era
of progressive reform with normalcy. As MacKaye retreated from government service and radical politics, the alternative futures he had envisioned for both federal conservation policy and the public domain took
a different form. In the spring of 1921, MacKaye sought refuge at the
New Jersey home of Charles Harris Whitaker and began the process of
reinventing himself as a regional planner.

"A Retreat From Profit": The Appalachian Trail

It would be easy to characterize 1921 as a year of stark transition for Benton MacKaye. In many ways it was. But there were also important continuities buried by necessity in his less radical tone. Some of the changes
came as a result of falling in with a new crowd of architects, planners, and
urban reformers who encouraged him to develop his conservation
thought within a planning framework and connect it with the problems
of metropolitan growth. He also grew preoccupied with the forces that
shaped the spreading popularity of outdoor recreation after World War I.
Yet despite these changes, MacKaye remained interested in labor issues
and their relation to natural resource use. "Though the sharpest shift in
my career," he later recalled of this period, "it left me, as I have indicated,
with the same goal—habitability."[32]

MacKaye's central intellectual achievement during the early 1920s—
the Appalachian Trail (AT)—represented these continuities and changes
almost equally. While living with Whitaker, he began to outline a plan
for a walking trail that would serve as a backbone for the regional development of Appalachia. Whitaker, who was then the editor of the
Journal of the American Institute of Architects, was well connected with a

group of urban planners and architects that included Lewis Mumford, Clarence Stein, and Henry Wright. Together, this group made up the bulk of the American Institute of Architects' Committee on Community Planning (CCP). MacKaye's vision of a planned Appalachian domain held together by a trail struck these urban-centered thinkers as refreshing and well worth supporting. As historian John Thomas has pointed out, they saw in MacKaye's plan "a conceptual bridge between the conservation movement in which [MacKaye] had been trained and community planning with which they themselves had been almost exclusively concerned."[33]

MacKaye refined his Appalachian vision in two lengthy memos written during the summer of 1921. Like much of his writing during this period, these memos convey a sense of crisis emanating from the shock of war and the social strife that followed. He spoke constantly of "reconstruction" and "readjustment," as if the current order had reached an impasse. While this feeling of crisis may have distorted his sense of the possible, MacKaye was one among many Americans who saw the war as a transformative moment and its aftermath as a time of great possibility. Yet he also seemed to sense the complacency and conservatism of the coming decade. As a result, these memos, though strong calls for action, also gave hints of a strategic retreat.

In his "Memorandum on Regional Planning," MacKaye defined his newly adopted discipline in familiar terms. Regional planning, he wrote, is "the conscious deliberate working out of a systematic method for developing, as far as still possible, the natural resources of a region (or locality) so as to convert those resources into human needs and welfare."[34] He then discussed, at length, his desire to see the reorganization, along socialist lines, of the resource economies of the Appalachian region. Here, in even more utopian form, were direct echoes of his colonization plans. Only in an appendix did he describe the trail itself.[35]

With the encouragement of Whitaker and Clarence Stein, MacKaye reworked his ideas in another memo, "Regional Planning and Social Readjustment," which focused more on his recreational vision and its relation to regional planning. "Regional planning," he insisted, "must provide not alone for man's work but for his leisure." While he maintained his interest in the relationship between labor and natural resources, he simultaneously embraced the increasingly conventional notion that mechanization and industrial abundance could, if properly controlled, reduce drudgery and allow more time for what he saw as life's meaningful pursuits. Appalachia could be, he noted hopefully, "a place to live, work, and play on a non-profit basis."[36]

MacKaye's new focus on recreation was also tactical, as he made clear in a rare discussion of garnering support for his Appalachian plan. He envisioned the creation of both industrial and recreational communities in Appalachia. Labor organizations and radical political groups, he wrote, would provide the natural constituency for the industrial communities, while hiking clubs and other outdoor organizations would likely support the recreational communities. Given that the political backing for the recreational component seemed stronger and less charged, MacKaye suggested that recreation be used to publicize his plan, while the industrial communities could be embedded as a "flank attack." He clearly understood that relying on a middle-class recreational constituency for support would be both necessary and dangerous. The danger was that recreational users might excise the social components of his AT vision. In this fear, MacKaye proved prescient.[37]

MacKaye refashioned these memos into an article worthy of public consumption, and Whitaker published "An Appalachian Trail: A Project in Regional Planning," in the October 1921 edition of the *Journal of the American Institute of Architects*. The article was a transitional piece for MacKaye. It was his first lengthy public discussion of the trail itself, and of the need to preserve portions of the landscape for outdoor recreation. In tracing the genesis of MacKaye's attachment to wilderness, this article would be a good place to start, but it also bore the unmistakable imprint of his intellectual activities during the previous decade.

MacKaye began the piece by noting a new development, the "recreation camp," that was an outgrowth of the outdoor recreation boom after World War I. The point, presumably, was to contrast this sort of camp with the timber and mining camps that had previously drawn his attention. Tying the "recreation camp" to what he generically called "scouting," MacKaye made a case for outdoor living as a way of escaping the "fopperies" of modernity and extending leisure opportunities to all Americans. This was fairly standard fare for the era. But he pushed this argument one step further, suggesting that scouting should seek to instill not only the ability "to sleep and cook in the open" but also the ability "to raise food with less aid—and less hindrance—from the complexities of commerce." By connecting the recreational aspects of scouting to what he called "the problem of living," he suggested a larger purpose toward which outdoor recreation might be directed. He hoped to develop in Appalachia a recreational domain that would itself create opportunities for resettlement and employment in the region.[38]

The AT article was a turning point in MacKaye's thought on the subject of wilderness, though the turn was not a particularly sharp one. "Camping grounds, of course, require wild lands," he wrote, and for his plan to work those wild lands would have to be protected. In the West, he noted, the nation had already provided for "national playgrounds" by creating national parks and forests. But in the East, where most could not yet afford to travel to such distant preserves, there was a need for accessible space devoted to outdoor recreation. The Appalachian skyline, MacKaye thought, "would form a camping base strategic in the country's work and play."[39] For this reason, some of its wild country would have to be preserved. His AT article was no more specific than that.

MacKaye's creeping interest in preservation did not mean that he was abandoning other possibilities for the region. He bemoaned the steady decline of the nation's rural population and the subsequent growth of cities—concerns at the core of the interwar period's pronounced regionalist sentiment.[40] Preservation of nature in Appalachia would provide a large portion of the nation's populace with an accessible recreational escape, but, again, MacKaye hoped for more. "There are in the Appalachian belt probably 25 million acres of grazing and agricultural land awaiting development," he concluded. "Here is room for a whole new rural population. Here is an opportunity—if only the way can be found—for that counter migration from city to country that has so long been prayed for." Such a migration required what MacKaye referred to as a "*new deal* in our agricultural system." And to this new deal in agriculture he added a potential new deal in forestry that would replace "timber devastation and its consequent hap-hazard employment" at a time when Appalachian timber production was in decline. He still hoped that his colonization ideas would find a home, this time within the broader confines of his AT proposal.[41]

In one of the more telling sections of MacKaye's important article, he asked the reader to envision a giant traversing the Appalachian ridgeline from north to south. For MacKaye, this giant was a metaphor for the regional perspective, and what better place to get a perspective on the region than from atop the mountains that flanked the eastern seaboard.[42] As the giant strode south, it took account of the surrounding countryside, but this was less a scenic journey than a survey of resources. When the giant finally reached the trail's southern terminus, it counted "the opportunities that yet await development." There were a number. First, there were opportunities for recreation; "secluded forests" and "pastoral lands" could serve the recreational needs of easterners, and workers in particular. Then there were opportunities for health and recuperation. In speculating that this mountainous realm could care for the "sufferers of tuberculosis, anemia, and insanity," MacKaye was part of a long tradition of ascribing to nature such healing powers.[43] But beyond these restorative qualities, there were "opportunities in the Appalachian belt for employment on the land" and possibilities for the redistribution of population. MacKaye saw recreation and recuperation as preliminary to what he hoped would be a reorientation in American economic culture toward nature and a direct working relationship with the land. "Coming as visitors," MacKaye wrote of the trail's users, "they would be loath to return [home]. They would become desirous of settling down in the country—to *work* in the open as well as to *play*."[44]

But how would such a transition be achieved? What was it about hiking the Appalachian ridgeline that would lure Americans back to the land? Americans, MacKaye thought, yearned for a more direct connection with the natural world, economically as well as recreationally. But what he really hoped the AT would encourage was "perspective," the essential link between mere recreation and a more thorough re-creation of modern living. From atop the ridges of Appalachia, MacKaye wrote:

> Industry would come to be seen in its true perspective—as a means in life and not as an end in itself. The actual partaking of the recreative and non-industrial life—systematically by the people and not spasmodically by the few—should emphasize the distinction between it and the industrial life. It should stimulate the quest for enlarging the one and reducing the other. It should put new zest in the labor movement. Life and study of this kind should emphasize the need of going to the roots of industrial questions. . . . the problems of the farmer, the coal miner, and the lumberjack could be studied intimately and with minimum partiality.[45]

This was a lot to hope for from a trail, but it was the essence of MacKaye's vision.

A wild setting was crucial to achieving the perspective MacKaye sought. From the ridgeline, hikers could see the world they had created below. In the solitude of a natural space that lacked the sights and sounds of their modern and mechanized lives, they could consider their impact on the world. Thus, from the early 1920s on, MacKaye connected Appalachian wilderness with the potential to achieve perspective on the socioeconomic state of the nation. In perhaps the most telling sentence of the piece, he insisted that the AT was "in essence a retreat from profit."[46] With access to such a physical and ideological retreat, MacKaye hoped, Americans would come to see the wisdom of his resettlement plans and their broader ideological goals.

MacKaye concluded his landmark article by outlining four specific features of his "project in housing and community architecture." First, there would be the trail itself. MacKaye suggested that the trail be divided into sections and that each section "be in the immediate charge of a local group of people." Upon the skeleton of the trail, he proposed a series of other developments. There would be "shelter camps," which would serve meals and provide sleeping arrangements for hikers. As a third layer of development, he urged the creation of "community camps," which he expected would "grow naturally out of the shelter camps and inns." These camps would be built adjacent to the trail and would provide for recreational users by offering "summer schools and seasonal field courses" that would foster the perspective he thought so vital to the trail's success. "Each camp should be a self-owning community," he insisted, "and not a real-estate venture." As the final stage of development, Mac-Kaye envisioned a system of "food and farm camps" that would give employment to those who needed it and make the trail as self-sufficient as possible. Permanent forest camps of the type he had suggested in his Labor Department report might even evolve out of such provisioning, perhaps on the newly created national forests of Appalachia. The trail, in other words, would be but an anchor for a larger project in regional colonization and development.[47]

MacKaye's 1921 article was a landmark in American environmental history—not only because it proposed an innovative through-trail that has become a well-worn part of the American cultural landscape, but also because it presented this trail proposal within the context of an unusual effort to craft a socially informed brand of conservation. Most scholars of American environmental thought have seen MacKaye's 1921 article as the beginning of his important career as a regional planner and a wilderness advocate.[48] But for MacKaye, who was forty-two when his AT article appeared, it was also the culmination of years of work and thought as a forester, a planner, and a political radical. Although others would find it easy to dismiss the social components of MacKaye's AT vision as quirky addenda to an otherwise admirable plan, for him these components remained central to what he hoped the trail would achieve.

In the wake of the AT article, trail activity proceeded on a number of fronts. The Committee on Community Planning (CCP) was the major early sponsor of the trail. Its chairman, Clarence Stein, secured reprints of MacKaye's article, wrote a brief introduction to it, and had the piece circulated to potential supporters.[49] MacKaye also threw himself into assembling support. In a letter to Stein, he mentioned an interesting cross-section of potential candidates for pushing the social aspects of this plan, including Gifford Pinchot, Jane Addams, and Sidney Hillman.[50] MacKaye also cultivated friends in recreational circles. He met with Allen Chamberlain, a leader of the Appalachian Mountain Club (AMC) and one of the organizational forces behind the New England Trail Conference (NETC). MacKaye attended the NETC's annual meeting in 1921 to gauge their support. "My impression from attending the meeting of the Conference," MacKaye informed Stein, "is that it will be comparatively simple to push on the trail portion of our program. The main problem will be how to handle the community feature."[51] He was correct in his first prediction; support for the trail came cascading in. But support for the "community feature," not surprisingly, proved more difficult to muster.

In the December 1922 issue of *Appalachia*, the AMC's publication, MacKaye spelled out the progress that had been made on the trail in its first year. The NETC was the soundest influence, with overarching authority over an already well-developed system of trails. The New York–New Jersey Trail Conference and the Appalachian Trail Committee of Washington also had done much to organize their divisions, as had the newly created Appalachian Trail Committee of Charlottesville. The Forest Service had aided in locating, building, and maintaining trail sections on its lands. Although four major sections remained unorganized, about one-third of the trail was already in place. MacKaye did mention the need for further development of communities, but he spoke of these only in recreational terms, as places where vacationers might settle for longer stays. In the end, the article was largely a pep talk for trail workers.[52]

MacKaye also offered an aside in his *Appalachia* piece that foreshadowed battles to come. The trail, he wrote, is "a walking trail, of course—not a so-called automobile trail."[53] The distinction was an important one, for at that time a variety of highway and "trail" associations were busy marking out recreational automobile routes through the country.

As he warned hikers about the threats that highways posed to his trail, MacKaye continued to hammer away at his broader vision. The vision was of utmost importance, and his defense of it would increasingly involve countering the inroads of the automobile.

The Regional Planning Association of America

In promoting his trail idea, Benton MacKaye had the support and critical feedback of one of the most important intellectual organizations of the interwar period—the Regional Planning Association of America (RPAA). An outgrowth of the CCP, it was founded in April 1923, largely at the behest of Clarence Stein. Other members included Lewis Mumford, Henry Wright, Alexander Bing, Stuart Chase, Robert Kohn, Charles Whitaker, Robert Bruère, Frederick Ackerman, and, later, Catherine Bauer. From 1923 through 1933, the RPAA was a leading voice in urban and regional planning and the springboard for MacKaye's regionalist explorations.[54] The AT was the group's first sponsored project, with Stein allotting MacKaye an initial $1,000 and feeding him cash increments throughout the decade.[55]

Like MacKaye himself, the RPAA is a tricky group to assess. Its energy dissipated after only a decade, when in 1933 the New Deal brought a number of RPAA members back into the government. They were loosely organized at best, more of an intellectual circle than a formal association, and their philosophy was never precisely spelled out. In letters to MacKaye, Lewis Mumford repeatedly complained that the group lacked coherent solutions; their regional planning was more a critique of how things were than a substantive vision of how things ought to be, Mumford said.[56] Nonetheless, there were certain shared concepts and concrete achievements that collectively make up the RPAA's legacy.[57]

Many of the RPAA's founders were housing experts concerned with the industry's inability to provide decent living spaces for low-income urban workers, and all urged a more prominent government role in meeting this need. Although such concerns predated World War I, the exigencies of mobilization prodded the federal government to get directly involved in providing housing for war workers, giving future RPAA members both hope and concrete work experience. Henry Wright, Robert Kohn, and Frederick Ackerman, for instance, all worked for the Emergency Fleet Corporation, a federal housing provider during the war. One of the distinctive features of the RPAA's agenda, then, was the desire to build affordable housing for workers and others whose needs were not met by the private housing market. In some respects, this was an urban analogue to MacKaye's colonization plans.[58]

The goal of equitable housing, while central to the group's identity, did not distinguish them from other modern urban planners. What did set them apart was their focus on the region as a planning unit. Regional planners criticized urban planners for failing to understand that the city and hinterland must be seen as a single unit, and for insisting that a city's problems could be solved by planning only within its borders. For the RPAA, urban problems were regional problems. The regional focus required that planners pay more attention to geographical and ecological boundaries and barriers; physical geography mattered, and proper planning both revealed and worked within natural boundaries. Finally, the RPAA aimed, whenever possible, to contain the relations of production and consumption within the region, an effort that foreshadowed some of the tenets of bioregionalism.[59]

Also emphasized in the RPAA's vision was the need to plan for economic efficiency. Stuart Chase was perhaps the one most committed to a planned attack on waste.[60] Economic inefficiencies were seen by the RPAA as among the greatest obstacles to the creation of just and habitable communities and sound environmental relations. In their attachment to efficiency through planning, RPAA members shared a commitment that animated many of the era's reformers. But, again, it was their ultimate goals—social equity and decentralized, livable communities—that distinguished them from other movements enamored of efficiency and planning.

A number of historians have emphasized the echoes of Progressivism in the RPAA's agenda, and such echoes certainly were there, but it was World War I that provided the more substantive influence.[61] National planning during the war was a compelling example of how government could intervene in the economy to rationalize production for the public good. Although MacKaye and his colleagues generally abhorred martial endeavors and criticized wartime infringements on civil liberties, the centralized economic planning that occurred during the crisis tantalized them.[62] If such an effort could have been shunted in other directions—toward the cooperative settlement of public lands, for instance, or toward improving worker welfare and urban living conditions—centralized planning might have been a powerful tool in rectifying the injustices of an economy driven by private capital. The armistice brought an abrupt end to centralized economic planning, but the war effort showed that the means existed and that the American public would support their use under certain circumstances. While World War I left many Progressives disillusioned, the RPAA saw in the extension of government planning power a glimmer of hope.

Other cornerstone concepts of the RPAA were borrowed from European planning traditions.[63] One was Ebenezer Howard's "garden city." As Howard, a British town planner, envisioned and designed them, garden cities were preplanned communities that combined the best elements of urban and rural living. Decentralization was the goal, but Howard was no mere proponent of suburban subdivisions. His garden cities were small and self-contained, with populated cores giving way to large concentric rings of open country dedicated to industry, agriculture, forestry, and recreation. By using greenbelts, Howard thought sprawl could be controlled and access to the countryside maintained. He envisioned such communities not as genteel retreats for the urban upper classes but as decentralized communities of workers engaged in local industry. He also emphasized public or municipal ownership as a way of mitigating speculative abuses and maintaining collective control over community development. The RPAA appropriated Howard's garden city idea, and elaborations by architects such as Raymond Unwin, as a rough blueprint for resisting market forces and situating planned communities in a regional context. Two such prototypes—Sunnyside Gardens in New York and Radburn in New Jersey—were among the most important achievements of the RPAA.[64]

Another important European influence was Patrick Geddes, who contributed several terms and concepts to the RPAA lexicon. In his idea of the "valley section" he suggested an organic metaphor—the region as watershed—that American regional planners, particularly MacKaye, embraced. Geddes also insisted that proper planning began with surveys of a region's resources, an idea that had a significant effect on MacKaye's AT vision. Finally, Geddes explained that the industrial world was on the verge of a transition from a "paleotechnic" era, defined by coal, iron, and dense urban concentration, to a "neotechnic" era in which emergent technologies such as electricity, the automobile, and the telephone would help decentralize population and industry. Mumford and MacKaye were particularly taken by Geddes's conception of an impending neotechnic era; their thinking during the interwar years centered on the promise and peril of planning around neotechnic technologies—especially the automobile.[65]

If any one term encompassed the fears of the RPAA, it was "metropolitanism," an epithet used frequently by MacKaye and Mumford. Metropolitanism was urban sprawl, the unplanned oozing of the modern industrial and commercial city into its hinterlands, guided by an economic logic that took little account of the quality of life. The attention to region, then, was an attempt to counter the growing power of the metropolis to shape regional geography. Decentralization was a double-edged sword, and they hoped to keep one edge honed and the other dull. MacKaye and Mumford also used metropolitanism to describe the spread of an urban-based, manufactured mass culture. With fewer Americans living on the land and with those who remained on the land increasingly connected to urban patterns of production and consumption, MacKaye and Mumford were concerned about the viability of cultures rooted in nature and place. Metropolitanism not only overwhelmed the livability of urban spaces, but its cultural influences eroded rural and regional traditionalism. In short, metropolitanism was hostile to regional distinctiveness.

This cultural regionalism allied RPAA members with other regional thinkers of the era who bemoaned the influence of the commercial metropolis and celebrated intimate connections between folk cultures and regional landscapes. Yet unlike other contemporary regionalist groups—the Nashville Agrarians, for instance—the RPAA was dominated by urban-minded thinkers who were sanguine about the possibilities of planning. They were notable among regionalist groups for taking the city seriously, even though, as later critics such as Jane Jacobs contended, many of their planning ideas seemed hostile to big city living—to vibrant street life and those facets of urban culture that resisted control by modernist planners.[66] The RPAA also was less inclined toward (though not immune to) folk nostalgia; their regionalism was self-consciously progressive, not reactionary. Rather than preserve distinctive regional facades in an otherwise metropolitan world, they aimed to create conditions under which future folk societies might thrive.[67] They were folk modernists. But, like most other regionalists, RPAA members were alarmed by the demographic and cultural urbanization they witnessed, and concerned that those processes would wipe out connections with nature, traditional knowledge, and the past.

The closest the RPAA came to a manifesto was the May 1925 "regional planning number" of *Survey Graphic*. The idea for such an issue had been MacKaye's, but it was Mumford who, as editor, gave it form and focus.[68]

The first substantive piece was Mumford's "The Fourth Migration," a quasi-historical, Geddes-inspired explanation of the ebb and flow of internal migration in U.S. history. For Mumford, the first migration was the pioneer's, and the settlement of much of the continent was its tangible result. Its point of departure was a New England civilization that Mumford and MacKaye venerated. The second migration saw a backflow of people into industrial towns. If the covered wagon and the agricultural homestead were the consummate symbols of the first migration, then the railroad and the factory town represented the second phase. Both had their problems. "[I]f the first migration denuded the country of its natural resources," Mumford concluded, "the second migration ruthlessly cut down and ignored its human resources."[69] A third migration to major financial and industrial centers emerged in the late nineteenth century, draining people and goods from factory towns, rural villages, and farms. The fourth migration, Mumford suggested, was the neotechnic movement out of those oversized urban centers and back into the countryside. The purpose of regional planning, Mumford wrote, was "not to create the fourth migration—that [was] already under way—but to guide it into positive and fruitful channels."[70] This was a crucial distinction. Providing such guidance necessitated a cautious embrace of the technologies and trends driving decentralization. Mumford, like MacKaye, did not see neotechnic influences such as electricity and the automobile as inherently positive. Only through planning could they be made to serve public ends. Without planning, they were powerful allies of metropolitanism.

"Regional planning," Mumford concluded, in a summary of the discipline that clearly owed much to MacKaye's influence, "is an attempt to turn industrial decentralization . . . to permanent social uses." It "is the New Conservation—the conservation of human values hand in hand with natural values"; it "sees that the depopulated countryside and the congested city are intimately related"; it is "permanent agriculture instead of land-skimming, permanent forestry instead of timber mining, permanent communities dedicated to life, liberty, and the pursuit of happiness, instead of camps and squatter settlements."[71] MacKaye could not have said it better.

MacKaye's contribution to the regional planning number, "The New Exploration: Charting the Industrial Wilderness," gave a different name and flavor to Mumford's fourth migration. For MacKaye, regional planning was a form of exploration, one that was peculiar to the end of the frontier and the rise of the industrial metropolis. The "old exploration," he explained, had charted global geography and mapped the wilderness of nature; the "new exploration" aimed at charting "the labyrinth of industry"—what he called the "industrial wilderness."

In "The New Exploration," MacKaye argued for a reorientation of the planning perspective. If the old exploration had started from urban centers and proceeded up river valleys to the fastnesses of mountain ridgelines, the new exploration would begin from the opposite end of the region. And if the industrial capitalism that guided the third migration had been hostile to topography and the friction of space, the new exploration would put the contours of nature to good use. Reaching back to his early Forest Service days, and to Geddes, MacKaye plucked the metaphors of the watershed and stream flow to give shape to his vision. The "New Exploration," he insisted, traced the flow of raw materials from "source regions" to industrial centers. It sought to unravel and rationalize the complex webs of commerce and migration, and to shape the landscape as a way of controlling commodity flows and demographic trends. "Our trail follows the crestline," MacKaye wrote, "and the new exploration may be symbolized in terms of the Appalachian Trail which is already threading its way down the backbone of the eastern states. For the crestline is the new frontier."[72] This hinterland orientation was crucial to MacKaye's regional perspective, and it suggested the importance that he would soon ascribe to wilderness.

Nonetheless, it is important to note that MacKaye used the term "wilderness" in this article mostly in a pejorative way—as a synonym for the messy complexity that needed to be decoded and rationalized. As of 1925, he would not have considered himself a wilderness advocate, though that quickly changed. His use of wilderness in this piece also suggested a deeper reservation about the term that would persist even after he became a wilderness advocate—a reservation that reflected his planner's urge for tidiness and legibility.[73] Wilderness meant disorder, among other things, and disorder was the foe in MacKaye's increasingly ambitious efforts to master modern geography. Even as he put a positive spin on the term in the coming years, he never quite embraced the disorderliness and lack of control that wilderness implied.

MacKaye's "new exploration" involved a process that he described as reverse pioneering. The goal was to conquer a new wilderness of industrial relations by saving remnants of the old wilderness of nature. Untangling such a mess, he was convinced, required stepping across the border that divided the spreading metropolitan world from what re-mained of "indigenous" America, as he referred to it, and looking back on what civilization had wrought. MacKaye talked about reverse pio-neering within a specific intellectual context. He had long been a critic of the environmental and social consequences of pioneering, but under Lewis Mumford's influence he also came to see pioneering as a culturally barren process. In *The Golden Day* (1926), a celebration of the flowering of an indigenous American culture in nineteenth-century New England, Mumford lambasted the pioneer life as "bare and insufficient." In the wake of "the epic march of the covered wagon," Mumford saw only "de-serted villages, bleak cities, depleted soils, and the sick and exhausted souls that engraved their epitaphs in Mr. Master's Spoon River Anthol-ogy." "In short," Mumford concluded, "the pioneer experience did not produce a rounded pioneer culture; and if the new settler began as an unconscious follower of Rousseau, he was only too ready, after the first flush of effort, to barter all his glorious heritage for gas light and paved streets and starched collars and skyscrapers and the other insignia of a truly high and progressive civilization."[74] It was an unsparing critique.

Mumford was not alone in this catastrophist assessment of pioneer-ing—an assessment that he later admitted was not entirely justified. Re-jecting America's frontier past as culturally sterile was a major theme of interwar cultural criticism. Critics such as Van Wyck Brooks, Waldo Frank, and Vernon Parrington also saw the frontier experience and spirit as prelude to a modern civilization that was void of cultural depth. The pioneer of the nineteenth century, they suggested, had not developed a useful folk culture but had become the Main Street consumer of the 1920s. Pioneering had indeed forged an American character, but it was not the one that frontier romantics, including some Western regional-ists, celebrated. Pioneering, they suggested, had produced Babbitt.[75]

MacKaye was deeply affected by *The Golden Day*, so much so that he mentioned it as one of two major influences that helped him turn "The New Exploration" into a book-length study by the same name—a task that occupied much of his time between 1925 and 1928. (The other major influence was Thoreau, himself a participant in Mumford's golden day.)[76] Mumford's cultural critique also shaped MacKaye's evolving commitment to wilderness. For him, the preservation of wilderness memorialized the missed chances to build rich regional cultures based in the land. Much of his early career had been aimed at keeping such an opportunity alive, but he also increasingly thought that the wilderness environment was itself worth preserving. In this way, one can see some congruity between Mac-Kaye's seemingly incompatible desires both to preserve wilderness and to settle and transform remaining portions of the undeveloped public domain. Even as he became a wilderness advocate, he remained keenly interested in how Americans might resettle the landscape in an equitable, sustainable, and culturally rich way. Indeed, MacKaye came to see wilderness preservation as a tool for guiding resettlement.

Toward an Architectonic Wilderness

After the regional planning number of *Survey Graphic* appeared in 1925, MacKaye returned to his home in Shirley Center to devote himself to writing a synthesis of his conservation and regional planning ideas. Life in Shirley Center, he admitted, was "on a cave man standard, with wood and water and light and every other primal element to be extracted from its source."[77] While he took a certain Thoreauvian pride in this direct working relationship with nature, it was not one of self-sufficiency. Over the next few years he fell back on several sources of income. He did some consulting, including a plan for recreational areas in Massachusetts. He did his best to emulate Mumford and make a living as a writer, but met with minimal success. Mostly, he relied on the financial kindness of friends, as official RPAA employment slid into informal patronage. Mumford and Stein sent a stream of "loans" MacKaye's way so that he could concentrate on his opus, and they helped him edit the manuscript and find a publisher. By 1927, MacKaye reported good progress. "In the present book," he wrote to Mumford in a concise expression of intent, "I am trying to develop a single idea—to meet the metropolitan challenge by the development of an indigenous environment as a *synthetic art*."[78]

The mid-1920s were good years for trail building. In 1925, eastern hiking clubs united to form the Appalachian Trail Conference (ATC), precisely the sort of decentralized federation that MacKaye had hoped would one day run the trail. Moreover, new clubs such as the Smoky Mountains Hiking Club in Knoxville, Tennessee, gave the ATC a strong presence along the entire Appalachian range. After 1925, the ATC would be the major force, and a very effective one, in building the AT, though this shift in the trail's sponsorship meant that power over its form was increasingly in the hands of hikers who did not necessarily understand or value MacKaye's larger vision.[79]

The years between 1925 and 1928 were also important ones for Mac-

Kaye in making connections with future Wilderness Society colleagues. He read and was impressed by Aldo Leopold's flurry of wilderness writings in the mid-1920s. Indeed, he wrote Leopold praising his work, and received an enthusiastic response. "It will be particularly interesting," Leopold wrote in 1926, "to talk over with you the unsolved questions involved in giving the wilderness idea actual expression in the form of a program. From the governmental end I can pretty well visualize the action which is necessary, but the idea really goes a lot further than merely governmental action. It is a point of view, not a piece of land."[80] With Leopold's help, MacKaye began to see that wilderness preservation could be a distinct part of his own agenda. In a 1926 letter to naturalist Walter Pritchard Eaton, MacKaye expressed a desire "for preserving a series of what Aldo Leopold would call 'wilderness areas'" along the Appalachian ridgeline.[81] Although he had earlier talked about the need for wild lands to protect the trail, this was the first time MacKaye explicitly connected the AT with wilderness preservation.

Part of what prompted this gravitation toward wilderness was MacKaye's growing concern with plans for new Appalachian national parks. In 1926, Congress passed legislation that enabled the creation of Shenandoah and Great Smoky Mountains National Parks, initiating a process that promised to protect large portions of the trail's route in the region.[82] But all was not well according to MacKaye. In a letter to Stein, he suggested that the RPAA should get involved in the planning of these new parks to prevent Appalachia from becoming "a Coney Island," the favored epithet at the time for commercialized and mechanized leisure.[83] The crux of MacKaye's concern was a nascent movement to build skyline roads in and between the proposed parks—roads that would intrude upon his trail and allow for the spread of automobile-borne metropolitanism. Given these threats, Leopold's definition of wilderness as roadless and recreationally primitive resonated with MacKaye.

Leopold was not the only future Wilderness Society founder with whom MacKaye corresponded in the mid-1920s. In November 1927, he received a letter from Harold Anderson, who announced the formation of the Potomac Appalachian Trail Club (PATC).[84] Anderson would be one of MacKaye's strongest allies in defending the trail against skyline incursions. A year later, Stuart Chase recounted having "met a bird the other night by the name of Ernest C. Oberholtzer, who is fighting for the Rainy Lake watershed park, as against a damned old lumber king." "He is eager to get in touch with you," Chase told MacKaye, "and anything you can

do to help him along I would appreciate." A month later, Oberholtzer re-
cruited MacKaye to serve on the National Board of Advisors of the
Quetico-Superior Council, an honor that MacKaye accepted.[85] Oberholtzer's efforts to preserve his region's canoe country were of a piece
with MacKaye's work in Appalachia.

As the AT inched its way toward fruition, MacKaye spent more time
developing trail philosophy and reminding trail workers of the AT's
larger purpose. In April 1927, he published an article, "Outdoor Culture—The Philosophy of Through Trails," that was both an important
preamble to his forthcoming book and a crucial transition point in his
developing commitment to wilderness preservation. Originally delivered
as a speech before a meeting of the NETC, the article explained a component of the AT proposal that had yet to receive a full airing—the "culture" of the trail. Again, this new attention to culture reflected the influences of Mumford and the broader regionalist movement. But there
was also a more specific purpose to this discussion. Developments in the
burgeoning field of outdoor recreation—its popularity, its commercialization, and its reliance on the automobile and other mechanized gadgetry—pushed MacKaye to further define the recreational nature of the
trail, a subject he had taken for granted in the past. "By 'culture' I mean
a special thing," MacKaye wrote. "It is not transcendentalism; it is not
erudition; it is not necessarily 'cultivated.' It is a special kind of ability:
the ability to visualize a happier state of affairs than the average hum-
drum of the regulation world."[86]

For MacKaye, the crucial distinction between "outdoor culture" and
modern outdoor recreation was that the former was constructive and
consciousness-raising while the latter was escapist and consciousness-
numbing. While he was careful to explain that outdoor culture was not
meant to be genteel or highbrow, he also hoped to distinguish it from
modern forms of commercialized mass culture that he, like many of the
era's radicals, thought substituted the opiate of consumer accommodation for the tonic of political mobilization.[87]

MacKaye saw the boom in outdoor recreation as mostly escapist and
accommodationist, and he attributed this to metropolitanism:

The modern metropolis is the product, not of its immediate region, but
of the continent and the world. It is a nerve center in a world-wide industrial system. Less and less is it indigenous; more and more is it stan-
dardized and exotic. It depends on tentacles rather than on roots. The

effect of an unbalanced industrial life, it is the cause of an unbalanced recreational life. For its hectic influence widens the breach between normal work and play by segmenting the worst elements in each. It divorces them into drubbing mechanized toil on the one hand and into a species of "lollipopedness" on the other.[88]

While "lollipopedness" was not a particularly enlightening term, MacKaye's larger point was clear. A major symptom of metropolitanism, he thought, was a segmentation of labor and leisure that made the former meaningless and the latter escapist. In each case, metropolitanism corroded "indigenous" culture—culture that had developed in relation to nature and place. The metropolitan forms of outdoor recreation that he saw developing seemed derivative of a mass culture that had grown away from nature.

To counter metropolitanism and to instill a sense of "outdoor culture," MacKaye proposed a barbarian invasion to challenge the reigning notions of the escapist "Civilizee." His language here betrayed the influence of Oswald Spenger, whose *Decline of the West* had just appeared. "Our Civilizee," MacKaye wrote, "sees in the mountain summit a pretty place to play tin-can pirate and strew the Sunday supplement; our Barbarian sees in the summit the strategic point from which to resoundly [sic] kick said civilizee and to open war on the further encroachment of his mechanized Utopia." The AT, MacKaye insisted, was not to become the domain of the Civilizee, a realm of effortless escape facilitated by the automobile. "The main thing," he wrote, "is to capture these areas and to hold them from further inroads of metropolitanism. Aldo Leopold would call them 'wilderness areas.'"[89] This last line was of considerable significance, for MacKaye increasingly defined the AT as a wilderness trail. As he conceived it, wilderness was one of a number of indigenous environments that stood in contrast to, and was designed to resist, metropolitanism as a specific urban form.

As a basis for MacKaye's through-trail philosophy, however, "outdoor culture" was a vague notion. He was trying to define a sort of recreational folk culture, and in doing so he hinted at the regionalist critique of consumerism as standardizing, spectatorial, and erosive of folk tradition. But he never developed this angle as much as he might have. Instead, "outdoor culture" functioned as a way of defending his ideological goals against those who saw the trail as only a recreational space. "Outdoor culture" was an abstract attempt to rewed work and play in a meaningful way. But, given the overwhelming popular interest in what MacKaye saw as escapist recreation, it was also a folk culture in search of a folk.

Nonetheless, MacKaye's articulation of "outdoor culture" marked an important juncture along his path to wilderness advocacy. It was in the context of making such *recreational* distinctions that he first began using the term "wilderness," and it was here that he started seeing roads and automobiles as the prime threats to wilderness. In 1921, the ridgelines of Appalachia had seemed safe from the immediate danger of metropolitanism, and recreational development seemed like a simple way of steering hikers and outdoor enthusiasts into a broader program of economic and social reconstruction. But, less than a decade later, roads, automobiles, and the consumer attitudes that came with them had changed all that.

In 1928, Harcourt, Brace, and Company released MacKaye's treatise, *The New Exploration: A Philosophy of Regional Planning*. Despite a long life of writing, it was the only book-length work MacKaye was able to publish.[90] Much of the book, particularly the early chapters, restated ideas he had expressed in previous articles, but *The New Exploration* did lend new depth to a number of familiar themes. For instance, he gave greater attention to defining the three "indigenous" environments—urban, rural, and primeval—that metropolitanism threatened to overwhelm. *The New Exploration* painted a vivid picture of Shirley Center and the qualities he revered in such settings. Shirley's scale, its embeddedness in place, its tangible connections to the past, its civic spirit, and its resistance to commercialization—all gave it a cultural depth and distinctiveness that made it, in MacKaye's mind, a pure product of the region. *The New Exploration* was an effort to communicate that kind of respect for place, to outline the threats to it, and to offer solutions.

MacKaye's elucidation of regional planning technique, however, was what made *The New Exploration* a book of lasting importance to planners and preservationists alike. *The New Exploration* was his most thorough explication of how regional planners could use open space conservation—and wilderness areas in particular—architectonically. Wilderness became a part of MacKaye's solution to the problems of modern metropolitan society, not only as a "psychologic" resource and physical protector of his trail, but also as a tool for the "remodeling of an unshapen and cacophonous environment into a humanized and well-ordered one."[91] In *The New Exploration*, MacKaye first fully articulated his sense of what planning historian Kermit Parsons has aptly referred to as "the city-shaping possibilities of open space conservation."[92] For MacKaye, wilderness was not an antipode but a counterpoise.

He again fell back on the watershed metaphor to describe his planning motives. Depicting metropolitanism as a flood or backflow from urban centers, he increasingly pointed to the automobile and modern roads as the vectors spreading commercialism and unplanned sprawl. Billboards, restaurants, shops, filling stations, and residences—an infrastructural system that MacKaye called the "motor slum"—spilled out of the city, littering the wayside environment. The question was how to control this flow.

In *The New Exploration*, MacKaye proposed two broad answers, using Boston as a model. His first suggestion was to create a system of open space "dams" to block and divert metropolitan flow. The Appalachian ridgeline was but the most distant of a series of natural barriers that, if preserved and put into public ownership, could function literally as dams against the spreading city. Second, he proposed to limit development along roadsides through zoning and wayside conservation. Such measures would provide a system of roadside "levees," preferably in public ownership, designed to prevent the motor slums he derided.[93] Under such a system, motorized traffic would flow across the landscape, connecting the urban, rural, and primeval environments and contributing to regional decentralization without the side effect of commercial sprawl. At one end of the region would be the contained regional city; at the other, the ridgeline world of wilderness; and in the middle, communities such as Shirley Center separated from main corridors of traffic.

Within this explication of regional planning technique, MacKaye gave a fuller description of wilderness and its planning role. In controlling the "metropolitan invasion," he noted, the regional planner should look for topographic features surrounding urban areas that might work as "embankments." Hills and mountains served this function well. Such areas, MacKaye continued, "could be reserved as a common public ground, serving the double purpose of a public forest and a public playground," by which he meant that they could supply recreation, watershed protection, and resources. Such a preserve, he concluded, "might be called a 'wilderness area.'" These wilderness areas would not only stop the spread of metropolitanism, but they would provide a form of counterdevelopment by creating "a linear area, or belt, around and through the locality, well adapted for camping and primitive travel (by foot or horseback)." MacKaye continued:

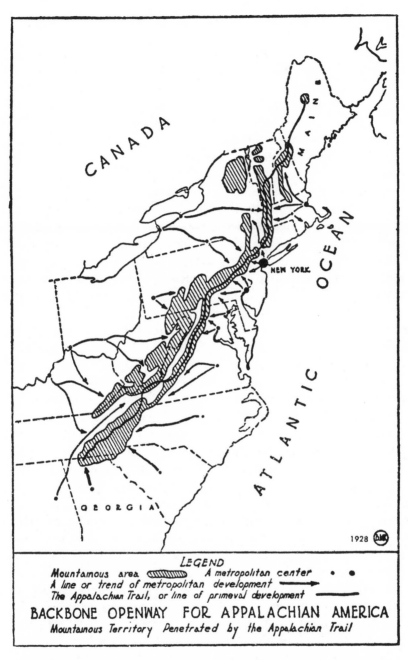

LEGEND
Mountainous area ⬭ A metropolitan center • •
A line or trend of metropolitan development ➝
The Appalachian Trail, or line of primeval development ━━━

BACKBONE OPENWAY FOR APPALACHIAN AMERICA
Mountainous Territory Penetrated by the Appalachian Trail

Backbone Openway for Appalachian America. By the late 1920s, as he grew more concerned about spreading metropolitanism, Benton MacKaye reconceptualized the Appalachian Trail as a dam set against metropolitan flows. (MacKaye, *The New Exploration* 1928, p. 199)

Overnight, weekend, and vacation trips could be made from the central city and from the adjacent villages by way of a number of varied circuits. This series of open areas and ways would form a distinct realm: it would be a primeval realm (or near-primeval)—the opposite realm of the metropolitan. These open ways (along the crestlines) mark the lines for *developing the primeval environment*, while the motor ways mark the lines for extending the metropolitan environment.[94]

Two aspects of this discussion of wilderness areas seem worth emphasizing. First, MacKaye's wilderness ideal, as a specific response to metropolitanism, was a direct reflection of the physical, technological, and cultural growth of the American metropolis during the interwar period. Roads and the automobile were a particular concern; pristine conditions and the absence of resource development were not. Second, wilderness areas were not to be preserved simply as exceptions to the metropolitan rule, as vestiges of a mythic past landscape, but as routes into, through, and around urban areas. They were meant not only to save nature but to save cities. For MacKaye, wilderness was a modern planning construct.

In the decade between *Employment and Natural Resources* and *The New Exploration*, MacKaye's environmental vision had become grandly architectonic. While he remained convinced about the need for resource communities along the lines he had sketched in the late 1910s, the influence of his RPAA colleagues and the landscape changes that had accompanied the explosion of automobile ownership during the 1920s had prompted MacKaye to widen his scope. By the late 1920s, his environmental vision encompassed the entire regional continuum, from resource hinterland to urban center, and it gave equal time to work and play. In expanding his view of the landscape, MacKaye embraced wilderness preservation as a tool of the regional planner. As Mumford later wrote in an introduction to a 1962 edition of *The New Exploration*, MacKaye's dams and levees "were a natural translation on his part of familiar wilderness terms to the new conditions for durable urban settlement."[95]

Although his wilderness thinking was indebted to Leopold's, MacKaye put his own stamp on the idea by stressing connectivity. In 1929, he published an article in *Landscape Architecture* defining what he called "wilderness ways." "A wilderness way," he wrote, "is a string of open spaces leading from somewhere to somewhere. But what is an 'open space'? An open space is an area of land, usually in public ownership, which is dedicated to some conservation use." The "open *way*," he continued, "is a somewhat new idea: it is a strategic method of distributing open spaces so as to gain greater access to the countryside in general."[96] The wilderness way was a method of connecting open spaces with through-trails to enable forms of travel other than the automobile.

In a suggestive conclusion to "Wilderness Ways," MacKaye intimated that, aside from its planning function, wilderness was also an ideal space to explore the ecological aspects of what he had earlier called "the development of an indigenous environment as a synthetic art." In explaining this function of wilderness, he made a persuasive distinction between productive and escapist recreation, giving greater substance to what he had earlier called "outdoor culture." "We can restore the primeval forest," he wrote:

> We cannot "make" a primeval forest any more than we can "make" an antique, but we can make a restoration. We can make it frankly and deliberately, like any other work of art. To do this we must learn to make it. . . . We must study the life history of the forest primeval from the lichens and the aquatics to the complete roof and canopy of hardwood, pine and spruce. In developing the forest, we are really developing ourselves; and after all it is ourselves which concern us rather than the forest. Here is the real strength of the wilderness way as a dam set across the streams of the metropolitan invasion: here is the ultimate pursuit of the real camper: to unfold the story of evolution on the out-of-doors horizon.[97]

We learn about nature to restore it and our relationship to it, MacKaye intimated, and we restore nature as a way of learning about it.

MacKaye was only beginning to give substantial attention to the "nature" of wilderness by the end of the 1920s. As a trained forester, he was well equipped to deal in ecological concepts, as his thoughts on restoration reveal. But his definition of wilderness did not rely on what some have called the pristine myth. Frankly, he was less concerned about the absence of human imprints than he was about keeping the modern world at bay. For MacKaye, wilderness was a place where nature dominated, but its preservation and management was a form of artifice that unified a deep respect for natural processes with a planner's modernist sensibility.

Controlling the Automobile

One of the central themes of *The New Exploration* was controlling the automobile through open space preservation. In the years after the book appeared, MacKaye devoted much time and thought to this problem. In the late 1920s, he worked on a design for the "Boston Bay Circuit," a two-mile-wide belt of parkland encircling Boston, anchored by a trail in the middle, and flanked by parkways on each edge. He envisioned the Bay Circuit as a hedge against metropolitan sprawl, hoping it would contain urban Boston and divide it from indigenous towns such as Shirley Center. His plan was never realized, though Boston did eventually get a similar loop road—Route 128—that became a magnet for just the kinds of development that MacKaye had hoped to thwart. Perhaps some of the features in his Bay Circuit design could have prevented this result, but he also miscalculated the developmental implications of even the most thoroughly planned transportation corridors.[98]

In *The New Exploration*, MacKaye had talked of designing what he called "intertowns" to control roadside commercial sprawl. By 1930, he had developed this notion into a more detailed plan for "townless highways," prototypes meant to complement the "highwayless towns" such as Radburn, New Jersey, that his RPAA colleagues were busy planning and developing. (These "towns for the motor age" strictly separated pedestrian and vehicular traffic.) MacKaye saw "townless highways" as a way of connecting such towns with each other, with contained urban centers, and with open space preserves like the AT. They were, in other words, key features in his vision of a planned regional landscape.[99]

The design of roads, MacKaye noted in a 1930 *New Republic* article titled "The Townless Highway," had changed little with the coming of the automobile. There had been advances in surface durability and other technical areas, but as a space the road was still treated as it had been during the horse and buggy era. MacKaye argued that highways, like railroads, were spaces for rapid and efficient transport more appropriately segregated from residential and commercial development. It was time, he thought, to redesign the road to make it safer, more efficient, and less likely to yield to the "motor slum." His design was for a limited access highway, without grade crossings, which preserved public waysides as open space and separated through-roads from living spaces. The townless highway was also an essential part of his broader vision of utilizing the automobile to achieve population redistribution.

Although MacKaye's "townless highway" design had an aesthetic component, it was much more than a plan for recreational parkways or highway beautification. It embodied his notion of roadside "levees," allowing people and commodities to flow from town to town and city to city without permitting the spread of metropolitanism. "Regional planning with these ends in mind," MacKaye concluded, "will in turn save both the local community and the open wayside environment, and give proper access to the wild places, instead of insidiously wiping out all these precious assets together."[100]

Some have suggested that MacKaye's "townless highway" was the seed of the parkway movement or of the interstate idea. Neither of these characterizations quite fits. Prototype parkways such as the Bronx River Parkway already existed, and, while MacKaye's plan did presage important elements of interstate design, the interstate system developed in ways that diverged from MacKaye's vision. What was most important about MacKaye's townless highway idea was not its technical design but its integration in a broader landscape planned, in theory, to be free from metropolitan sprawl. The success of the townless highway relied on other planning successes. As he told his congressional ally, Representative Robert Crosser of Ohio, the townless highway was "no mere anti-billboard scheme. It involves the control of the land (and its values) being used for further development, and so it is but another aspect of the colonization problem on which we worked together fifteen years ago."[101] This was a telling reference, and there were certainly connections. But it is equally important to note how preoccupied with the automobile MacKaye had become by the late 1920s.

A few members of the hiking community, such as the PATC's Harold Anderson, also were growing uncomfortable with the influence of the automobile. In 1930, Anderson sent MacKaye a copy of an essay he had written, "The Recreational Value of Hiking," that was about to appear in *American Motorist*. The piece made several important observations about the automobile's impact on the hiking community. The automobile, Anderson wrote, gave Americans easier access to the countryside and therefore to a wider array of recreational opportunities. But it also drove hikers off the country lanes (to which they were accustomed) and into the forests and mountains. Thus while it may have improved access to hiking venues, it tended to push such places further into the hinterland. "But if the automobile has made hiking harder," Anderson noted, "it has forced many a hiker to take to the woods and thus to get into closer contact with nature." Although he may not have realized it, this was a crucial point. The automobile had hardened the distinction between the worlds of road and trail by propelling hikers into wilder portions of the landscape. And "closer contact with nature" was not only a result of being pushed to the margins of the landscape; closeness to nature itself was increasingly defined by the automobile's absence.[102] Like MacKaye, Anderson, as he came to value this separation of hikers from the automobile, increasingly associated the domain of the hiker with wilderness.

In his letter to MacKaye, Anderson also mentioned a recent article in *Scientific Monthly* on wilderness preservation that he had immediately associated with MacKaye's thought on the subject. "I note that your gospel of the need for wilderness tracts," wrote Anderson, "is being preached by Dr. Robert Marshall of the Johns Hopkins University." The article, of course, was Marshall's "The Problem of the Wilderness," and Anderson and MacKaye would later prove key players in the creation of "the organization of spirited people" that Marshall called for, though only after the conflict between roads and trails came to a head during the mid-1930s.[103]

In 1931, MacKaye received a letter from another hiking enthusiast, a member of the Smoky Mountains Hiking Club named Harvey Broome. It was the beginning of a long friendship and a remarkable correspondence. Broome, an avid explorer of the Smokies, was particularly taken with MacKaye's broader plans for the AT, as they were laid out in *The New Exploration*. "I am profoundly, *vitally*, interested in the philosophy which underlies the 'New Exploration,'" Broome wrote at the conclusion of a long letter to MacKaye.[104] Broome was an ally of tremendous energy who over the next few years would remind MacKaye that his vision mattered and that it needed persistent restatement. Like Anderson, Broome became an indefatigable proponent of preserving the wilderness conditions of the trail. Moreover, as Anderson was an important observer of developments in the Shenandoah National Park, so Broome would be a vigilant steward of the Great Smoky Mountains National Park's wilderness condition.

Broome was involved in organizing the 1931 ATC meeting in Knoxville and was intent on getting MacKaye to attend. In the end, MacKaye could not, but he sent Broome a message, to be read to the gathering, that took a strong stance against automotive intrusions. The automobile, he wrote, should "be a servant in going to and from the trail, but never . . . a trespasser upon the trail itself." "Beware of the innocently insidious!" he implored trail activists. "Whatever instruments we use let them not get started toward the cause of our undoing." Although his wilderness advocacy was defined largely in opposition to the automobile, MacKaye also understood—perhaps more explicitly than other advocates of the era—that there was an important symbiosis between automobiles and wilderness. Wilderness preservation would have made little sense prior to the proliferation of automobiles, not only because the very essence of wilderness was its resistance to mechanized transport but also because mechanized transport was itself essential to wilderness access. There is a measure of irony in this, of course, but it is also worth noting that MacKaye's emergent wilderness advocacy was never simply a rejection of the automobile as a technology. For him, wilderness preservation was about shaping and restricting the automobile's presence, not denying it.[105]

In an April 1932 article, "The Appalachian Trail: A Guide to the Study of Nature," which appeared in *Scientific Monthly*, MacKaye returned his attention to trail aesthetics. If the first decade of the trail's history had been devoted to putting together the physical trail, he wrote, then the second decade, or stage, would be devoted to learning to read the nature of the places it traversed. "Primeval influence," he added parenthetically, "is the opposite of machine influence."[106] Learning to read the primeval, he implied, involved digging beneath the machine civilization to reveal nature's substrate. Most of the article was devoted to a detailed reading of the geological and biological processes that defined the region and its scenery. Gone, for the moment, were calls for sustainable communities based in reorganized labor regimes. Benton MacKaye was sounding much more like Robert Sterling Yard.

Why this change in the stated purpose of the Appalachian Trail? There were a number of reasons. Although MacKaye had described fairly clearly the cultural purposes and importance of his other two indigenous environments—the rural and the urban—he had a harder time explaining the details of "outdoor culture." As modern roads proliferated in previously isolated areas, and as driving itself became a form of outdoor recreation, MacKaye felt a greater need to define the AT in opposition to motor trails and scenic drives. But how was he to put into words the positive side of the trail's recreational experience? Nature study was one answer. His commitment to studying the trail's nature grew in direct proportion to the threats posed by roads and automobiles. The more the mechanized world intruded upon Appalachia, the more adamant he became in defining a nature that stood apart from, and in opposition to, such intrusiveness.[107]

The following summer saw a flurry of letters between MacKaye and Anderson about road building in Shenandoah. In June 1932, Anderson reported ruefully that the skyline drive through the park had been extremely disruptive of the trail. "The road and the Trail," he wrote, "on top of this part of the Blue Ridge at least, are incompatible."[108] Later that year, Anderson mentioned that he and fellow members of the PATC had divergent opinions on the impact of the road. "Myron has not yet grasped the idea of the incompatibility of road and trail," he wrote, referring to Myron Avery, a PATC member, president of the ATC, and the most tireless blazer of the trail. Anderson thought it best to relocate the trail away from the road, a process that would have sacrificed its ridgeline position. Avery, on the other hand, thought that maintaining the scenic views was more important than minimizing the intrusions of the road. In Anderson's mind, the wilderness quality of the trail was more important than scenery. He urged MacKaye to write to Avery emphasizing that the AT "should above all things be a wilderness footpath . . . and that we should as far as possible seek the primitive environment for the trail even if we have to sacrifice the skyline route."[109] MacKaye agreed: "It all narrows down, as you say so well, to whether 'it is worth while to preserve the primitive.'" "As to the Appalachian Trail," MacKaye concluded, "its whole point is to preserve the 'primeval environment.'"[110] These conflicts—between road and trail, and between competing concepts of trail design—gave further form and detail to a specific conception of wilderness wielded with increasing aggressiveness by MacKaye and his hiker allies.

During the early 1930s, MacKaye remained in close contact with Clarence Stein and, to a lesser extent, Lewis Mumford. Throughout 1931 and 1932, MacKaye and Stein traded information about the development of skyline drives, and about the chances of Franklin Delano Roosevelt becoming president. As governor of New York during the early Depression years, Roosevelt had shown himself willing to use government as a planning force. In 1931, Stein met with Roosevelt, ostensibly to get him to attend a conference at the University of Virginia on regionalism.[111] But part of Stein's purpose was to assess Roosevelt's openness to national planning. Just weeks before the election of 1932, MacKaye wrote to Stein urging him to impress on Roosevelt certain core convictions of the RPAA. His letter provides a good look at what regional planning meant to him at that moment. In clipped form, he insisted that Stein emphasize four specific premises:

(a) Build not houses but communities; shape them not like Roadtown but like Radburn. (b) Proclaim a "moratorium on road building" (as the saying now is) until the people can be taught that a road is a highway and not a hog trough; and that a *through* road is a "railroad" and not an endless alley. (c) Save the wilderness from Coney Island; interest the folks in the inexpensive joys of nature in lieu of the jarring jams of jazz. (d) Build up (not down) the country's forests, gaining thereby the maximum of jobs and the minimum of taxes.[112]

In animated fashion, MacKaye insisted that regional planning involved the combination of town planning, townless highways, open space preservation, resource conservation, and labor reform.[113]

MacKaye's growing attachment to wilderness must be seen in this larger planning context. But wilderness also was, as his comments made clear, an important haven from the commercialization of leisure and the cacophony of the jazz age. In making this point, MacKaye showed an inability or unwillingness to see much that was positive in mass culture. His distaste for jazz, a musical form that was itself a stunning example of the sort of indigenous cultural fluorescence for which MacKaye so eagerly hoped, was a particularly unfortunate by-product of his Yankee crustiness. But it is also important to realize that MacKaye often used the word "jazz" in a more generic way, to signify the literal and metaphorical noisiness of modern America—a din that was an important counterpoint to his conceptualization of wilderness.

A Return to Government Service

Franklin Delano Roosevelt's inauguration in 1933 brought Benton MacKaye to Washington in search of a place within an administration that seemed willing to entertain regional planning ideas. The New Deal ended a twelve-year period in which MacKaye and his RPAA colleagues operated largely outside government circles, and with little hope that government would be a planning force. Harding and Coolidge had let business take the reins during their administrations, and even Hoover, the engineer, had preferred to craft an associative state built on business-government cooperation.[114] Early signs were that the New Deal would be different.

Initially, MacKaye fixated on the proposed Civilian Conservation Corps (CCC), a program that he hoped would resuscitate his integration of resource and labor concerns. He went so far as to draw up a memorandum for the Secretary of Labor on that subject.[115] When he failed to get a position with the CCC, he turned his attention to a congressional bill seeking to create a regional planning authority in the Tennessee Valley. Here was legislation that regional planners could get behind.

In early 1933, MacKaye wrote a series of articles about the Tennessee Valley Authority (TVA) imparting a sense of his hopes for the agency. In "Tennessee—Seed of a National Plan," he lauded the potential for public development of the region. He envisioned a river basin plan, complete with a system of reservoirs, dams generating electrical power, and the conservation of watershed forests. He remained confident about these technologies and the democratic possibilities of such planning, and made a strong appeal for prioritizing forest jobs over those in dam building and power generation, believing this a better way of absorbing the unemployed and anchoring regional development in a long-term relationship with renewable resources. He also warned Roosevelt against allowing the TVA to become an engine for regional real estate speculation. In short, he revisited the prospect that Appalachia might be developed as an alternative economic realm.[116]

Unfortunately, MacKaye was unable to secure work immediately with the TVA. He had accepted a position doing emergency conservation work in Connecticut's state forests when he received an offer from the Bureau of Indian Affairs (BIA). Sensing there might be a possibility of planning "forest communities" on Indian lands, he accepted the offer. To prepare for his job with the BIA, he composed yet another memo, "Memorandum to the Commissioner of Indian Affairs," which argued, predictably, that the preservation of Indian communities and cultures should be integrated with a program for conserving Indian natural resources.[117] All this occurred in the context of John Collier's reorientation of the federal government's Indian policy. Particularly attractive to MacKaye was Collier's repudiation of the Dawes Act and its emphasis on assimilation through individual ownership of homesteads. The Indian New Deal's encouragement of communal ownership of land resonated with MacKaye's commitment to such a form of land tenure. Moreover, the respect for native cultures that Collier brought to the Indian Service complemented his own commitment to "indigenous" culture. Although he found himself in unfamiliar regions amid cultures with which he had little experience, the ideological reorientation of the Indian New Deal was comfortable, at least in the abstract.

MacKaye's stint with the BIA was brief. In early June of 1933, he headed for the Dakotas to inspect the emergency conservation work on the Rosebud and Pine Ridge Sioux Reservations. He was quickly transferred to the Southwest, where he met Aldo Leopold for the first time. (Leopold was doing summer inspection work prior to assuming his new duties as a professor of game management.) The two spent several days together discussing the "erosion question," and they eventually co-authored a plan for a flood control project utilizing Indian labor.[118] Such work took up most of MacKaye's energy, though he did give some thought to a regional plan for the Navajo Reservation.[119] In a few instances, including at the end of the report he co-wrote with Leopold, MacKaye mentioned preserving wilderness on Indian reservations, an initiative that Bob Marshall pushed more forcefully in his work with the Indian Service. Such suggestions were largely a response to his fears that "scenic sections are in special jeopardy of invasion by development of the Coney Island type." Wilderness areas would protect regions from tourist, not resource, development.[120]

On August 31, 1933, MacKaye received a letter from John Collier relieving him of his duties with the BIA. His dismissal, Collier insisted, was due solely to budget cutbacks. MacKaye seemed not to mind too much. He headed back east through Knoxville, where he hoped to meet Harvey Broome and "size up the Gov't Tenn. Plan and especially Nat'l Park and A.T. affairs." At a stopover in Florence, Alabama, the site of the TVA's Wilson Dam, MacKaye received a letter from Harold Anderson. "I have been giving a little consideration, "Anderson wrote, "to a plan to start some sort of organized movement here in Washington to advance the cause of those who are opposed to the unwarranted invasion of our wilderness areas by automobile roads with their train of desecrating abominations."[121] Anderson's was one of the earliest suggestions for what would eventually become the Wilderness Society.

MacKaye arrived in Knoxville in mid-September and immediately began assessing the TVA. He was prepared to be skeptical. He already had expressed certain reservations about the New Deal: work relief seemed to take priority over planning, immediate results were at a premium, and coordination was often the casualty. All were concerns quite similar to Leopold's. Mumford also had sent MacKaye a scathing brief on TVA and New Deal efforts to date. "Morgan [Arthur E. Morgan, chairman] apparently is an egoist who runs a one-man show," Mumford wrote; "he is also, apparently, an antiquarian of the Henry Ford school who wants to keep the Tennessee valley merely quaint and primitive: so he can go to hell as far as I am concerned." "Good God what a mess Washington is," he wrote of the entire New Deal effort:

[O]utside the Housing Department and Mr. Ickes it all looks to be like the most hopeless lot of amateurs and bunglers and well-meaners who ever got together for the purpose of effecting a half-baked revolution which will neither break with the past nor lead us into the future, but will combine the worst feature of both capitalism and socialism without deriving the benefits which either may claim.

Roosevelt, Mumford concluded, was a political Mary Baker Eddy: "As a faith healer he is all right. As a surgeon he is useless, because he doesn't believe in operations."[122]

Within days of receiving Mumford's letter, MacKaye began suffering from what turned out to be an acute case of diverticulitis. He underwent an emergency operation at the end of September. After a six-week hospital stay in Knoxville (paid for by Clarence Stein since MacKaye was all but destitute), MacKaye returned to Shirley Center for the winter. As he recuperated, he continued to pursue a position with the TVA.[123] He also fielded a barrage of letters from Anderson warning of other skyline drive proposals and the general threat of New Deal projects.[124]

The New Deal admixture of work relief, road building, and recreational development seemed on the verge of obliterating large sections of the AT in favor of automobile access to the Appalachian ridgeline. The most immediate issue was the skyline drive nearing completion in Shenandoah National Park. In December 1933, MacKaye got wind of an exchange between Myron Avery of the ATC and Arno Cammerer, the new head of the National Park Service. Avery had contacted Cammerer to discuss the disruption of the AT caused by the Shenandoah road; Avery was concerned for the trail, but he also sensed that the disruption of the AT could be turned into financial assistance to rebuild and improve the trail. Moreover, government-acquired rights-of-way for such skyline roads could be used by the AT, thus circumventing one of the trickiest parts of bringing the trail to completion. Avery was bent on finishing the trail and was less concerned with its integrity or its philosophical undergirding. He saw the federal government as a powerful ally, one to be dealt with diplomatically. In a letter to another trail worker, Avery wrote: "It behooves us to be on the alert for there is much happening which can be turned to the Trail's advantage." Then he sounded a note of concern about those who might undermine this advantage. "It is, of course, fruitless," Avery wrote, "to offer a vain, ineffective opposition to the road projects."[125] Such opposition, he implied, would only hinder federal co-operation.

Avery asked that the Park Service, in its construction of the skyline drive, also include an adjoining portion of the AT in its plans. In re-sponse, Cammerer assured Avery that he had sought "recognition for a trail along the proposed highway." He even suggested that the trail be made wide and smooth enough that it could serve as a bicycle path.[126] When MacKaye heard about this, he was apoplectic. It was entirely un-acceptable, he wrote to Cammerer, to have the AT skirt the skyline drive within its 250-foot right-of-way, and to have the trail widened and graded for bicycle traffic. MacKaye insisted that there were two funda-mental qualities that defined the AT: "that it be a footway and not a wheelway" and "that it be a secluded way (at a maximum distance from sounds and influences of town and highway)." The Avery-Cammerer plan compromised both of these conditions. The AT, MacKaye insisted, was to be a "real wilderness footpath."[127] A week later, MacKaye got a let-ter from Anderson expressing similar dismay at the Avery-Cammerer scheme. "I am to have luncheon with Robert Marshall," Anderson noted, "to discuss ways and means of combating the drive to put a road over the large section of the A.T. If we could only get the sentiment which exists against such a project organized!"[128]

Meanwhile, MacKaye continued to push for a TVA job. In December 1933, Ned Richards, the Chief of the TVA's Forestry Branch (and, inci-dentally, a forester who shared some of MacKaye's socialist leanings), asked, as a precursor to potential employment, that MacKaye spell out his ideas regarding an appropriate TVA forest policy. MacKaye responded with a five-page treatise in which he insisted that any TVA forestry pol-icy be part of a broader regional plan. The challenge of regional plan-ning, he continued, was to create an environment worth working in, liv-ing in, and recreating in. With this in mind, he projected three types of forests for the Tennessee Valley. First there would be the "protection for-est," designed "for conserving the *flow of streams* (water streams and soil streams)." Next, there would be the "production forest," designed to conserve "the growth of timber and *flow of timber products*." MacKaye again suggested that planned forest communities should be nurtured. Fi-nally, there was the "wilderness forest," designed "for conserving the wilderness environment (by control of the local highway system and its *flow of population*)." Warning of the destructiveness of "the unplanned highway" to both community and wilderness values, he argued for the coordinated planning of highways in the region that were "townless," "insulated," and "flankline" in orientation. In essence, he was advocating his townless highway proposal, but with a flankline component to distinguish it from skyline roads.[129] It was another grand vision.

In early 1934, MacKaye finally got his wish when Earle Draper, head of the TVA's Division of Land Planning and Housing, hired him as a regional planner. MacKaye moved to Knoxville in April 1934, and began what he would later refer to as the "two damndest years" of his life.[130] His early letters from Knoxville were buoyant. "Knoxville is a little buzzing side-edition of Washington," he told Clarence Stein, and he compared the atmosphere at TVA to the idealistic early years of the Forest Service. Two months later, he still sounded upbeat. "I've got the best job in the whole outfit," he told Stein.[131]

MacKaye was charged with designing a comprehensive regional vision and philosophy for the TVA. There were few jobs more perfectly suited to his strengths. But, as planning historian Daniel Schaffer has pointed out, MacKaye was also in the tricky position of being a "resident philosopher" among "builders and bureaucrats."[132] During his twenty-six months with the TVA, MacKaye composed a number of his characteristically lengthy memos on the philosophy of regional planning and its relation to the Tennessee Valley, but such philosophizing did not mesh well with a practical agency bent on action.[133] Part of the problem was MacKaye's abstract thinking and his inability to communicate the applied aspects of his philosophy. But his vision also was at odds with the dominant ethos of TVA: to plan for economic development and to leave quality-of-life issues—the ends of planning—for private citizens to define. TVA leaders, Harcourt Morgan and David Lilienthal in particular, saw economic development as an appropriate goal, while MacKaye thought this approach would enable the very sorts of social inequality and metropolitan sprawl that regional planning sought to alleviate. The TVA, MacKaye soon concluded, was in the business of facilitating growth, not planning it.[134]

This difference of opinion was particularly evident in the area of recreation. By and large, the TVA Board saw recreation as an industry; their goal was to use it as another engine for regional economic development. MacKaye, on the other hand, saw recreational planning as a process of controlling the very developments that his superiors sought to enable.[135] "Recreational development," he suggested in a 1935 memo, "like power development, should be removed as far as possible from the motives of commerce."[136] Wilderness preservation, he thought, was a way of doing this.

Despite MacKaye's growing sense that the TVA would not fulfill his hopes, his life in Knoxville was stimulating. In the evenings and on weekends, he served as visionary and guru for a group—composed of fellow TVA employees and members of the Smoky Mountains Hiking Club— that called itself "The Philosopher's Club." Among its members were Harvey Broome and Bernard Frank. "Philosopher's Club" gatherings combined intense discussions of political, social, and environmental issues with singing, square dancing, and outings to the Smoky Mountains. The group epitomized the sort of communitarian spirit and participatory culture for which MacKaye pined. In a letter to his old friend Judson King, he compared the "Philosopher's Club" to their former affiliation in the "Hell Raisers."[137] It was, in part, the spirit of this group that got the Wilderness Society going.

By 1934, the Skyline Drive in Shenandoah National Park was a done deal, and, as a prototype, it spawned numerous offspring. One was a plan for a Blue Ridge Parkway to connect the Shenandoah and Great Smoky Mountains National Parks. While there were conflicting reports about the route it would take, almost every option involved considerable interference with the AT. Even more troubling was a proposed skyline road through the Great Smoky Mountains National Park. Meanwhile, in New England, there were plans afoot for a Green Mountain Parkway.[138] Public enthusiasm for skyline roads and the willingness of the Roosevelt administration to fund them in the name of work relief spelled disaster for the AT's wilderness qualities.

In response to these various proposals, and in an effort to craft a more constructive position, MacKaye promoted design alternatives to skyline roads. In 1934, *Appalachia* published MacKaye's manifesto, "Flankline vs. Skyline," which served as a rallying cry for those who opposed skyline development. He began the article by dividing the users of the outdoors into four groups. First there were those for whom "the primeval environment is a natural resource in itself." For this class of users, "the essence of wilderness is solitude," and road development threatened this quality. He called this group the "campers." Second, there was a group for whom "wilderness is *not* a serious asset." The outdoors was a place to play, a "pleasant background" rather than an essential environment. These were the autocampers and recreationists "seen along the highways on any summer Sunday afternoon." MacKaye referred to them as the "frolickers." A third group were the "motorists," those who stuck to the roads and their cars, and for whom scenery was nature's most important asset. Finally, there was a group "composed of persons who (as hotel-keepers, food-vendors, or contractors) hope to capitalize the wilderness by means of selling something." This group, by far the smallest in Mac-Kaye's estimation, he called the "salesmen," and he heaped most of the blame for the promotion of skyline roads on them.[139]

MacKaye admitted that each of the first three groups deserved their areas. He made it clear that he was not opposed to all scenic roads. But he thought that, though skyline drives were being sold as developments catering to frolickers and motorists, they were being pushed by the sales-men who would profit from them. In MacKaye's mind, skyline road de-velopments were less about access and more about sloganeering and "A-1 publicity."[140] Like Yard and Leopold, MacKaye indicted boosters and commercial interests while recognizing as legitimate the desires of all recreational users. This did not mean that he saw all such uses as equal; he still privileged wilderness recreation. But the article was a frank admission that, from a political standpoint, the success of wilderness preservation would necessitate a broader recreational planning effort aimed at meeting various public needs. The only group whose desires should not be taken seriously, said MacKaye, was the profiteers.

The upshot of the article was a proposal to abandon skyline roads in favor of scenic routes following the flanks of the Appalachian range. Such roads, augmented by occasional spur roads to the ridgeline for views, were equally scenic and afforded motorists greater variety. More-over, MacKaye suggested (without much evidence, it should be noted) that neither frolickers nor motorists were particularly attached to the skyline routes. By developing flankline roads—perhaps even a single route extending the same length as his trail—the needs of these two groups could be met admirably without compromising the needs of the campers, for whom the ridgeline wilderness was a vital and tenuous re-source. MacKaye saw the flankline road and skyline trail as complemen-tary, and he concluded his article by calling for an "entente" between the first three groups of recreational users as a way of thwarting the plans of the salesmen. He hoped that his flankline proposal would appeal to all recreational users in a way that a single-minded focus on wilderness did not.[141]

These various skyline threats generated new interest in a wilderness organization. In a letter to MacKaye in August 1934, Harold Anderson again mentioned creating a group to oppose skyline road development, this time suggesting it take the form of a rogue committee of hiking group representatives. Myron Avery and the ATC, Anderson reported, had made few efforts to oppose road building. Indeed, a dramatic schism was developing between MacKaye and Avery over whether the ATC should publicly protest skyline drives. Avery thought protest fruitless and politically damaging; MacKaye saw such protest as an essential part of protecting the core values of the trail. In turn, Avery accused MacKaye of armchair philosophizing while ignoring the physical trail work. "I doubt if any of you realize or know the smallest part of the labor of getting these links put in," Avery wrote to MacKaye. He insisted that the AT needed to be "more than a squirrel track."[142] Throughout this debate, Avery portrayed himself and his supporters as the true workers, and criticized MacKaye, Anderson, and Broome as out-of-touch complainers and purists. Although Avery was right that MacKaye had drifted away from trail building, he was wrong to assume that MacKaye did not know the work of putting the trail together. He had done much of that work during the AT's early years. But MacKaye defined that work differently.

The MacKaye-Avery debate revealed two very different conceptions of the AT. Avery's goal was a wide, graded, and aggressively marked trail; MacKaye, Broome, and Anderson were much more inclined to err on the side of underdevelopment—they saw nothing wrong with a "squirrel track" as long as it was surrounded in most places by a continuous strip of undeveloped land. To Avery, the engineered trail was everything; MacKaye remained committed to the entire realm. In the end, Avery successfully controlled the ATC and quelled any public statements opposing road building. For MacKaye, who had never sought to control the ATC, the schism with Avery was a deathblow to his efforts to define the trail's philosophy, at least from within the ATC. The Wilderness Society was a result, in part, of the departure of MacKaye, Broome, and Anderson from the ATC and their search for an alternative approach to opposing skyline drives.[143]

Avery was not MacKaye's only concern. The federal government, after all, was the party building the roads. MacKaye tried to get Arno Cammerer, whose agency was in charge of most recreational road building, to appreciate the value of wilderness and the threat skyline roads posed to it. But Cammerer was unwilling to fully support the wilderness program; indeed, he did not quite get it. "A parkway with a right of way of some 200' to 1000'," Cammerer wrote to MacKaye, in an effort to offer some solace in the face of the Blue Ridge Parkway's disruptiveness, "at least assures for all time the retention of certain wilderness conditions in that width."[144] Cammerer, it seems, assumed that wilderness preservation simply meant preserving trees from the ax.

To MacKaye, this inability of the head of the National Park Service to fathom what he meant by wilderness was startling. In a reply, he tried to clarify the wilderness idea. Wilderness, MacKaye wrote, had one common denominator: "the *absence of city sounds*." He mentioned the automobile and the radio as specific intrusions. He then took Cammerer to task for suggesting that the edges of parkways could be wilderness: "This appears offhand like a thoroughly amazing statement,—especially from the Guardian of American-owned wilderness. What you mean, I suppose, is that these spots may retain primeval forests." But such parkways, MacKaye argued, created the very metropolitan conditions that ruined wilderness.[145] In a brief reply, Cammerer simply invoked the specter of timber cutting in those areas if they were not opened up to recreational traffic. Ironically, the Park Service, under the guise of vista thinning and view enhancement, would take considerable liberties with roadside trees.[146]

By the fall of 1934, the groundwork had been laid for a wilderness organization, with twin centers of activity in Knoxville and Washington. Harold Anderson continued to push for a group to oppose skyline drive proposals, and he pulled Marshall, a Washington resident, into that orbit. Yard also joined this Washington cadre. When Marshall visited Knoxville in August 1934 to scout out the southerly route of the Blue Ridge Parkway, Broome mentioned to him Anderson's proposal for an organization to protect wilderness along the AT. Marshall's response was enthusiastic (he and Anderson had been talking, so the idea was nothing new), but he suggested that the group extend its purview to seek "the protection and preservation of wilderness wherever it might occur."[147]

In the weeks that followed, the groups in each city—Anderson, Marshall, and Yard in Washington, and MacKaye, Broome, and Frank in Knoxville—met and corresponded as a prelude to formal organization. On September 12, 1934, Anderson fired off a letter to MacKaye that recounted a meeting with Marshall, during which they discussed potential founding members. Many of the candidates were involved in hiking clubs, but Marshall also included friends such as Elers Koch and Olaus and Margaret Murie (westerners all) as well as Roger Baldwin of the ACLU, John C. Merriam, and John Collier. Marshall and Anderson also

suggested two outsiders who would lend the group national stature—
Aldo Leopold and Ernest Oberholtzer. Anderson included with the letter
a draft "Statement of Principles" to get the definitional ball rolling; it was
mostly an indictment of the automobile and the "sights and sounds of
the machine-age world." "We are opposed," he wrote, "to the invasion
of our few remnants of primeval environment by those three abominations of our present civilization—noise, ugliness, and congestion." Mac-
Kaye replied with his own list of potential members that included Stuart
Chase, Lewis Mumford, Clarence Stein, and Sinclair Lewis, and he expressed regret that Anderson would not be able to attend the meeting of
the American Forestry Association (AFA), to be held October 17–20 in
Knoxville.[148] Bob Marshall was headed down, however, and his return
sparked the Society's formation.

In October 1934, when the Wilderness Society had its beginnings along
the side of a road near Norris Dam, MacKaye was still a relative newcomer
to the TVA. Although he sensed the limitations of the agency from the
start, his feelings about the TVA soured as time passed. In a letter to Stuart Chase, he recounted the vacuousness of the attention given to outdoor recreation. "There's a lot of guff about aesthetics and beautification," he grumbled, "and non-industrial planning is in the hands of
'landscape artichokes' who would cope with the problem of life's setting
and background by planting a few pansies." As for TVA's industrial engineers, they showed too little concern about issues of habitability; in his
mind they were "merely grown up men who like to play tin engine and
choo choo car."[149] By 1936, MacKaye was a marginalized visionary whose
utopian memos received scant attention; Earle Draper asked him to leave
that June.[150]

MacKaye returned to Shirley Center and a financially tenuous life as a
writer and consultant. He continued to defend the AT against metropolitan incursions, though with the growing clout of the Wilderness Society
to back him. MacKaye was fifty-seven years old when he left Knoxville in
1936. His career as a wilderness advocate was just beginning.

Conclusion

What prompted Benton MacKaye—the trained utilitarian forester and
progressive conservationist, the political radical, the regional planner—
to become a leading advocate of wilderness preservation? In part, his
turn to the wild was a defensive one. He had failed to get the federal gov-

ernment to adopt his ambitious colonization schemes in the teens; the communitarian components of his AT proposal had been lost in the recreational tenor of the ATC; the broad critique of the RPAA had only a few applied successes; and New Deal conservation agencies had not lived up to their planning promise. The importance of wilderness grew as MacKaye's more sweeping goals foundered. As the metropolitan world spread ever outward, MacKaye retreated to higher ground and into a defense of wilderness.

MacKaye admitted as much in his contribution to the first issue of *The Living Wilderness*. In a reprint of a speech he had given at the 1935 meeting of the ATC, titled "Why the Appalachian Trail," he returned to what he insisted was "the precise nature and conception" of the trail. This time, however, the message was notably different than it had been in 1921. He wrote:

> One function of true wilderness is to provide a refuge from the crassitudes of civilization—whether visible, tangible, audible—whether of billboard, of pavement, of auto-horn. Wilderness in this sense is the absence of all three. Just so of the wilderness footpath; it is unadorned; it is foot-made; it is noise-proof. Such are its qualities in essence. The advertising sign (whether board or edifice), the graded way (known as "Grade A"), the auto-horn (or its refrain the radio)—all of these are urban essences; all are *negations* of the wilderness. No true Appalachian Trail can follow within the influences of any of these invasions, for the Appalachian Trail is a wilderness trail or it is nothing. Such is the original, and never abandoned, conception of the thing which the Appalachian Trail Conference was founded to preserve.[151]

Few quotes are clearer in suggesting that MacKaye defined wilderness in response to what he saw as the essences of the modern metropolis and their steady drift into wild landscapes. But MacKaye, like Yard, was guilty of a bit of revisionism, for his original AT proposal had contained very little about the importance of the AT as a "wilderness trail." It took the developments of the intervening years to tease out, to create, this wilderness commitment.

Like Aldo Leopold and Bob Marshall, Benton MacKaye began his career as a forester, and he retained his commitment to reforming the exploitative aspects of the modern relationship between humans and natural resources. His robust social critique of Progressive conservation policy was indeed a vital nurturer of his wilderness advocacy. The connections between his social conservation and his preservationist sentiment suggest that we have overdrawn the ideological differences between utilitarian conservation and preservation. MacKaye did not reject use in favor of preservation. He developed particular and complementary visions of how each should be done. His notion of wilderness was not a romantic ideal of recreational nature detached from labor and social problems. Quite the contrary, it was a reaction to the creation of just such an ideal at a time when he thought conservationists should have been more attentive to such problems. Wilderness was a recreational alternative, just as colonization was an economic alternative. Both were, in essence, retreats from profit.

By 1935, after more than a decade of trying to plan the automobile into the landscape, MacKaye came to advocate wilderness as a refuge from this new and troubling machine in the garden. In his identification of the automobile and road building as the major threats to wilderness, and in his critique of the consumer relationships with nature that they fostered, he was of a mind with the other founders of the Wilderness Society.

But MacKaye's wilderness idea was also unique, particularly in its role within his broader regional planning vision. That vision was utopian and thus full of contradictions. He often underestimated the tensions between planning and democratic choice; he was not a particularly keen observer of cultural complexity or the richness of popular culture; and he almost always overestimated the ability of planners to control the technological and economic forces of the era. As a planner, he tended to substitute the simple order of artifice, what he would come to call "geotechnics," for the complex order of life.[152]

Nonetheless, MacKaye's grand planning vision made his wilderness idea a bold one. For him, preserving wilderness was never just about sequestering large areas of wild nature and holding them against the depredations of an otherwise unreformed world. Rather, wilderness preservation was a reformist tool, a modernist attempt to reshape American geography. MacKaye saw wilderness as a design choice. But it was more than that, for MacKaye also came to realize that wilderness preservation meant saving a world that was profoundly nonhuman, a world in which Americans could immerse themselves and realize the folly and hubris of their efforts to grow beyond nature. Such efforts, MacKaye thought, were more than environmentally destructive. They were also dehumanizing.

CHAPTER 6

The Freedom of the Wilderness:
Bob Marshall

Each of the four founders of the Wilderness Society discussed in this study brought unique concerns and strengths to wilderness advocacy. Aldo Leopold brought an impressive and rounded conservation career— an accretion of experience molded into a profound environmental philosophy. Robert Sterling Yard brought the preservationist values of an older generation retrofitted to match the forces that shaped interwar American culture. Benton MacKaye brought his creative thinking on the spatial role of wilderness and its relationship to the modern urban form.

Bob Marshall's contributions were both more and less obvious. For those who have written about Marshall, the force of his personality and his voracious appetite for traversing wild spaces have seemed his defining qualities.[1] Marshall was energy and enthusiasm personified, an achiever who quickly grew restless when snared in the organizational webs of early twentieth-century America. While other founders theorized about the threats to and importance of wilderness, Marshall's impulse was to explore, chart, and preserve. Indeed, for Marshall more than any of the other founders, wilderness was a place of masculine physicality, of direct bodily engagement with the natural world.[2] As a result, he knew the nation's remaining wild lands better than anyone of his generation. He also had the luxury of wealth, which allowed him to dictate the trajectory of his career and to support the Wilderness Society's activities, even after his premature death in 1939 at the age of thirty-eight. Together, Marshall's practical knowledge, wealth, and charisma made him the most politically effective of the Wilderness Society's founders.

There were other important facets of Marshall's wilderness advocacy that have garnered less attention. Although his writing was neither as compelling as Leopold's nor as original as MacKaye's, his arguments for the freedom of wilderness and wilderness as a minority right were vital to interwar advocacy. And they reflected his broader political agenda: along with his wilderness advocacy, he was a strong supporter of socialism and civil liberties. While scholars have tended to downplay the interconnection of these political commitments, they were, in fact, quite strong. Marshall's wilderness advocacy must be understood in the context of these two other commitments, and within the broader matrix of interwar political and social thought.[3]

That said, there are some good reasons why Marshall's radical politics have seemed separate from his wilderness advocacy. He devoted most of his working career to conservation, whereas his advocacy for socialism and civil liberties was largely extracurricular. Thus the records he left are tilted toward conservation. There was also more at stake in vocally supporting these other two causes. Marshall's involvement in organized socialism eventually landed him on the enemy list of Martin Dies, the early ringleader of the House Un-American Activities Committee and a leading voice in the backlash against New Deal liberalism.[4] Although outwardly cavalier about public charges of "un-Americanness," Marshall was guarded about airing his radical views. As a member of a prominent Jewish family, he also was a natural target for the strong nativism of the era, and, although he recorded few incidents in which anti-Semitism affected him, his Jewish identity was central to his commitment to civil liberties.

There are also historiographical reasons why Marshall's integrated political commitments have been perceived as separate. Historians of U.S. environmental thought have tended to see a dichotomy between utilitarian conservationists and preservationists. Because Marshall was such a high-profile wilderness advocate, most scholars have portrayed him as a preservationist hero.[5] Yet throughout his career, even as his commitment to wilderness grew, he remained wedded to utilitarian public resource stewardship. Indeed, like Benton MacKaye, he attempted to radicalize utilitarian conservation. To fully understand Marshall's wilderness advocacy, then, we need to buck the logic of this dichotomy and appreciate how he integrated the two commitments.

Understanding the full scope of Marshall's wilderness advocacy also involves reexamining his place in the history of American preservationist sentiment. This has become tricky terrain as recent scholarship has revealed the mixed motives and results of some preservationist activities. Marshall's example suggests the need for caution in caricaturing wilderness advocates. Like his Wilderness Society colleagues, he was less interested in saving pristine and unworked nature than in protecting large areas from road building, automobiles, and the various forms of modernization that characterized the interwar period. Sensitive to perceptions that wilderness recreation was elitist, he devoted considerable energy to defining wilderness as a social as well as an environmental ideal. A trained forester with a Ph.D. in plant physiology, he was acutely aware of what wilderness did and did not preserve ecologically, and he used that knowledge to his advantage in defining a wilderness aesthetic and defending it against the arguments of foresters. Finally, Marshall was a pioneer in thinking about how wilderness preservation and subsistence resource use could coexist.[6] His wilderness advocacy had its unexamined assumptions and biases. He was not, for instance, particularly sensitive to how gender shaped his advocacy, and there were unresolved tensions between his collectivist and libertarian leanings. But Marshall did craft a socially informed brand of wilderness advocacy that deserves closer attention.

Forever Wild: The Adirondack Backdrop

Bob Marshall was born on January 2, 1901, in New York City. His father, Louis, was a well-known constitutional lawyer, a champion of civil liberties, a leader in the Jewish community, and a conservation activist. By all accounts he had a tremendous influence on his son. Marshall attended Felix Adler's Ethical Culture School in New York City, an educational environment that nurtured independent thinking and a commitment to social justice.[7] Among other initiatives, Adler's Society for Ethical Culture endeavored to extend outdoor vacation opportunities to the urban working classes—a model, perhaps, for Marshall's later efforts to make public recreation grounds accessible to all classes.[8] Such access was no problem for the Marshall family, however, for they owned a share in the summer camp "Knollwood" on Lower Saranac Lake in the Adirondack State Park. Adirondack camps were popular among New York City's upper classes at the time, and Knollwood had many of the trimmings. During his youth, Marshall spent summers in the Adirondacks, exploring the region and reading about its natural and human histories.[9]

The politics of preservation in the Adirondack Park formed an important backdrop to the development of Marshall's environmental thought. In northern New York State, the vast area of the Adirondack Mountains had remained relatively unsettled well into the nineteenth century. Between 1836 and 1840, New York's Natural History Survey, under the leadership of Ebenezer Emmons, explored and charted the area. Although some settlement came in the wake of this survey, the rising tide of sportsmen was a more significant outcome of increased knowledge of the region. Joel T. Headley's *The Adirondacks; Or, Life in the Woods* (1849) helped fuel antebellum recreational interest in the region, while William H. H. "Adirondack" Murray's popular *Adventures in the Wilderness; Or, Camp-Life in the Adirondacks* (1869) prompted a postbellum "recreational revolution."[10] Commercial logging also came to the Adirondacks in the late 1800s, often followed by reversion of abandoned cutover lands to state ownership through tax default.[11] Together, these changes nurtured a growing, though confusing, preservationist movement in New York State.

The political process that led to the famous constitutional clause declaring that the Adirondack preserve would "be forever kept as wild forest lands" has long perplexed scholars tracing the history of preservationist sentiment in America. Nationally (and, in fact, internationally), extensive commercial logging in the late 1800s led to a growing perception of the need for timber and watershed protection.[12] New York's Forest Preserve Law of 1885, which set aside 681,000 acres of forest lands in the Adirondacks to be managed by a new state Forestry Commission, was a response to these concerns more than it was a gesture of recreational or aesthetic preservation.[13] Even when, in 1892, the Forest Preserve was subsumed within a 2.8 million acre Adirondack State Park, timber and watershed concerns remained paramount, with timber cutting still permitted on state-owned lands.[14] "Forever wild" largely meant forever public. Two years later, however, during the 1894 constitutional convention, delegate David McClure introduced a constitutional amendment prohibiting timber cutting on the state lands within the park. Riding a wave of antilumbering sentiment, distrustful of the Forestry Commission, and concerned about a serious drought, convention delegates, Louis Marshall among them, passed McClure's article and ended timber cutting on the state-owned lands in the park.[15] The "forever wild" clause took on a new, more modern meaning. In the future, the clause would be a strong tool for preserving wilderness within the park, though that was not necessarily its original intent. Political motives other than a public commitment to wilderness as a higher preservationist ideal were at work in closing off the state lands to logging. There was not, in other words, a simple lineage of preservationist sentiment linking the creation of the Adirondack State Park in the late nineteenth century to the wilderness advocacy of the interwar period. Each was the product of its own context.[16]

The convoluted politics of preservation in the Adirondacks were matched by the region's heterogeneous and confusing landscape. The park's public-private patchwork was one aspect of this complexity, but there were also other changes under way in the late nineteenth century. Creation of the park encouraged the consolidation of private game preserves in the region, which came to control about 800,000 acres within the park, an area roughly equivalent to the state's holdings. Infamous for their exclusivity (the Marshalls would not have been welcomed by most), the private clubs that ran these reserves provided for elite recreation while restricting customary local access to resources. Locals who relied on the Adirondack commons for their subsistence thus found themselves squeezed between state lands suddenly off limits to timber cutting and enclosed hunting preserves guarded by private security forces. There were violent incidents as a result, and some feared a class war in the region.[17]

The park's creation also coincided with property consolidation by timber companies such as International Paper, which bought up land and began a second, more destructive logging boom in the region. Driven by new technologies permitting the manufacture of paper from wood pulp, Adirondack logging reached its historic peak between 1890 and 1910. This new round of logging, all of which occurred on private lands, resulted in a series of catastrophic slash fires in the decades leading up to World War I. Although the logging boom was fairly brief—it collapsed around 1910—it resulted, ironically, in a profound transformation of the park in the years immediately after its protection. By World War I, when Marshall was entering his teens, the Adirondack Park was a patchwork of private game reserves, cutover and charred timber company land, protected and regenerating state forests, and various permanent and seasonal settlements.[18]

There were, after 1894, continual efforts to reopen state lands in the park to timber cutting, but none met with success.[19] In 1915, the state held its first constitutional convention since the fateful 1894 meeting. The Conservation Commission (formerly the Forestry Commission) and timber interests made a last-ditch effort to pass amendments to open public lands to timber harvesting. Again, Louis Marshall opposed these efforts, making a memorable speech against an industry-sponsored proposal that was ultimately defeated. His father's defense of the "forever wild" clause made a profound impression on fourteen-year-old Bob Marshall.[20] But the 1915 convention passed amendments that allowed the leasing of campsites and construction of a highway from Saranac Lake to Old Forge, harbingers of a new era in Adirondack Park politics.[21] Bob Marshall came of age on the cusp of this important shift.

Where his father had exposed Marshall to the politics of protecting the Adirondacks, the writings of Verplanck Colvin introduced him to the romance of exploring the region. Colvin, who spent the post–Civil War decade surveying New York's north woods, published a multivolume study, *Report of the Topographical Survey of the Adirondack Wilderness,* during the 1870s. Marshall discovered Colvin's writings during his early teens, and came to share many of his obsessions: an addiction to discovery and charting, a love of climbing mountains and surveying areas from peaks, and a hard-driving quest to cover ground as a way of accumulating knowledge. Whether Colvin's example produced or reinforced these attitudes in Marshall is hard to say, but there were startling similarities between the two.[22] In 1915, Marshall climbed his first Adirondack peak, one of many ascents that would make him, his brother George, and family guide Herb Clark pioneers in climbing all of the Adirondack peaks known to be over 4,000 feet.[23]

Marshall's early hiking records reveal an Adirondack Park in transition. Beginning in the mid-1910s, Marshall kept a series of hiking notebooks, illustrated with photographs, which both reveal his preoccupation with statistical achievement and give a strong impression—literary and visual—of what hiking in the region was like during this transitional period. Marshall gave his first "Adirondack Notebook" the subtitle "Mostly Road Walks," indicating that the region's roads were its major hiking thoroughfares. Before the widespread use of the automobile, there was little need to have both roads and trails, or to distinguish between the two.[24] Marshall's notebooks also recorded the extent to which the region had been transformed by human activities. He noted and photographed frequent clearings, including one for a golf course under construction, and he commented on the extent to which the region's forests were in early stages of regeneration.[25] While Marshall's biographer, James Glover, has written that the "howling wilderness described by Colvin was right outside his back door," in actuality Marshall passed through a more complex mosaic of natural and human landscapes.[26] The region's roads were themselves at a transition point. As more tourists and recreational users accessed the park via automobile, the roads that Marshall relied on for his early feats of endurance grew less appealing. Indeed, by the early 1920s, hikers had begun developing and documenting a separate trail system, a process to which Marshall made important contributions.[27] Although he lauded the "forever wild" protection, and would help craft for it a new set of meanings after World War I, the Adirondack landscape of his youth was much more complicated than that phrase suggested. Marshall's wilderness advocacy would be similarly complex.

Becoming a Forester

To his considerable education in Adirondack nature and history, Bob Marshall soon added a professional degree. After graduating from the Ethical Culture School in 1919 and spending a year at Columbia University, he transferred to the New York State College of Forestry in Syracuse—a school that his father had played a crucial role in creating.[28] Marshall spent his winters in Syracuse studying and his summers exploring the Adirondacks. During the summer of 1922, he was among a group of forestry students and faculty who made camp in the Cranberry Lakes region to do research. On weekends, he explored the area, much of which had been cut and burned. The region elicited mixed feelings in Marshall. It was a remote portion of the park, and a sense of solitude came often and powerfully. He reveled in exploring uncharted (or minimally charted) areas and being alone in places that lacked the signs of civilization. But the human imprint on the land was considerable, and more noticeable because of his forestry training. Finding tracts of old-growth timber was thus a revelation. For Marshall, these twin feelings—a desire for solitude and a special appreciation for "virgin timber"—were the cornerstones of his nascent wilderness sentiment.[29]

Despite his utilitarian training, Marshall took an increased interest in Adirondack recreation during the early 1920s. The year 1922 saw the founding of the Adirondack Mountain Club (AMC), an organization devoted to the building and maintenance of trails and the diffusion of knowledge about hiking in the park. Both were important functions as more people came to the region in automobiles and then sought to escape their presence. Marshall was a charter member of the AMC, and he prepared a guidebook for the group, published in 1922, entitled *The High Peaks of the Adirondacks*.[30] Through the AMC, he made contacts with important members of the hiking community such as Russell Carson, an explorer and writer who shared his zeal for the region.[31] After Marshall left upstate New York, Carson and other AMC members would keep him abreast of developments within the park.

During these years, Marshall also published a student essay in which he argued for prioritization of recreation in the Adirondacks. "Recreational Limitations to Silviculture in the Adirondacks," which appeared in the *Journal of Forestry* in 1924, provides a telling glimpse of Marshall's thought as he embarked on a forestry career.[32] His basic point was that recreational preservation and timber cutting were not compatible in the Adirondacks, and thus publicly owned forest lands in the park should not be opened to timber cutting. He also made a strong case for the aesthetic value of "virgin forest," though in doing so he did not repudiate his new profession. Rather, he grounded his thesis in forestry arguments. "From a forestry standpoint," he conceded, "virgin forests are undesirable." The productivity of such stands is hampered by their age, and valuable wood goes to waste through decay. But in certain cases, he argued, "the grandeur of primeval woodland" was of greater value than the "finest formal forest." The public utility of such preservation, he contended, outweighed the money lost to forestry.[33]

Marshall had to respond to the familiar forestry argument that recreation and silviculture could coexist, and he did so by addressing the specifics of forestry practice in the region. Under certain silvicultural conditions, recreation and timber exploitation could coexist, he admitted. But in the Adirondacks, which was dominated by four forest types—swamp, spruce flat, mixed hardwood, and spruce slope—the preferred methods for timber harvesting were either clear-cutting or heavy selection. Both methods, he maintained, were aesthetically undesirable because they left a great percentage of the forest in either cleared or immature states. Even if virgin stands were preserved intermittently, the resulting checkerboard would not suffice for those who sought the aesthetic of a vast, mature forest. It was not silviculture in the broadest sense to which Marshall objected, but the intensive silvicultural practices most foresters thought appropriate to a region whose recreational importance was growing. The Forest Preserve embraced only twelve percent of the total timberlands of the state. Surely, Marshall thought, this amount of land preserved for recreation was appropriate.

As was characteristic throughout his career, Marshall balanced his commitment to scientific forestry with his desire to see areas preserved. Indeed, a key if somewhat understated tenet of his stance in this case was that careful scientific forestry would make preservation possible by limiting the amount of productive acreage needed to meet resource demands. One goal of efficient production, in other words, could be recreational preservation. And he placed his argument skillfully within both the political context of the "forever wild" clause and the intellectual context of scientific forestry. Although he conceded that "the right to do limited cutting in the Forest Preserve would be beneficial," he sympathized with the reluctance of the state's residents to change the constitution to allow such cutting. In the process, he taught fellow foresters about the perils of stereotyping the opposition:

> Perhaps when the exponents of opening the Forest Preserve have satisfied the justifiable apprehension . . . the so-called "absurd conservationists" will be more amenable to changing the constitution. For these conservationists have seen in the past the most outrageous exploitation of the public domain. . . . They are not the selfish, sentimental type of conservationist. But they know that the only thing which has prevented the utter destruction of the Adirondacks has been Article VII, Section 7 [the "forever wild" clause], and so they are naturally quite wary of giving it up.[34]

Marshall urged foresters to appreciate the public's justifiable suspicion of scientific experts. Before foresters could hope to convince the public of the need for even the most surgical cutting, they must prove themselves the true protectors of a public resource and cease their dismissive treatment of opponents as uninformed sentimentalists. "Recreational Limitations to Silviculture in the Adirondacks" was an admirable piece of mediation between two arguments that had been speaking past each other. It was also a testament to how a forester might support preservation within the bounds of professional logic. It was a strong hint of things to come.

Despite his love for the Adirondacks, Marshall longed for wilder country. Upon completing his undergraduate forestry degree, his greatest hope was to be assigned a Forest Service position in Alaska. While he was intrigued by the unique opportunities for public resource stewardship there, his most compelling interest was Alaska's sheer wildness.[35] He had played wilderness explorer in the Adirondacks; now he wanted to experience real wilderness. But the Forest Service was unable to accommodate his desire, so he took a summer position at the Wind River Experiment Station near Carson, Washington, where he worked under Richard McArdle, who went on to become Chief of the Forest Service from 1952 to 1962.[36]

Marshall returned east in the autumn of 1925 to continue his forestry training at Harvard, which offered a one-year master's program at its experimental forest near Petersham, Massachusetts (where Benton Mac-Kaye had studied and taught two decades earlier). Each student in the program was required to complete a research project, and Marshall chose a fascinating one. He noticed that a section of the Harvard Forest had grown back in a stand of softwoods, mostly hemlocks, rather than in the less valuable deciduous stands that characterized most of the region by the early 1900s. Figuring out why this had happened proved to be a neat study in environmental history. He discovered that the area had been logged selectively rather than clear-cut, as had most of the region's forests. Loggers had removed the merchantable old-field white pine, but they had left a suppressed understory of hemlock that thrived in the absence of competition. While Marshall drew numerous conclusions from the study, most of them technical and specific to the site, he came away with a stronger appreciation of the value of selective cutting as a forestry measure.[37]

Marshall completed his master's degree in the spring of 1925, returned to the Adirondacks for a brief vacation, and then reported to his new Forest Service assignment at the Northern Rocky Mountain Forest Experiment Station in Missoula, Montana.[38] As formative as his youth in the Adirondacks had been, it was his time in and around Missoula that impressed him with the need to preserve wilderness. But this was not the only lesson he learned during these years. He worked with loggers and fire fighters, witnessed the conditions under which they labored, and absorbed powerful lessons about labor and natural resources that pushed his conservation thought in a radical direction. In this regard, he again followed in MacKaye's footsteps, but he did so in a very different era of federal forest policy. During his three years in Montana, Marshall's interests in wilderness preservation and socialized forestry developed in tandem.

A Liberal Forester in a Conservative Age

Forest Service policy drifted in a conservative direction after World War I, as foresters traded a strong regulatory approach to the timber industry for a more cooperative stance. William Greeley, who became Chief of the Forest Service in 1920, was the primary architect of this drift. A pragmatic administrator with a sanguine view of industry, he focused Service efforts on removing obstacles to achieving sustainable timber harvests on private lands. He also preferred to emphasize the few sincere industry efforts to crop timber, rather than criticize industry for its slow embrace of sustained-yield practices. Greeley was convinced that fire, not industrial cutting, was the chief threat to the nation's timber supply, and he pushed a cooperative program of fire suppression on public and private lands. For his efforts he earned the wrath of Gifford Pinchot and other foresters who, on ideological grounds, opposed his cooperative approach to industry. Marshall, too, would chafe under the reins of this policy direction.[39]

Marshall's main research at the Experiment Station focused on the dynamics of forest reproduction after fires. But before he could get his research under way, he faced more pressing concerns—fires themselves. A July 1925 storm produced "the most severe lightning discharge ever seen in this region" and set off over 150 fires in Idaho's Kaniksu National Forest alone. "The fight against them," Greeley wrote in his annual report, "necessitated the largest concentrated organization and supply of men, subsistence, tools, and motor-truck and pack-train transportation ever thrown into a single field by the Forest Service."[40] Marshall was in charge of supporting and provisioning one of the crews. As he recalled, "I am working 18 to 20 hours a day as time-keeper, Chief of Commissary, Camp Boss, and Inspector of the fire line."[41]

In carrying out those duties, Marshall came to know the men fighting the particular fire to which he was assigned—the Mount Watson fire. In letters home, he paid uncommon attention to the social realities of fighting fires and protecting the public domain in the West. Most of the fire fighters were from Spokane, and were the type of men, Marshall recalled, "who could not hold a permanent job."[42] Slowly, he got to know these men, "their good traits and the tragedies that reduced many of them to their present extremity." He was particularly disturbed by the way they had been plucked from the streets of Spokane, thrown into a summer-long battle to which they reacted mechanically, and then deposited back on the city's streets. The nation's efforts at protecting and sustaining its forests, Marshall thought, stood in marked contrast to its lack of interest in these human lives.[43] The 1925 fires remained a touchstone for Marshall throughout his time in the northern Rockies. He frequently returned to assess the scientific and aesthetic implications of their damage, and he continued to think through their social implications.[44]

During his official travels through District One, Marshall often found himself in the lumber camps of the region and was quick to comment on their condition. In October 1925, he visited a camp in Idaho and noted that conditions seemed good. "The camp serves uniformly high class meals nowadays," he observed; "they are well-cooked, clean and in limitless quantities. From all accounts, this desirable condition is in marked contrast to the situation before the war, when the food was poor, dirty, and mostly fried." "That this tremendous improvement has taken place so suddenly," he continued, "can fairly be attributed to the I.W.W., or Wobblies, as they are universally called here. They also forced a revolution in sleeping quarters and today the dirty wooden bunks of yore have given place to iron cots and clean bedding."[45] Marshall was impressed with such results.

But, as sympathetic as Marshall was with timber workers and the IWW, his social background and managerial position filtered his observations. He was a privileged twenty-four-year-old expert from the East who had traveled to Montana imbued with both frontier romanticism and a civic commitment to help protect federally owned resources. He had not expected to supervise roughs from Spokane or dine among timber workers. In the abstract, he maintained an ideological affinity for the workers' cause, and their plight had a strong influence on his conservation thought. But he also found these interactions jarring. There was a dissonance between his ideal of strenuous labor in nature and the decidedly less romantic conditions under which these men worked. Although he absorbed the intellectual and moral lessons of his encounters and pushed foresters to recognize the social aspects of their profession, he would always be more comfortable alone in nature, without the distractions of this mediating social reality.

Marshall's reactions to the wild character of the region were more comfortable. In August 1925, he wrote a letter to his family about a trip to Idaho's Clearwater National Forest. It was one of his first statements on wilderness and the threats to its existence.

In days to come, when all of the once wild places of the country are dissected with highways and the honk of the auto horn on one road can be distinctly heard on the next, I will probably adjust my false teeth, and tuning up my trembling falsetto voice, tell my grandchildren about this past week. We were living for a few days 14 miles from the nearest habitation, 41 miles from the closest post office and settlement (population 50), 69 miles from the nearest railroad . . . 78 miles from the first pavement, electric lights, or doctor, and 125 miles from the first real town with a thousand or more inhabitants. We were camped on the north fork of the Clearwater 14 miles above Bungalow Ranger Station, well into the 28,000 square mile wilderness which, as I have mentioned several times before, forms the largest region left in the country undissected by roads.[46]

It is not clear when Marshall learned about the Forest Service's early experiments in wilderness preservation, though it was probably either just before his move west or not long after.[47] The Forest Service, after preserving the Gila Wilderness in 1924, set aside several other areas the following year—including a couple in the northern Rockies. Marshall could not have remained ignorant about the policy, or those areas, for very long. In fact, his letter may well have described one of those areas.[48]

Marshall had developed some strong preservationist views by the time he arrived in Missoula. But it also appears that Aldo Leopold's thinking significantly influenced his conceptualization of wilderness. Among Marshall's personal papers there is a copy of a 1925 letter from Leopold to Arthur Ringland that dealt extensively with Forest Service wilderness policy. The letter stressed the primacy of roadlessness and Leopold's belief that wilderness designation was not necessarily incompatible with limited resource use. "I would rather see cut-over lands than Fords any day," he tersely noted. Leopold's letter was also strident about the social utility of wilderness. "The prime purpose," he said of wilderness preservation, "is to take care of the citizen who cannot afford to travel far or expensively." Thus, relative proximity to population centers was vital. "The superficial thinker," he added, "cannot disassociate a wilderness trip from expensive guides and pack outfits. He concludes that the wilderness areas are for the idle rich. We must kill this conclusion or it will kill the wilderness idea."[49] All of these points resonated with Marshall's own experience and commitments. Nine months to the day after he composed his letter to Ringland, Leopold penned a brief note to Marshall expressing regret that they had missed each other on a visit Leopold had made to District One. How they knew each other is not clear, but Leopold's note marked the beginning of a fruitful alliance.[50]

Marshall remained quiet through the early stages of Forest Service wilderness discussions, as foresters debated the policy and took stock of wilderness holdings.[51] But there was evidence of his concern. A 1926 hike through the Lolo Pass on the Montana-Idaho border invoked in Marshall romantic thoughts of Lewis and Clark, who had followed the same Nez Perce hunting trail through the pass more than a century earlier. But his reverie was disrupted when he realized that a road would replace the remnant trail. "In a few years, the road from Lolo Hot Springs will push through this country and cut the last great wilderness in two," he lamented. "I am certainly glad I had the chance of standing at its edge in mid-winter before the wilderness is ruined forever by a highway."[52] Lewis and Clark were childhood heroes of Marshall's, and he rued the fact that a road along the trail would ruin the chance for others to experience the pass as they, and he, had.

As his veneration of Lewis and Clark suggests, Marshall's wilderness ideal romanticized exploration. He admitted as much. But he stuck to his position on the importance of wilderness in the face of others' arguments for modern recreational access. "'Think of how many more people,' they would always say, 'can enjoy the woods if you open them to autos than can ever benefit from them by this Daniel Boone stuff of yours,'" Marshall later wrote, aping the arguments of his critics. Yet for him, such a question missed a crucial point. Wilderness recreation involved escaping "the strangling clutch of a mechanistic civilization," and road building spread the reach of that civilization. Converting the Lolo Trail into a motor road made it easier for people to travel the route of Lewis and Clark and the Nez Perce, but it did so at the expense of being able to imagine how they had experienced the country. With motorized access, Marshall wrote, "certain doors of sensation were entirely shut."[53]

Developments in the Adirondacks also got Marshall thinking about how wilderness preservation might apply there. In 1927, the New York State Assembly passed an amendment to the state constitution allowing construction of a road up Whiteface Mountain as part of a broader initiative to modernize recreational facilities in the region.[54] As a letter from Russell Carson intimates, Marshall had suggested the wilderness idea as a way of opposing such developments. "While I have held the same sentiments unconsciously," his friend Carson replied, "your words crystallized these sentiments into a definite thought. I feel more and more that the word 'wilderness' is a word that conveys a wider shade of meaning than any other, and that the Adirondack Mountain Club would do well to adopt that word and emphasize it again and again in the Club's phraseology." In "Recreational Limitations to Silviculture in the Adirondacks," Marshall had not used the word "wilderness" to describe his recreational vision. By the late 1920s, it had become a keyword.[55]

Marshall was soon more directly involved with Forest Service wilderness politics. In September 1927, he wrote to Leon Kniepp asking for specific information on the remaining roadless areas in the national forests.[56] Thus began more than a decade of wilderness fieldwork that culminated with an impressive 1936 map, prepared with Althea Dobbins, that charted the nation's remaining roadless areas.[57] A year later, Marshall joined the public debate over Forest Service wilderness policy with a response to Manly Thompson's *Service Bulletin* article that charged wilderness advocates with elitism. Leopold, who also responded to Thompson's tract, had raised the question of "whether we think human minorities are worth bothering about," but this was only one of a number of points he made. Marshall's rebuttal, "The Wilderness as Minority Right," dealt with this issue almost entirely.[58]

Thompson accused wilderness advocates of pushing a policy whose intent was to preserve huge expanses for an elite group who sought only to escape the "hoi polloi." Marshall took issue with two aspects of Thompson's characterization. First, he thought Thompson's estimate that only 0.5 percent of recreational users sought wilderness conditions was "pure guesswork."[59] The figure could well have been much higher, he thought. But whatever the percentage, he identified an even more fundamental issue raised by Thompson's objections: "the old problem of minority rights." Preserving wilderness areas in the national forests for the use of less than one percent of the American population, Thompson argued, was an egregious violation of the utilitarian credo of "the greatest good for the greatest number." "Democracies," Marshall countered, "which are founded on the principle that the will of the majority shall govern, have a tendency to ignore the prerogatives of minorities. The outstanding champions of democracy, Voltaire, Mill, Paine, Jefferson, all appreciated this danger, and their works are interjected with eloquent pleas for the rights of the few." Public funds had long gone to the maintenance of cultural institutions that a small fraction of the populace chose to visit, yet few complained about them. "A small share of the American people have an overpowering longing to retire periodically from the encompassing clutch of mechanistic civilization," Marshall intoned. "To them, the enjoyment of solitude, complete independence, and the beauty of undefiled panoramas is absolutely essential to happiness." This minority, he concluded, had "a right to a minor portion of America's vast forest area for the nourishment of this peculiar appetite."[60]

Although brief, this response was one of the most important pieces Marshall wrote on wilderness. Two themes are particularly worth noting.

First, Marshall played upon the Jeffersonian notions of individual rights and the pursuit of happiness, even quoting the Declaration of Independence. In future wilderness writings, he often invoked this tradition of individual rights, which he saw staggering beneath the organizational burdens of an increasingly corporate society. While these notions formed one cornerstone of his defense of wilderness, his reading of Mill's utilitarianism formed the other. In exploring the mechanics of achieving "the greatest good for the greatest number," Mill had raised the specter of the "tyranny of the majority." In its management of the national forests for recreation, Marshall argued, the Forest Service had to take this possibility into account. Utilitarian management, Marshall insisted, compelled that the rights of minorities be protected from the will of the majority, which in this case was for mechanized recreational access. This was a crucial argument, for it made room for wilderness preservation within the Forest Service's utilitarian ethos.

Marshall displayed his liberal pedigree proudly in "Wilderness as a Minority Right." But, like Leopold before him, he also recognized the danger in making such an argument. Politically, he would have preferred widespread support for a wilderness policy; the minority rights argument was a strategic choice in the absence of such support. But it was also true that if there had been widespread support for a wilderness policy, and if such support had translated into significant wilderness use, tighter controls on access would have been needed (as they are in many such areas today). Because wilderness, by definition, could absorb only so much use, it was easy to construe it as elitist and antidemocratic. The wilderness idea was hostile not to the principle of equal access but to mechanization and the politics of mass consumption. Unfortunately for wilderness advocates, Thompson and other critics were easily able to conflate the two.

Throughout his stay in Missoula, Marshall had entertained thoughts of graduate school. By 1928, he had decided to pursue a Ph.D. in botany at the Johns Hopkins University.[61] But before he left the northern Rockies, he got in a few last words on the importance of wilderness. "There are only 15 areas left in the country of one million or more acres," Marshall wrote to his colleague Meyer Wolff, showing that he had done his homework, "and when all those damn roads are built only seven will remain."[62] That same month, Marshall attended a district meeting at which wilderness policy was discussed. "There was little sympathy shown for the idea of leaving a few large undeveloped areas for the enjoyment of those who might delight in recreation amid entirely primitive surroundings," Marshall lamented. "The general consensus of opinion," he wrote, sensing the compromises that would define Regulation L-20, "seemed to be that wilderness areas should be kept as long as there wasn't any other demand for them, which means precisely nothing." Achieving permanence would drive Marshall's advocacy in the decade to come. And increasingly he felt that such permanence could not be realized without fundamental forestry reform.[63]

The Freedom of the Wilderness

From the late 1920s on, Bob Marshall's career and thought oscillated between the increasingly defined poles of wilderness advocacy and socialized forestry. In the autumn of 1928, he began his studies at Johns Hopkins under plant physiologist Burton Livingston. His dissertation, which was eventually published in the first issue of *Ecological Monographs,* looked at how evergreen seedlings reacted to the drying of their soil.[64] While in Baltimore, Marshall also embarked on a more rigorous exploration of socialism. He became a prominent member of the Johns Hopkins Liberal Club, and he donated money and time to groups such as the League for Industrial Democracy (LID), an alliance of socialist intellectuals and one of the most important American leftist organizations of the interwar era. LID's motto, "Education for a new social order based on production for use and not for profit," struck Marshall as an apt description of his own conviction that rational discourse could achieve radical results. Among LID's prominent members during this period were Florence Kelley, John Dewey, and Stuart Chase. Norman Thomas, a perennial Socialist Party presidential candidate, and Harry Laidler were the group's executive directors.[65]

Of LID's members, John Dewey seems to have had the strongest influence on Marshall's thought. Dewey's philosophical explorations of democratic collectivism spoke to Marshall's intellectual struggle to fit his commitment to civil rights and liberties within the parameters of a collectivized society. Indeed, squaring individualistic principles with the dictates of a mass society was perhaps the central intellectual problem of the era. Dewey offered Marshall a pragmatic socialism stripped of authoritarian tendencies. Moreover, Dewey's liberalism retained, according to historian Robert Westbrook, the core values of "liberty, individuality, and freedom of inquiry" while sloughing off "the adventitious connection between liberalism and the legitimation of capitalism."[66] America's parallel libertarian traditions, Dewey argued, had collided in the modern age, revealing profound contradictions. The "ideal of equality of opportunity and freedom for all" had crashed up against the divisive concentration of power that was the product of economic freedom.[67] Like Dewey, Marshall became convinced that salvaging individualism in modern America involved keeping social and economic freedom distinct, and rooting the former in a sense of community responsibility. At the same time, however, Marshall was suspicious of collectivism and not entirely content to scrap what Dewey called the "old individualism." While he saw democratic socialism as the only just response to the socioeconomic realities of the era, he also yearned for a sense of freedom from collective entanglements. His affection for wilderness was very much related to this desire for freedom.

Marshall's misgivings about collectivism must be understood within the logic of a new brand of civil libertarianism that came of age during the interwar era. Although his commitment to civil liberties had initially been nurtured by his father's efforts in this area, Marshall was also greatly affected by ideological reverberations from the federal assault on the civil liberties of workers, socialists, and antiwar activists during and after World War I. Out of that crisis was born the American Civil Liberties Union (ACLU), which got its start defending conscientious objectors and protesters during the war, and which turned its energies to supporting the besieged labor movement after the war. The ACLU's approach to civil liberties reflected a profound postwar disaffection with majoritarianism that stood in marked contrast to the faith in the people that had animated reformers prior to the war. Postwar civil libertarianism was a direct result of a concrete historical sense that majority rule, particularly when it was achieved through the sort of consensual propaganda behind calls for "100 percent Americanism," could be a threat to individual rights and freedoms. Marshall became a strong supporter of the ACLU during this period, and he and Roger Baldwin, its founder and head, became fast friends. More than that, he shared the ACLU concern about majoritarianism, particularly in terms of its threat to the health of the labor movement and organized socialism in America. Marshall also came to fear that wilderness would be sacrificed to majoritarian recreational impulses. When he defended wilderness recreation as a minority right, then, he did so within this specific intellectual context.[68]

Alaska remained central to Marshall's fantasy of escape, but it also became a crucial vantage point from which to launch his social critique of modern America. During the summer of 1929, after his first year at Johns Hopkins, he finally fulfilled his dream of going to Alaska. The ostensible purpose was to study tree growth at the northern timberline, but the real reason was exploration pure and simple. "The section which I chose for this purpose," Marshall later explained, "covered some 15,000 square miles at the headwaters of the Koyukuk River, a major tributary of the Yukon, in the neighborhood of the Arctic Divide in the Brooks Range, north of the Arctic Circle."[69] That summer, Marshall spent a month exploring the Brooks Range and what became Gates of the Arctic National Park. Indeed, Marshall gave the park, and the region, that name. "Fortunately," Marshall recalled of his first view of the valley, "this gorge was not in the continental United States, where its wild sublimity would almost certainly have been commercially exploited. We camped in the very center of the Gates, seventy-four miles from the closest human being and more than a thousand miles from the nearest automobile."[70] Where his adventures in the Adirondacks and the northern Rockies had required that Marshall pepper his wanderings with imaginings of past explorers, in Alaska he was secure in his conceit that he was the first person to explore portions of the territory. Marshall was quick to assume the role of the great wilderness explorer, an archetype that has stuck with him and defined his environmental legacy.[71] But what ultimately fascinated him most about the Alaskan wilderness were the people who lived there. He returned to Alaska a year later, and the product of that stay was a community study that far surpassed in originality his hackneyed narratives of exploration.

Back in Baltimore after his summer in Alaska, Marshall got more deeply involved in a growing debate about Forest Service policy.[72] In 1928, a forester named Major George Ahern had fomented the debate by publishing a scathing booklet on the failure of forestry in America.[73] Not only had federal forest policy gone in a pro-industry direction, Ahern maintained, but most professional foresters had become conservative technocrats whose work was uninformed by any social vision. Marshall's initial foray into the debate occurred when *The Nation* published his article "Forest Devastation Must Stop" in August 1929. He began with a revealing interpretation of the history of public forestry in the United States. "The Pinchot-Roosevelt theory had been a militant one," he insisted, though "their crusade attracted to its banner people who under ordinary circumstances condoned all damage to public welfare when it stood in the way of private profit." He revered the earliest practitioners of Progressive forestry, but he recognized that capitalists had used the program for their own purposes. Then, in the early 1920s, a "remarkable change in attitude" occurred, as foresters turned to the "more facile path of cooperation and persuasion." Ahern's treatise attempted to shatter the complacency that had built up around the cooperative approach by showing that it had not changed industry behavior. "Then what is to be done about it?" Marshall asked. "Sit by hopefully and pray that cooperation may bring better times? Or go militantly forward to procure those better times by compulsion?" Ahern proposed the latter course, as did Ahern's greatest supporter, Gifford Pinchot.[74] For Marshall, this militancy, not the gospel of efficiency, was the true legacy of Progressive forestry. With the help of Ahern, Pinchot, and a number of other foresters, Marshall sought to resuscitate that spirit.[75]

In his *Nation* article, Marshall proposed a series of solutions to the twin problems of industry malpractice and regulatory acquiescence. He resigned himself to the decline of old growth. "We can only reserve a certain share of the remaining primeval forest for recreational purposes," he wrote, "and make up our minds that from a commercial standpoint the ancient woods will soon be a thing of the past." While this was a depressing prospect, Marshall also realized that the easy returns from the harvesting of "ancient woods" stood in the way of responsible cropping. The sad truth was that only with the disappearance of commercially available old-growth timber—either through exhaustion or preservation—would industry be forced to practice sustainable forestry.[76]

The problem of fire was more complex. Marshall accepted the industry argument that if private timber companies were to practice sustainable long-term forestry, their timber would have to be protected from fire damage. But he also wanted timber companies held accountable for their cut-and-run tactics and the resultant slash fires that retarded regrowth and degraded land.

Finally, Marshall suggested that capitalist motivations had to be blunted. "Cutting the forest for the future public welfare," he concluded, "implies in almost all cases a sacrifice of present private profits." "The government must step in and compel the private timber owner to leave the forest which he exploits in a productive condition," he demanded. "If the operator protests that economic conditions make this impossible, then his lands should be taken over by the government." This was the gist of Marshall's program, and the spirit, he insisted, of Progressive forest conservation since its inception.[77]

In joining this debate so forcefully, Marshall caught Gifford Pinchot's eye. In January 1930, Pinchot invited Marshall to have lunch with him and discuss the formation of an organization to counter forest depletion.[78] It was the first time they had met, and Pinchot immediately impressed Marshall. "Governor Pinchot is one of the most amazing men I have ever met," he told his friends Gerry and Lilly Kempff.[79] The upshot of the meeting was "A Letter to Foresters," a public statement issued in February 1930 and subsequently published in the *Journal of Forestry*. The letter, which echoed many of Marshall's recently published sentiments, also drew upon Pinchot's long-held concerns about timber famine. "The destruction of the forests of America has been a long-drawn out tragedy of waste," the letter began. "Now we face the danger of a moral tragedy also: that the foresters of America will accept that destruction and by silence condone it."[80] The letter charted the decline of forestry's "high ideals and great purposes" and suggested that forest devastation resulted from nothing less than a loss of professional will.

Marshall had always seen forestry as more than just a technical pursuit. Unfortunately, an exclusive focus on technique was rife within the profession, a product, perhaps, of professionalization itself. Evidence of this came in a reply to the "Letter" by forester Royal Kellogg. The Society of American Foresters, he wrote:

> should not be an institution for propaganda nor used for the advancement of personal theories of public policy. To the extent that it is so used, the Society will be torn by internal dissensions and its proper functions obscured. Propaganda should come from organizations publicly known to be subsidized for such purposes. It has no rightful place in a scientific organization.[81]

"A Letter to Foresters" and such responses revealed an ideological divide between those in the forestry community who saw themselves as disinterested scientists and those who insisted that the profession necessitated a particular political approach to industry. While Kellogg pointed out to these rebels that their anti-industry stance compromised the objectivity essential to scientific management, Marshall and others insisted that the forestry profession, in its retreat to the high ground of disinterestedness, was implicitly choosing sides. Unlike their more conservative colleagues, Marshall, Pinchot, and their co-conspirators were skeptical about whether the timber industry ever would, or could, serve the public interest.

Marshall's next task was to define that public interest more completely. In "A Proposed Remedy for Our Forestry Illness," which appeared in the March 1930 issue of the *Journal of Forestry*, he urged his colleagues to see forests as a vast public utility.[82] Forests, he wrote, performed four important functions: they furnished timber, regulated stream flow, curtailed erosion, and provided for recreational and aesthetic needs. Like Aldo Leopold, Marshall argued that public interests inhered in private lands, and that the mismanagement of private forests had public impacts. Management of all forests thus required that "four classes of people" be protected: the owners of the forest, the people who make their living harvesting forest crops or manufacturing forest products, the consumers of wood, and "all those who are part of the community in which the forest industry functions." According to Marshall, only the owners and consumers had been thus served, and he criticized industry for failing and provide workers with "a stable home life," regular work, and safe conditions. "Abandoned towns and deserted farms," he concluded, sounding much like MacKaye, "have typified the unregulated pursuit of private profit in every forest region in the United States." Industry had also failed, for obvious economic reasons, to provide properly for forest recreation, and most private forestry was negligent in the protection of watershed values. Marshall concluded that "voluntary private forestry has failed to meet any requirement of an adequate forestry policy, except providing cheap wood to the present-day consumer." If forests were to be seen as utilities, as he suggested, the ability of private landowners to use their timber resources as they pleased had to be sharply curtailed.

Although the solutions Marshall proposed in this article involved a complex mix of cooperation, regulation, and public ownership, he was moving toward the latter as a comprehensive solution.[83] Indeed, that same year LID published Marshall's *The Social Management of American Forests* as part of its pamphlet series on "the movement toward the social ownership of industry." Although Marshall cribbed much of the booklet from previous articles, one theme emerged more clearly in this piece: "With such a clear record, the rational conclusion seems inescapable that commercial privately owned timber lands should be socialized."[84]

Marshall's attention soon swung back to wilderness. In February 1930, *Scientific American* published his most important article on the subject, "The Problem of the Wilderness."[85] In many ways, it was a conventional piece of thinking. His basic historical model was one of wilderness giving way to civilization, the timelessness of nature succumbing to human transformations. There was also a fair dose of masculine rhetoric, as Marshall used terms such as "virility" in describing the therapeutic value of wilderness. He also alluded to William James's famous 1911 essay, "The Moral Equivalent of War," to suggest how wilderness could be an outlet for the instinct for adventure that he, like James, thought was buried alive by modernity and exhumed only in outbursts of martial violence.[86] "One of the most profound discoveries of psychology," Marshall observed, noting another of the era's intellectual developments, "has been the demonstration of the terrific harm caused by suppressed desires."[87] Marshall thought that wilderness could play a part in releasing those desires. Indeed, by briefly alluding to this psychological argument for wilderness, Marshall revealed another facet of his disquiet with mass society.

Marshall's language here was consummately antimodern. In a world full of enervating artificialities, modern Americans often sought out experiences that put them in touch with the primal realities of life. His search for a "real" wilderness experience certainly fit this mold. But as revealing as "The Problem of the Wilderness" was of Marshall's own cultural and psychological needs (and they were considerable), his argument for wilderness was more than just a recreational prescription for a flaccid urban elite. Marshall's reformist agenda kept his wilderness advocacy from becoming merely a consumer gesture. There was a certain amount of escapism in his idealization of wilderness as a space of freedom, but that escapism never lapsed into a therapeutic accommodation to consumerism.[88]

"The Problem of the Wilderness" begins with a definition of wilderness quite similar to Aldo Leopold's. A wilderness area, Marshall wrote, is "a region which contains no permanent inhabitants, possesses no possibility of conveyance by any mechanical means and is sufficiently spacious that a person in crossing it must have the experience of sleeping out." The dominant attributes of such an area were "first, that it requires any one who exists in it to depend exclusively on his own effort for survival; and, second, that it preserves as nearly as possible the primitive environment. This means that all roads, power transportation, and settlements are barred." Like Leopold, he defined wilderness primarily in opposition to roads and "power transportation."[89]

Mixed in with his celebration of romantic exploration and strenuous recreation, Marshall made an interesting case for wilderness as a space of forced self-sufficiency. In part, this was to separate the wilderness experience from modern forms of outdoor recreation, but he had other goals as well. "As long as we prize individuality and competence it is imperative to provide the opportunity for complete self-sufficiency," Marshall wrote. Wilderness was not just an escape from "the effete superstructure of urbanity"; it was also a space that resisted the economic entanglements threatening the ideals of individual autonomy in modern America. He reached back, implicitly, to a tradition of self-reliance in American political and environmental thought, a tradition that had grown in New England amid the early strains of the industrial revolution. Emerson and Thoreau had each connected individual freedom and self-reliance.[90] Marshall, who found himself in the midst of an advanced stage of industrial development, also invoked the need to preserve "the freedom of the wilderness." Like MacKaye, he saw wilderness as a place where one might envision a different future. Marshall was more inclined to see wilderness as a vestige of freedoms past, but both saw it as an integral part of the modern landscape.

Marshall devoted much of "The Problem of the Wilderness" to aesthetics, making a case for distinguishing wilderness from conventional scenic notions. A wilderness experience involved much more than the "ordinary manifestations of ocular beauty."[91] Wilderness recreation was not about seeing sights, but about being encompassed by the natural world. One looked at scenery from a distance, but one experienced wilderness from within. Rather than privileging the visual, wilderness engaged all of the senses. By definition, then, the wilderness experience could not be had in an automobile, or in the midst of a designed landscape that shaped how people experienced the nature around them. The expression of such an aesthetic was not new, but it took on new meaning in relation to the automobile.

Marshall had a penchant for fairness, and in that spirit he addressed the supposed disadvantages of wilderness preservation. First, there was the threat of fire, which he thought necessitated compromise: "[C]ertain infringements on the concept of an unsullied wilderness will be unavoidable in all instances." Trails, telephone lines, and lookout cabins would thus be tolerated. "But even with these improvements," he maintained, "the basic primitive quality still exists: dependence on personal effort for survival." He thought that the second supposed drawback, economic loss, was mostly a red herring: "It is time we appreciate how little land need be employed for timber production, so that the remainder of the forest may be devoted to those other vital uses incompatible with industrial exploitation."[92] Careful forestry, he argued, was the ally of wilderness protection. The final drawback was that wilderness would "automatically preclude the bulk of the population from enjoying" it. "Far more people can enjoy the woods by automobile," he admitted, and far more "would prefer to spend their vacations in luxurious summer hotels set on well-groomed lawns than in leaky, fly-infested shelters bundled away in the brush." To counter this argument, he resorted to minority rights. Motorists had plenty of roads "traversing many of the finest scenic features of the nation. . . . But when motorists also demand for their particular diversion the most insignificant wilderness residue, it makes even a Midas appear philanthropic." Quoting from Mill's *On Liberty*, Marshall asked: "Why should tolerance extend only to tastes and modes of life which extort acquiescence by the multitude of their adherents?" Wilderness areas were open to all, but they were still vital to only a few. Nonetheless, those few deserved the right to pursue their happiness unfettered by the preferences of the majority—at least until that majority could be brought around to the wilderness cause. Indeed, Marshall closed "The Problem of the Wilderness" with a call for the "organization of spirited people who will fight for the freedom of the wilderness." "If they do not present the urgency of their view-point the other side will certainly capture popular support. Then it will be but a few years until the last escape from society will be barricaded."[93]

Within a month of publication, Marshall received a letter from Aldo Leopold complimenting him on his article, and a month after that, similar praise from Harold Anderson. "I am glad to see," Anderson wrote, "that Benton MacKaye is not the only one who is preaching the benefits to be derived from contacts with wilderness tracts."[94] Marshall's article proved to be a pivotal expression of the need for wilderness in a nation being transformed by the automobile. But Marshall also made clear that wilderness preservation was about perpetuating a tradition of individual rights and liberties that seemed to be running out of steam. More than any of his Wilderness Society colleagues, Marshall defended the freedom of the wilderness, even though that freedom was only recreational. Social freedom demanded a different set of reforms.

An Arctic Middletown

Bob Marshall finished his Ph.D. in the spring of 1930 and began planning a return to Alaska. "I am contemplating going back to the north country for a whole year," he told his friend Lincoln Ellison, "to make a study of the economic and social conditions in the far north as well as to continue my ecological observations which I made up there last summer."[95] "In almost all economic and social discussions of our present civilization," he told another friend, "the question seems to arise whether we are happier than our preindustrial ancestors."[96] Marshall's guiding purpose in returning to Alaska was to see if simpler living surrounded by wilderness made Alaskans happier than modern Americans.

Leaving in August 1930, Marshall spent the next fourteen months in Wiseman, Alaska, testing his social, political, and environmental beliefs. Committed to an idea of freedom rooted in the nation's past, he was looking for a community type that had been lost in the lower forty-eight. Dedicated to an ideal of social and racial equality, he also went to Alaska to demonstrate that Western civilization was a bankrupt standard against which to judge the achievements of other races and cultures. And with the rest of the nation in a prolonged depression, he wanted to prove that living in a wilderness granted its residents an important measure of economic independence. Marshall's results were convincing on these points in a limited way.

Arctic Village, the book that resulted from this visit, was a mix of sociological and ethnographic examination, a hybrid of then-fashionable scholarly methodologies.[97] Recognizing its similarity to another well-known community study of the era, one reviewer referred to *Arctic Village* as an "Arctic Middletown."[98] But Wiseman was no Muncie, Indiana. It lacked many of the problems, such as class differences and status anxieties, of the Middle American town that Robert and Helen Lynd made famous. Wiseman also lacked overt tensions between racial and ethnic groups, despite the community's diversity.[99] Indeed, these qualities were, for Marshall, part of Wiseman's allure as a subject of study. He also was influenced by the assumptions and methodologies of a new school of American anthropologists. Franz Boas, and students of his such as Margaret Mead and Ruth Benedict, embraced cultural relativism as their guiding ethos.[100] Beyond that, this interwar generation of anthropologists used close study of other cultures not to bolster assumptions of Western superiority but to reveal significant shortcomings in modern society.[101] Marshall too had the specific intent of providing a critical perspective on modern America. But where these anthropologists often relied on purer cultural manifestations for their critiques, Marshall used the cultural hybridity of Wiseman to make one of his major points: it was the structural relations that prevailed in Wiseman, particularly the direct economic relation with nature, that made the town a model.

Wiseman's civility, according to Marshall, had everything to do with the environmental setting.

For all its apparently utopian qualities, Wiseman was a weird place in ways that Marshall was not quite willing to admit. The community had a short and not particularly glowing history as a boom and bust gold-mining town, and most of the miners who had made anything from the experience had left. The population was a mix of indigenous peoples and marginally successful white prospectors who had lived there for at most three decades. The white residents who remained did so partly out of fail-ure and partly because of the freedom Wiseman afforded them—includ-ing a freedom from wives, many of whom either had remained in the lower forty-eight or had fled Alaska when it became apparent that their husbands were irreversibly attached to the place. Residents were mostly male and single; the eligible bachelors were usually older white prospec-tors while the few eligible women were younger Eskimos. There was some intermarriage, but the demographic trends pointed to a younger Eskimo community destined to outlast an aging white male population. In "The Happiest People," her review of *Arctic Village* for *The Nation,* Ruth Bene-dict noted that most of Wiseman's residents had sacrificed a stable fam-ily life to live the way they did.[102] That Marshall felt so comfortable in this frontier bachelor society was perhaps more revealing of his predilec-tions than it was of a widely held ideal of independent, communitarian living in close contact with nature. Most Americans would not have rec-ognized Wiseman as a utopia.

But despite his flawed romanticization of this unorthodox community, Marshall's study worked as a critique of modern American civilization, and it spoke in important ways to his various political commitments, in-cluding wilderness preservation. He knew that most Americans were well past the point of being able to revert to living as the residents of Wise-man did, and that most would not want to even if they could. The real point of the book was to reflect the apparent happiness of Wiseman's res-idents back on modern Americans. Marshall saw this happiness in a num-ber of qualities peculiar to Wiseman: the relative tolerance of the town's residents, the lack of class or status divisions, the independence and self-reliance that Wisemanites developed and prized, the enforced coopera-tion required by living under such harsh conditions, and the ready avail-ability of adventure. Marshall thought these qualities were shaped by a simple economic system closely connected with nature. He saw in Wise-man an ideal organic socialism in which individual liberty and a communitarian ethos coexisted. Although most Americans would never have the opportunity to live this way, Marshall hoped that by preserving wilderness as a recreational space, they might gain the same sort of perspective on their lives that his year in Wiseman had given him. If wilderness disappeared, he feared, there would be no more imagining a freer, more equal, more tolerant, and more adventurous world.

Back into the Forestry Fray

Marshall returned to Baltimore in September 1931, rented a room, and wrote *Arctic Village*. He also jumped back into forestry politics and renewed his friendship with Gifford Pinchot. Still a regulator at heart, Pinchot resisted Marshall's growing commitment to public ownership of the nation's forests, but in their zeal for reform the two remained united.[103]

During the summer of 1932, Marshall received the most important assignment of his career to date. At the age of thirty-one, he had established himself as an important voice in forestry and conservation circles, even though he had not developed much of a career. Earle Clapp, head of the Forest Service's Branch of Research, contacted him about a study of the nation's forests just commissioned by Congress at the request of New York Senator Royal Copeland. Clapp, the study's director, wanted Marshall to write the recreation sections of what was to become *A National Plan for American Forestry*, known to most as the Copeland Report. From the beginning it was clear that the report would repudiate the cooperative policy of the 1920s in favor of a more aggressive government program. It might also influence the policy of a new president with a strong interest in forestry. Marshall accepted the offer, moved to Washington, and immersed himself in a study of the recreational uses of the nation's forests.[104]

The Copeland Report did indeed suggest a revised course for American forest policy. It blasted industry for contributing to the major problems of overproduction, wasteful utilization, land devastation, community instability, fire-related destruction, and tax delinquency. Although the report acknowledged structural impediments faced by industry, it concluded that cooperative policies had changed industry behavior very little. It also pointed to two other significant problems. First, there remained a vast amount of public land left behind by the homesteading process, some of it forested but not controlled by the Forest Service, which demanded public management. Second, "tax delinquency [was] creating a new public domain not of forested land but largely, instead, of devastated forest land, and of such size that it promises to be a heavy burden." The report thus recommended a two-pronged effort involving a "large extension of public ownership" and "more extensive management of all publicly owned lands." This "next big step in American forestry" was to be part of the larger New Deal effort to contend with a permanent—and growing—public domain.[105]

Marshall's major contribution to the Copeland Report was a twenty-five-page section entitled "The Forest for Recreation." The national forests, he began, had seen huge increases in recreational use since World War I. Visitation had risen from three million in 1917 to thirty-two million in 1931.[106] In explaining this surge, he pointed to the "entirely new recreational habits" inculcated by the automobile. The most urgent need, he suggested, was a program for preserving a variety of areas that would meet user demands without having the preferences of some intrude on those of others. Most of his report was devoted to explaining these specific types of preservation.

Three important contextual points need to be made here. First, the Copeland Report was about the nation's forestry problem as a whole. Though the national forests were a focus of the inquiry, the authors also considered private timberlands and forests within other federal, state, and local designations. Second, Marshall's contribution was a strong statement on the need for recreational planning at the national level, something that the Forest Service had long resisted. In this sense, the wilderness idea was part of a broader effort to differentiate and zone recreational land uses. Third, recreational use of the automobile drove the need for careful recreational planning. As Marshall put it, the "differentiation between forms of recreation employing mechanical transportation and those employing natural transportation is of fundamental significance."[107]

Marshall recommended that seven types of recreational areas be preserved. The first was the "Superlative Area," a locality "of unique scenic value." These areas could vary in size depending on the scenery, and would be protected against both resource and recreational development. Trapper's Lake fit this category. The second type was "Primeval Areas," a category Marshall saw as synonymous with "natural areas." These were "tracts of virgin timber in which human activities have never upset the normal processes of nature." This type was congruent with the sort of preservation sought by the Ecological Society of America, but Marshall added an aesthetic rationale to the ESA's program. "Primeval Areas" certainly were necessary for studying natural processes, and as a standard against which to measure forestry practices. "The importance of the primeval in the more subtle aspects of forest recreation," he added, pointing to the aesthetic value of such areas, "is much less generally recognized." He recommended the preservation of a number of these areas in every important forest type (he listed twenty) in the country. He also argued that such areas should be at least 1,000 acres, but preferably on the order of 5,000 to 10,000 acres. Because he thought that "Primeval Areas" served an important aesthetic function, he argued for preserves larger than scientists generally had asked for, though they were still smaller than wilderness areas.[108]

The third category was "Wilderness Areas." These Marshall defined as "regions which contain no permanent inhabitants, possess no means of mechanical conveyance, and are sufficiently spacious that a person may spend at least a week or two of travel in them without crossing his own tracks." The definition had changed little since his 1930 article. He recognized that wilderness areas could contain primeval areas—that the primeval should be a subset of wilderness preservation—but he also assumed preservation of primeval conditions would not be possible within all wilderness areas. They were compatible though not synonymous forms of preservation. "The difference between primeval and wilderness areas," he wrote, "is that the primeval area exhibits primitive conditions of growth whereas the wilderness area exhibits primitive methods of transportation." The chief function of wilderness was not to preserve the primeval, "but rather to make it possible to retire completely from the modes of transportation and the living conditions of the twentieth century."[109] "Cattle and sheep grazing is not incompatible with wilderness use," Marshall wrote, echoing Leopold, and he recognized that limited timber cutting might occur in wilderness areas. In terms of size, he suggested a 200,000-acre minimum. There were still a number of these areas left unprotected, and "almost no sacrifice of economic values would result from preserving these forest areas as wilderness." "The only sacrifice involved," he suggested, "would be in barring tourists," by which he meant those who sought motorized access. He ended this section by listing sixty-five areas, some established and others candidates for wilderness preservation, totaling almost twenty-seven million acres.[110]

The remaining four recreational types were "Roadside Acres," "Camp-Site Areas," "Residence Areas," and "Outing Areas." All were designed to cater to motorists. "Roadside Areas" were strips of timber along highways used for recreational travel. Such an aesthetic gesture, Marshall wrote, would go a long way toward appeasing motorists. "Camp-Site Areas," or developed campgrounds, were primarily for "the benefit of the many automobilists and boat travelers who spend their nights in camp." Citing E. P. Meinecke's concerns and his innovative campground design, Marshall suggested that these areas would minimize the damage autocampers were doing to the forests. But he also hoped such development would occur only in the most heavily used areas. Beyond the pale of mechanization, such developments would "give the recreationalist an unnecessary impression of the very regimentation and artificiality which he is seeking to avoid."[111] "Residence Areas," developed under the auspices of the Term Permit Act, allowed concentrated construction of "private homes, hotels and resorts, group camps, sanatoria, and stores and services of one sort or another." Finally, he suggested the need for "Outing Areas," where motorists could stop, get out of their cars, and enjoy "more intimate contact with the woods." These did not require the acreage necessary for wilderness areas, the scenic magnificence of superlative areas, or the "virgin" conditions characteristic of primeval areas. "Outing Areas" would allow people "to get away from the sounds of the highway" without necessitating strict conditions of preservation. Such areas could be compatible with forest utilization, and many could be created near cities. Thus did Marshall hope to meet the recreational needs of all Americans while also accommodating conflicting desires.

One of the most fascinating sections of Marshall's report addressed the natural foes of forest preservation, particularly the "one natural enemy against which the primeval forest can not in the long run be protected": "senility." "Sentimental conservationists," he continued, "talk glibly about setting aside virgin timber tracts to be preserved in all their natural glory forever," though relatively few forest trees lived more than 400 years. "What is a beautiful virgin forest today may in 40 years be a very ragged stand in which most of the old trees are dying and in which the understory will require a century or more to attain the size and beauty of the former forest." Such "overmature timber" was also susceptible to insect epidemics and fire. These observations alone were not particularly novel; foresters such as Howard Flint often used a temporal understanding of old growth in arguments against "sentimental conservationists."[112] What was unique here was Marshall's way of arguing for a more extensive commitment to preserving the primeval. He concluded:

The primeval forest, though it is a self-perpetuating unit, is bound to go through cycles of deterioration and upbuilding. Since deterioration is inevitable, and many years or even several centuries may elapse before the beauty of the primeval is restored, sustained-yield principles must be applied to primeval areas as well as to lands which are being logged. It is necessary, in other words, to maintain *in a primitive state* a complete rotation of age classes, so that when the overmature forest decays a mature stand will be growing up to take its place and a stand of reproduction will be advancing toward maturity.

Marshall understood that forests were dynamic and that senescence might confound scenic preservationists. Today's "overmature" primeval areas would last only so long in their present state. But instead of using this logic to urge harvesting and intensive management, as most foresters did, he used it to urge a sweeping program of preservation—what he called primeval rotation—that included not only mature primeval forests but also a variety of less mature age classes. Marshall noted, as Leopold had, that nature too harvested the forest, and he used that knowledge to argue for a more sophisticated brand of preservation.[113]

In a final section on the relation between forestry and recreation, Marshall reiterated his simple but crucial contention that without "the practice of forestry on the lands devoted to timber production, the best values of forest recreation would be doomed." If, he warned, "our physical forest needs cannot be met on the areas devoted to commodity production, it is almost certain that the aesthetic and inspirational forest values will be sacrificed."[114] Preservation and utilitarian forestry were, for Marshall, symbiotic commitments, not antithetical worldviews. The former required the latter, and both required a government willing to thwart the liquidation of old growth and the destruction of wilderness.

Marshall had another remarkable opportunity during this period to shape federal policy. After he became president, Franklin D. Roosevelt wrote to Gifford Pinchot for advice on forest policy. Pinchot, feeling out of touch with the Forest Service, summoned Marshall to his Pennsylvania home for a briefing. When Marshall arrived, Pinchot told him about Roosevelt's request and asked if he would be willing to ghostwrite a policy letter. Marshall accepted, crafting a statement that predictably stressed "large scale public acquisition of private forest lands." "Neither the crutch of subsidy nor the whip of regulation," Marshall wrote above Pinchot's signature, "can restore [private forestry]." He then recommended that, in combination with a program of land acquisition, the president begin a large-scale public works project to restore devastated forest lands.[115] Roosevelt considered this plan—and the recommendations of the Copeland Report—but opted for a more cooperative program consonant with the National Recovery Administration's industrial codes. However, he did embrace the recommendation that the unemployed be put to work improving the nation's forests, though even that victory would come to have its dark side.

In early 1933, Marshall began work on *The People's Forests,* a book for popular consumption that advocated the complete socialization of the nation's timberlands. At the same moment, he was growing more critical of organized socialism in the United States. He wrote scathingly that winter to Norman Thomas, leader of America's Socialist Party, reneging on a pledged $1,200 contribution: "[I]f the party still stands chiefly for Soap-Box Socialism, then I feel that there are much more valuable ways to invest the money I have available."[116] He also wrote a lengthy "Letter on Economics" in which he admitted that he was moving away from the "mildly socialist" views he had developed under the "impact of *The Nation*" and from "the frank discussion of western lumber jacks." By 1933, he realized that "general socialization will not come through learned discussion, soap-box orations, or idealistic capitalists." The "main tools" would have to be "militantly organized workers and farmers who will use mass pressure for bringing about the change of conditions." This change of heart was largely the result of Marshall's growing fear that farmers and workers would choose the emotional appeal of fascist movements over the intellectual high-mindedness of American socialism. The New Deal had also pushed him in this more strident direction. "The present dying economic order," he lamented, "will be pumped full of stimulants and kept alive" instead of being fundamentally reformed.[117]

The People's Forests was shaped by this growing disillusionment, though it contained little not found in Marshall's previous writings. His two chapters on recreation, for instance, reprinted his Copeland Report contribution broken only by an inserted section from "The Problem of the Wilderness." He did include a chapter, "A Biological Interlude," in which he argued that forests should be respected as complex organisms whose management required constant scientific research. The point was that economic and social reforms in ownership were not enough; protecting the public interest involved gaining a much greater knowledge of how forests work. But beyond that, *The People's Forests* was little more than a reaffirmation of Marshall's beliefs, though radicalized by the events of 1933 and packaged for popular consumption. "The time has come," he wrote in the book's last sentence, "when we must discard the unsocial view that our woods are the lumbermen's and substitute the broader ideal that every acre of woodland in the country is rightly a part of the people's forests."[118]

Marshall's critics were many. In a departure from protocol, Franklin Reed, editor of the *Journal of Forestry,* reviewed *The People's Forests* himself. He did so, he explained, at the author's request. True to his belief in the importance of open discourse, Marshall sought the opinion of a respected member of the forestry community who almost certainly would not like what he had written. Reed did not. He excoriated Marshall for writing a "dangerous book," and for making the process of socialization sound easy. Reed saw Marshall's proposals as a threat to cherished economic liberties, and he suggested that socialization would substitute totalitarian for democratic methods.[119] Marshall clearly was no friend of totalitarianism, and to paint him as such was unfair. Reed's critique would have hit closer to the mark by recognizing the fundamental weakness in Marshall's scheme: rather than ditching democracy in favor of centralized planning, Marshall was attempting the difficult feat of combining the two. His plan for socializing the nation's forests exalted democratic ends while fudging democratic means. (Marshall's prescription also ran counter to Aldo Leopold's concurrent contention that the future success of conservation in the United States lay in the hands of private landowners.) He would run into a similar problem in his next major endeavor.

Indian Wilderness

With a successful year of publishing under his belt, Marshall hoped to land a position as a recreational planner with the Forest Service. There had been talk of such a position materializing during the early New Deal, but by July of 1933, the position had fallen through due to lack of funding.[120] Instead, on August 1, Marshall began official duties in one of the most important New Deal agencies. John Collier, the new Commissioner of Indian Affairs, hired him to head the BIA's forestry division. This position was a fine second choice, for it allowed Marshall to travel throughout the West and conduct wilderness reconnaissance. It also gave him another context in which to assess his economic, social, and environmental ideas.

Marshall was well aware of Collier's history as a defender of Native American sovereignty and rights. In fact, he had been an ardent supporter of Collier's for the position of Commissioner of Indian Affairs, writing to the new Secretary of the Interior, Harold Ickes, soon after Roosevelt's inauguration to recommend Collier for the job.[121] Collier's major achievement with the BIA was the passage of the Indian Reorganization Act of 1934, which created mechanisms to rebuild tribal sovereignty and to return lands, both individual allotments and remaining public domain within old reservation boundaries, to tribal ownership. Collier was committed to rebuilding the cultural lives and communal economies of Indians, and in Marshall he found someone with similar goals. Over the next few years, Marshall became an important adviser to both Collier and Ickes.[122]

One of Marshall's most curious initiatives in the BIA involved setting aside portions of Indian reservations as wilderness areas. During his summer reservation tours, he took time out to explore the large roadless areas remaining under Indian control. As early as 1934, he sent policy memoranda to Interior Department officials suggesting wilderness preservation for such areas. By the end of his tenure with the BIA, he had convinced Collier and Ickes to protect sixteen Indian wilderness areas, mostly in the West, ranging from 6,000 to 1.59 million acres. His rationale revealed the particular contours of his wilderness thinking.[123]

It might seem strange, even arrogant, to contemplate wilderness preservation on what was then a shrunken and marginal Indian estate.[124] But Marshall's wilderness policy was not meant to dispossess Indians or lock up their resources. Rather, he aimed to protect Indian economic and cultural autonomy from what he saw as a new set of threats. In an early memorandum on roads and truck trails on the Navajo Reservation, Marshall wrote: "highways bringing the interior of the reservation into intimate contact with the outside world should be eliminated, in spite of their convenience for administrative purposes." Such roads, he continued, "will suddenly thrust Twentieth Century white civilization into the entirely different Navajo culture."[125] In a press release announcing the adoption of Marshall's wilderness policy on a number of reservations, Collier reiterated this theme: "The new order is a fulfillment of one of the promises implicit in the Indian Reorganization Act of 1934, the promise to permit the Indian to follow his own way of life. Consequently, I am establishing the policy that existing areas without roads or settlements on Indian reservations should be preserved in such a condition, *unless the re-*

quirements of fire protection, commercial use for the Indians' benefit or actual needs of the Indians clearly demand otherwise."[126]

Rather than preserving a romanticized pristine nature against any human use, Marshall's policy sought to prohibit roads and other modern developments as a way of protecting both Native Americans and wilderness. He hoped to preserve Indian reservations from the outside world in the same way he wanted to protect a town like Wiseman. Indeed, when later given an opportunity to shape the recreational development of Alaska, he recommended that "all of Alaska north of the Yukon River, with the exception of a small area immediately adjacent to Nome, should be zoned as a region where the Federal Government will contribute no funds for road-building and permit no leases for industrial development." The native population, he wrote, "would be much happier, if the United States experience is any criterion, without either roads or industries." Here was the germ for a unique Alaskan preservationist regime—what historian Theodore Catton has called "inhabited wilderness"—in which native populations have subsistence rights to resources within otherwise preserved parks and wilderness areas.[127]

Such noble intentions, however, did not mean that Marshall's Indian wilderness policy was without its problems. Like many Indian New Deal initiatives, the wilderness policy was a paternalistic directive foisted on Indians without consulting them. Moreover, Marshall's policy idealized pre-industrial, pre-mechanical Indian economies and cultures at a moment when the land and resources needed to support them were wanting, and when many Indians found participation in modern civilization a necessary—and even an attractive—option. Marshall's wilderness policy mapped Indians as primitive when the reality was much more complex.[128] Marshall was also too sanguine about the possibility of Indians earning a living as wilderness guides as an alternative to off-reservation wage labor or the industrial extraction of reservation resources.

All of these shortcomings were informed by a deeper misperception on Marshall's part: he mistook tribalism for organic socialism; he assumed that Native Americans were single-minded avatars of his critique of modern America.[129] He clearly did not intend his Indian wilderness policy to be dispossessive, but he was guilty of offering his own romantic conceptions of the best interests of native peoples instead of seeking a truly democratic expression of native interests. Had he sought such input, he likely would have encountered the sort of messy pluralism that characterizes most democratic societies and often foils the master plans of well-intentioned radicals. Not surprisingly, Indian needs and desires would eventually abrogate most of these wilderness designations.[130]

New Deal Connections

During his years as chief forester with the BIA, Marshall made several important connections that allowed him to become a key player in emerging wilderness battles. One such acquaintance was Ernest Oberholtzer, who, it turns out, was also a friend and advocate of the Native Americans of the Quetico-Superior lake country, particularly the Ojibwa. Like Marshall, he thought wilderness preservation could help maintain a way of life that would vanish if the area were opened to modern recreational and industrial uses. In the 1920s and early 1930s, a number of plans threatened the sanctity of the canoe country, including one frequently proposed by a lumberman named Edward Backus to develop the region's waterpower. But equally challenging to the wilderness character of the region was a series of planned roads. Local booster groups, seeking to cash in on the meteoric rise of the region as a recreational getaway, pushed these roads vociferously. As much as the plans advanced by Backus, these efforts—which also included calls for developed portages, term permit developments, and the facilitation of motor boat travel— threatened an end to the region's isolation. In the early 1920s, Arthur Carhart and others had successfully opposed Forest Service plans to build roads into the region, and Oberholtzer soon joined the battle.[131] From then on, the Superior National Forest was an important battleground in the effort to define and preserve wilderness.

Boundary Waters politics were also crucial to bringing together future founders of the Wilderness Society. In 1934, Franklin Roosevelt formed the Quetico-Superior Committee and charged it with creating an international preserve linking the canoe country of the Superior National Forest with a vast chain of lakes in Canada's Quetico Provincial Park. Roosevelt named Oberholtzer, Sewell Tyng, and Charles Kelly, all regional activists, to the Committee, and he asked the Departments of Interior and Agriculture to each name a representative. Harold Ickes chose Marshall to represent the Interior Department. Among the Committee's Advisory Board members were Leopold, Yard, and MacKaye. Oberholtzer's cottage on an island in Rainey Lake proved a focal point for the coalescence of modern wilderness advocacy.[132]

Although Marshall's job with the BIA put him in the Interior Department, he managed to stay on top of Forest Service wilderness issues. He also continued his assault on the forestry profession. In a 1934 letter to the editor of the *Journal of Forestry,* he again blasted the profession for seeking to serve private interests rather than public welfare.[133] He was also growing impatient with New Deal conservation policy. As early as March 1933 he had expressed concern that the considerable funding the Forest Service was getting for work relief might lead to poorly conceived and inadequately supervised projects. That the Forest Service was cutting back its trained staff as the emergency work began was likely one source of his concern.[134] Another was the primacy of road building. In July 1933 he wrote to conservationist Willard Van Name about "the prospect of needless road building into the few wilderness areas which remain through the use of emergency funds." He suggested calling a conference in the autumn to bring attention to the issue.[135] Ten days later, Marshall received a letter from Harold Anderson on the same subject. Anderson's concern was the building of skyline drives along the Appalachian crest at government expense. "I presume you are very busy at this time," he wrote Marshall, "but I wish we could get together at some time in the near future to discuss a plan proposed the other day for furthering to some extent the ideas expressed in your article."[136]

As an Interior employee, Marshall was also pondering the question of wilderness and the national parks. In the early 1930s, he joined the National Parks Association and became an at-large member of the NPA's board. He took a liking to Robert Sterling Yard, who by this time had solidified his reputation as a thorn in the side of the national park lobby. In his 1933 letter to Willard Van Name, Marshall cited Yard as an important defender of wilderness conditions: "I know you don't like him, but he has done a splendid piece of practical work in fighting the road-building and water power developing proclivities of the Park Service."[137] As early as 1932, Marshall and Yard were comparing notes on threats to wilderness and strategizing about how to toughen NPA stands. By 1934, both were keen to organize a group to promote the preservation of wilderness and oppose needless New Deal roads, many of which were being completed in or planned for the national parks.[138]

Marshall's most powerful ally on park policy was not Yard but Harold Ickes. Having initiated a relationship with Ickes in the course of recommending Collier for the BIA job, Marshall became Ickes's trusted adviser, and the Secretary soon turned to him for advice on a variety of road-building proposals. In August 1934, Ickes wrote Marshall asking his opinion about road development in Sequoia National Park.[139] That same month he dispatched Marshall on a tour of the Blue Ridge Parkway's southern terminus, which was supposed to go right through the Great Smoky Mountains National Park (GSMNP). That Ickes sought Marshall's opinion when he knew him to be a vocal critic of such road building was either a testament to his faith in Marshall's objectivity or, more likely, a sign that he wanted to hear what Marshall was inevitably going to tell him.

The Blue Ridge Parkway assignment proved a fateful one. It was certainly impeccably timed. Anderson and MacKaye had been pushing for the leaders of the Appalachian Trail Conference to oppose planned skyline drive projects, but the ATC leadership proved unwilling to act. So Anderson wrote to MacKaye suggesting that an offshoot group organize such a protest. On the same day that MacKaye received this letter from Anderson—August 11, 1934—he also received a telegram from Marshall announcing his visit to Knoxville. MacKaye and Broome, unaware of the reason for Marshall's visit, met him at the Andrew Johnson Hotel to discuss a potential advocacy group. When they discovered that Marshall of all people had been assigned to scout the southern terminus, they were beside themselves. "We almost forgot our organization project," Broome recalled, "in our eagerness to press upon him the reasons for a low-level, valley routing." They saved their discussion of a wilderness group for the following day's inspection of a spur road being built to Clingman's Dome.[140]

When Marshall returned to Washington, he wrote a memo to Ickes urging him to stop construction of the skyline drive through the GSMNP. Ickes then urged Park Service Director Arno Cammerer, for whom Ickes had notoriously little tolerance, to reconsider the project. But Cammerer stuck to his plans.[141] In 1935, Marshall submitted yet another memo on the proposed road and trail programs in the park in which he stressed the efficacy of low-level roads, just as Broome and MacKaye had urged, and criticized Park Service plans to grid the park with wide, graded "celluloid" trails. Since "the finest primitive areas occur in the center of the Park," Marshall wrote to Ickes, "it does not seem unreasonable to cater to non-primitive values around the Park's periphery." "My own feeling," he concluded, "is more strong than ever that a skyline drive, or any additional fraction of it, would be indefensible."[142] Later, Ickes wrote to Cammerer insisting that before he submitted any more proposals for roads in the park, he pass them by Marshall.[143] This unusual diversion in the chain of command shows how important Marshall's opinion had become to Ickes. That the GSMNP today remains free of a skyline road bisecting the park is, in part, a result of Marshall's influence with Ickes.

In October 1934, Marshall was back in Knoxville for the annual meeting of the American Forestry Association. He was scheduled to speak to the meeting, and when his turn came he made an impassioned case for keeping skyline drives off the Appalachian ridgeline. Cammerer, who was in the audience, took these remarks as "an improper and ungracious attack."[144] But Marshall's case was apparently convincing, as the AFA passed a resolution that read: "the invasion of this [Appalachian] belt through the greater part of its length by motor roads which would bring to it the sights and sounds of our machine civilization would constitute a serious and irretrievable blunder."[145] It was during this meeting that Marshall, MacKaye, Broome, and Frank had their roadside conference leading to the founding of the Wilderness Society.[146]

Bob Marshall had been the first to call for an organization to promote the preservation of wilderness, and it was his wealth and energy that carried the Wilderness Society through its early years. It would have made sense for him to serve as the Society's first president. But as a government official, he thought it best to check with his superiors. It is not clear whether Marshall asked John Collier for permission to assume this duty, though he had tried to get Collier to sign on as a founding member. But Marshall did go to Ickes to ask permission. "[A] Wilderness Society is sorely needed in this country," he wrote to Ickes, "to counteract the propaganda spread by the Automobile Association of America, the various booster organizations and the innumerable Chambers of Commerce, which seem to find no peace as long as any primitive tract in America remains unopened to mechanization."[147] Although Ickes sympathized, he thought it better that Marshall, as a government official, not take the post.[148] He could be a member, but not the president. After Leopold refused the presidency, Yard assumed the duties, but Marshall continued as the leader of the group in all but name.

Marshall's success in getting Ickes to table skyline drive proposals was soon matched by successes more palatable to the Park Service. Irving Clark, a member of the Mountaineers, a Seattle-based hiking club, urged Marshall and his group to support the creation of an Olympic National Park, to be carved out of national forest land. As mostly foresters and Park Service critics, the Wilderness Society founders were often partial to

Forest Service preservation. The Park Service, they thought, was too willing to build roads and other modern developments. Marshall, however, insisted that the Wilderness Society take a nonpartisan approach and recognize that the Park Service was more adept at protecting primeval forests, one of Olympic's greatest selling points. "If an act were passed to set it up as a National Park," he told Clark, "I would only be for it if it was specifically stated in the act that this would be a wilderness National Park and that no roads could ever be built within its boundaries." "Unless this were done," he continued, "I do not feel that it would be any safer as a National Park than as a National Forest."[149] Eventually, with mediation by Marshall and the Wilderness Society, the Olympic area became a national park on a new wilderness model.[150]

With the Park Service paying more attention to roadlessness, Marshall saw an opportunity to push the Forest Service to expand its preservationist efforts. "I ran into Cam last night, who is apparently turning yearning eyes towards every pretty spot in the National Forests," Marshall wrote to Ferdinand Silcox in September 1935, baiting the new head of the Forest Service. Marshall still thought it important that each bureau have a recreational role, "but my God! Sil," he continued, "it is hard to put up an argument for the Forest Service in view of the total disregard for any form of recreational planning which seems to have occurred in many of your regions."[151] As this letter makes abundantly clear, Marshall had become a skilled political operative. Having gotten Ickes to oppose parkways and support wilderness national parks, he moved into position to help shape recreational planning on the national forests.

The following year, Silcox offered Marshall the recreational position he had sought three years earlier. Ickes initially refused to let him go, but he relented in 1937 and allowed Marshall to take charge of the Forest Service's newly created Division of Recreation and Lands. The Forest Service controlled the bulk of the nation's remaining de facto wilderness, and Marshall convinced Ickes that it was there that he could do the most good. By all accounts, he was right. During his brief tenure, he worked on two major initiatives. The first was an effort to extend national forest recreational opportunities to more people by facilitating use by lower income groups and dismantling discriminatory barriers against minority groups. His second effort was to preserve more wilderness within the national forests, and to craft a system ensuring that such preservation had permanence. These twin efforts provided a fitting conclusion to Marshall's career.[152]

Conclusion

In June 1937, Catherine Bauer, a housing expert who had been a prominent member of the RPAA, wrote to Marshall accepting an invitation to join the Wilderness Society. She included a five dollar check in support of the cause, but she also took the opportunity to point out two criticisms of the wilderness program she had heard among her reformist colleagues: "(1) that the enjoyment and appreciation of wilderness areas is a snobbish form of recreation, touching only a few people, and, much more important, (2) that wilderness areas are relatively inaccessible, which means not many average working people could take advantage of their recreational possibilities even if they realized what they had been missing." "While both of these criticisms are essentially factual at present," she continued, "I do not believe they constitute a valid criticism of your movement." Like Marshall, she thought that workers would find wilderness appealing and worth supporting if they could overcome the barriers of cost and distance. As a result, she stressed the need to subsidize transportation to such areas. While she understood that the Wilderness Society did not have the resources to undertake such a program, she encouraged Marshall and his colleagues to make alliances with labor groups and to make working-class access central to their program. "There is still enough wilderness left in America to make it a great national playground," she concluded. "And it won't be preserved unless the urban wage-earners feel that way about it."[153]

Marshall thought Bauer's letter important enough to send on to the Society's co-founders. Both Yard and Broome wrote back expressing concerns. "Natural areas cannot be applied to social uses and stay natural," Yard replied crankily; "See National Parks."[154] For his part, Broome, though more inclined toward Marshall's political sympathies than Yard, saw "no gain in identifying *our* aims with those of any group or class." Broome thought that subsidizing and promoting working-class use might lead to the very pressures for development that they all sought to avoid. Both Yard and Broome feared, in other words, that working-class tastes would undermine wilderness values.[155] Marshall responded vigorously, rejecting the notion that the lower classes were culturally predisposed to commercialized and mechanized recreation. "If we can enlist them on our side instead of on the side of endless auto roads and Coney Island developments," he insisted, "it will be a tremendous gain. If we can't enlist this mighty force, if we let it escape to the other side, I am afraid we are beaten."[156] For Marshall, the purpose of the Wilderness Society was to give voice to the wilderness option and to persuade people that it was worth supporting. As he saw it, wilderness advocacy was a battle for the masses, not against them.

In his wilderness advocacy, Bob Marshall combined traditional frontier romanticism with a brand of radical politics not usually associated with the preservationist movement. Indeed, he made social equality and civil rights central components of his advocacy so that wilderness preservation would be more compelling in both moral and logical terms. Moreover, his sophisticated understanding of forest ecology allowed him to defend the new ideal of wilderness against his forestry colleagues who saw it as yet another manifestation of mushy sentimentalism. He was adept at building an aesthetic argument for wilderness grounded in forestry science and distinct from other preservationist models. Unwilling to pit preservation and conservation against each other, he insisted that the national forests were big enough for all sorts of uses; they would have been even bigger if he had his way. But Marshall also urged foresters to recognize that if and when they invaded remaining wilderness areas for their timber resources, their actions would be an admission that sustainable public forestry had failed. His fight for a more permanent and hands-off wilderness policy that strongly emphasized primeval qualities as well as roadlessness was thus a challenge to foresters to practice what they preached. His wilderness advocacy and his critique of forestry were inseparable in this regard.

Marshall saw himself as part of a radical Progressive tradition. For him the scientific and civic principles of his forestry education were rooted in an ideology that was decidedly egalitarian and anticorporate. Although the dominant wing of the forestry profession had, by the 1920s, accepted cooperation with industry and management for efficient production as appropriate and sufficient goals, a cadre of foresters still stressed an ideal of aggressive public regulation or, in Marshall's case, public ownership. This revolt of foresters suggests, contrary to conventional wisdom, that there was a complex and contentious professional forestry culture during the early twentieth century, and that many Progressive conservationists had ideological concerns that extended well beyond the gospel of efficiency.[157] Indeed, understanding the professionalization of forestry during the 1920s may be the key to rethinking the legacy of Progressive conservation, and to explaining why an adversarial group of foresters, Marshall among them, became wilderness advocates. It was not just that Marshall was able to juggle supposedly disparate commitments to conservation and preservation, or that he chose the latter over the former. Rather, his sense that forestry had drifted from its core values fed his wilderness advocacy. The same might be said about Leopold and MacKaye.

Yet, like his Wilderness Society colleagues, Marshall was most concerned about the extent to which roads were carving up the nation's remaining wilderness areas. It was roads, and not the threat of resource extraction, that prompted him to define and defend modern wilderness. Indeed, he was willing to countenance resource use—and in certain cases whole communities of resource users—within wilderness areas so long as modernization was not part of the equation. Saving wilderness was thus a way of keeping a portion of the American landscape free from modernizing forces, and retaining a remnant of freedom that was rapidly disappearing in modern America.

Like many Progressive thinkers, Marshall faced a dilemma. On the one hand, he revered basic American precepts such as political freedom, civil liberties, and social equality—the values that had supposedly defined what was best about the "island communities" of nineteenth-century America. But he also recognized that in an era of market expansion, corporate consolidation, specialization, urbanization, and mass culture, it was foolish to hang on to such an antiquated set of values, at least in an unreconstructed form. The organizational and technological changes of the early twentieth century required a new ideology, one that could refurbish old values for new circumstances.[158] In promoting socialism and in pushing for the socialization of natural resources, Marshall tried to save cherished aspects of traditional America, like community and a civic commitment to the public good, from an economic juggernaut that was making a mockery of them. He tried, in other words, to save one strand of America's liberal tradition from the ravages of another. His fight for "the freedom of the wilderness" was based in the same diagnosis of what ailed America.

In a recent essay comparing American and Canadian views of nature, historian Donald Worster asks why the wilderness idea, a powerful part of the U.S. environmental tradition, has not been as resonant with Canadians. The answer, he says, is that Americans have tended to fuse their attachments to wilderness and freedom, particularly through the frontier myth, into a national environmental tradition with a fatal flaw:

The Achilles heel of American environmentalism is the fact that, despite all their calls for government activism and regulatory power, environmentalists in their heart of hearts share the same ideology of liberty and self-determination that has created a degraded environment. The distance between the "wilderness freedom" of an Abbey or a Muir and the "economic freedom" of laissez-faire capitalism may at times not be very great. This confusing overlap of a liberty-seeking ideology with its enemies may constitute the greatest embarrassment the wilderness movement has, one that even its most thoughtful philosophers have never fully addressed or clarified.[159]

Perhaps more than any other figure in America's environmental tradition, Bob Marshall scrutinized and struggled with this "confusing overlap," trying to separate an individualism grounded in political equality and civil liberties from free market libertarianism. His celebration of wilderness as a preserve of liberty, as a space free from physical and psychological restraint, may have revealed a level of personal discomfort with modern collectivized society—a discomfort shared by many of his contemporaries. But it never translated into an unwillingness to limit economic liberty in the name of social justice and environmental protection. To the contrary, when Marshall stepped from the wilderness into modern America, he was an unabashed critic of the environmental and social results of economic liberalism. He knew that the negative freedom of the wilderness was not an appropriate reformist model for a society that, in his view, needed a strong infusion of positive freedom in the form of government intervention. But he was also unwilling to see the freedom of the wilderness disappear, and he saw no reason to think that a commitment to one sort of freedom meant giving up the other.

Epilogue: A Living Wilderness

Modern wilderness emerged as a new preservationist ideal during the interwar years because of the profound changes wrought by the automobile, road building, a growing leisure-based attachment to nature, and a federal government increasingly willing to fund recreational development on the nation's public lands. The founding of the Wilderness Society, which was triggered by New Deal developments but had its roots in the 1910s and 1920s, represented the coalescence of individual concerns about these forces. As such, it was an important moment in the history of the American environmental movement. But it was not a simple moment, for the founders brought to the new organization unique experiences and varied perspectives that shaped their wilderness thought. The wilderness idea that would guide the Society's political future, and that eventually found legislative validation in the Wilderness Act of 1964, had to be negotiated.

"One of the greatest obstructions to progress today in all land causes," Robert Sterling Yard wrote to Benton MacKaye in October 1934, "is the difference in meaning conveyed by essential words." Yard had just received the news from Bob Marshall of the Wilderness Society's provisional formation, and he was pleased to be among the founding members. But he also had reservations about the term "wilderness"; he preferred "primitive." As Yard understood it, wilderness and primitive were kindred concepts, but they were not synonyms. "Not all wilderness is primitive," he wrote, "but all primitive is wilderness."[1]

Both primitive and wilderness were terms with multiple meanings, even among this small group of advocates. For Yard, the defining characteristic of wilderness was that it was roadless and otherwise uninvaded by the sights and sounds of modernity. Primitive nature, on the other hand, was synonymous with primeval or pristine nature. A large roadless area could be second growth, selectively logged, or grazed; it could be wilderness, in other words, without being primitive. But a primitive area could not be affected substantially by such processes, or by roads for that matter. Because his advocacy was still strongly shaped by national park standards, Yard hoped to make preservation of the primitive the primary focus of the new organization. Indeed, just months before, he had proposed to John C. Merriam that they form an "Organization to Preserve the Primitive."[2] For Yard, wilderness was a concept that was useful primarily as a sanction against recreational development and the political, commercial, and rhetorical abuses that came with it. He wanted both sorts of preservation, but he privileged the primitive.

Not all of the founders agreed with Yard's priorities, let alone his definitions. Like Yard, both Leopold and Marshall thought that wilderness was largely defined by roadlessness, but both adhered to a definition that also required substantial acreage. In other words, not all primitive, as Yard used the term, was wilderness; smaller untouched areas, though deserving of preservation, did not qualify as wilderness for reasons of scale. Here again was the major difference between the preservation of natural conditions proposed by ecologists and wilderness preservation. Because Leopold and Marshall valued recreational solitude, the ability to lose touch with modern civilization, such scale was an essential part of what they sought to preserve. But beyond this, they did not concur with Yard about what the word primitive meant. In his letter to MacKaye, Yard had mentioned that some people used the term primitive to describe areas in which "people entering them would have to travel in a primitive way." Leopold, Marshall, and MacKaye all made such a quality central to their advocacy. For them, primitiveness was a recreational rather than an ecological condition.

As this wrangling about terminology suggested, providing a precise definition of what they sought to preserve was not an easy task for the founders of the Wilderness Society. In the simplest sense, the founders wanted to protect both large, roadless areas and areas of relatively pristine nature. Ideally these would come in the same package, but the founders would also seek the preservation of these qualities separately, though with an emphasis on roadlessness. That they chose to call themselves the Wilderness Society reflected this emphasis. But their choice of the word wilderness was also a pragmatic one. They needed to settle on a single term with political resonance, and wilderness seemed the best option. It was a concept with which Americans were familiar, and it did a better job in highlighting what was distinct about their platform—that they sought to preserve large areas free from all major development, recreational as well as industrial. In short, the founders settled on wilderness as a construct that they hoped would encapsulate and express their concerns. The founding of the Wilderness Society was not the result of the founders appreciating the value of an age-old notion of wilderness, as the traditional wilderness narrative would have it. Rather, it involved the intellectual work of reconditioning a term with common currency to reflect the developments of a new age.

At the Wilderness Society's first official meeting in January 1935, those in attendance—Marshall, Yard, MacKaye, Anderson, and Broome—created a four-page statement of principles that described the organization's purpose. Their platform made clear that the term wilderness would have to carry a lot of baggage. Erring on the side of inclusiveness, the document presented an expansive and multifaceted notion of wilderness, one that rang with the particular concerns of each of the individual founders. "Primitive America is vanishing with appalling rapidity," the statement began:

> Scarcely a month passes in which some highway does not invade an area which since the beginning of time had known only natural modes of travel; or some last remaining virgin timber tract is not shattered by the construction of an irrigation project into an expanding and contracting mudflat; or some quiet glade hitherto disturbed only by birds and insects and wind in the trees does not bark out the merits of "Crazy Water Crystals" and the mushiness of "Cocktails for Two."[3]

In this brief introductory paragraph, the founders quickly addressed the interwar era's most important threats to wilderness: road building, persistent efforts to develop water resources (particularly in national parks), and the consumer ethos guiding modern outdoor recreation. The inclusion of the water development threat probably came at Yard's insistence, as he had fought against such schemes throughout the late teens and twenties as part of his campaign for complete conservation. But it was the other two threats that were of more universal concern, and of more importance to defining what made wilderness unique.

The statement then turned to the argument for minority rights, made most coherently during the preceding decade by Marshall and Leopold:

We recognize frankly that the majority of Americans do not as yet care for these values of undisturbed nature as much as for mechanically disturbed nature. Cheerfully, we are willing that they should have opened to them the bulk of the 1,800,000,000 acres of outdoor America, including most of the superlative scenic features in the country which have already been made accessible to motorists. All we desire to save from invasion is the extremely minor section of outdoor America which yet remains free from mechanical sights and sounds and smells.

This statement made clear an important truth about the interwar argument for wilderness: although the founders sought to preserve wilderness from agricultural and industrial transformations, their claims on the natural world must be seen within the context of the era's complex and charged recreational politics. The modern wilderness idea was a product of battles between preservation and recreational development, not preservation and resource use.

The founders followed this introductory statement with an eleven-point platform that outlined the specific principles and goals of the Society. As the collective expression of the founders' desires, these principles said much about the paths that had converged to create the Wilderness Society.

First, the founders suggested that wilderness, as an "environment of solitude," was an increasingly scarce natural resource as worthy of conservation as timber. As they often had in preceding years, the founders made familiar utilitarian arguments serve recreational distinctions. By utilizing the logic of scarcity that had helped to define the Progressive conservation movement, they hoped such an argument would have a commonsense appeal to the land managers who had the power to preserve wilderness during the interwar years.

Second, the founders insisted that wilderness was "a public utility and therefore its commercialization should not be tolerated." This principle was crucial to wilderness advocacy during the period, and its various facets deserve a strong reiteration. The founders hoped that preserved nature would function, to use MacKaye's apt phrase, as "a retreat from profit," as an antidote to a society otherwise obsessed with all things pecuniary. This was not a new hope, for preserved nature in America had long symbolized escape from the capitalist ethos, though mostly it had stood in opposition to the forces of production. Consumerism reconfigured this tradition. By the interwar years, it was the commercialization of outdoor recreation and nature tourism that was undermining the "publicness" of natural spaces.

The modern wilderness idea was certainly the product of an emergent consumer society. Wilderness areas, after all, were designed for recreation, not production. But to see wilderness only as a consumer ideal is to miss the critical stance toward consumerism taken by these early wilderness defenders. Consumerism redefined outdoor recreation after World War I, but the result was a wide spectrum of recreational sentiment that deserves closer scholarly attention. Consumer relationships emerged most powerfully when the automobile, improved roads, and the culture's growing interest in nature combined to create opportunities for commercial profit. The interwar years saw the beginnings of what Edward Abbey would refer to, more than three decades later, as "industrial tourism."[4] Although Americans had profited from nature tourism prior to the interwar years, it was not until after World War I that the economic promise of national parks and other areas could match the returns from resource extraction. This economic argument was a powerful one in getting natural areas set aside, and it transformed environmental politics. But it also tended to make the development of such areas for tourists a fait accompli. In defending modern wilderness as a public utility, the founders were reacting to the beginnings of industrial tourism and the commercialism that it encouraged.

Understanding the centrality of this consumer critique to the birth of wilderness advocacy helps to make sense of one of the apparent paradoxes of this story: that the interwar era's most important preservationists were mostly foresters. Indeed, not only were Leopold, MacKaye, and Marshall trained and committed utilitarians, but they were among a cadre of foresters who pushed Progressive conservation in more radical directions. Unlike other foresters of the era, they refused to jettison the civic vision and economic critique that had attracted them to professional forestry in favor of a detached preoccupation with technique. Their examples suggest that scholars need to rethink the traditional interpretation of Progressive conservation as a movement motivated primarily by a "gospel of efficiency." To understand how these foresters became wilderness advocates, we need to rediscover the importance of the "public utility" argument to Progressive conservation. For Leopold, MacKaye, and Marshall, the salient political dilemma was not whether to preserve nature for recreation or to develop its resources efficiently under the guidance of apolitical experts. It was to figure out how the public good might be best protected in *both* processes. When the founders saw recreational development headed in a direction that would benefit commercial interests at the expense of the public, they responded in the same way that they did when they saw federal forestry headed in a similar direction. These foresters became wilderness advocates by following a resource-based critique of commercialization into the field of outdoor recreation.

This "public utility" argument also suggests the importance of seeing wilderness as a public lands ideal. The rise of conservation sentiment in the early twentieth century has often been portrayed as a heroic call for limits on economic growth and the abuses of capitalism. To a great extent it was, and the founding of the Wilderness Society was no exception. But conservation sentiment was also the result of a more prosaic reality: by the early twentieth century, the federal government had on its hands an extensive though marginal public estate that, it increasingly appeared, was going to remain in public ownership and demand public management. Modern wilderness politics emerged at a historical moment when federal recognition of long-term managerial responsibilities for this marginal public domain collided with an automobile-driven enthusiasm for outdoor recreation. Such concerns about "marginality" joined recreational politics as a defining quality of interwar conservation. Thus, when the founders spoke of wilderness as a public utility, part of what they meant was that wilderness preservation was an efficacious way of protecting the public interest in portions of a public domain whose permanence was a fait accompli. Interwar wilderness advocacy was less a call for a permanent public domain than a response to it.

Seeing the modern wilderness idea within this public lands context suggests another important conclusion. While some have seen the prominence of the wilderness ideal in American environmental politics as indicative of a persistent bifurcation of the natural and human landscapes, it was also a product of a solidifying distinction between public and private lands. The modern wilderness idea emerged as a claim on the public landscape, and it has remained a dominant environmental ideal in the United States precisely because preservationist politics have remained so focused on the fate of *public lands*. This does not mean that the founders were not concerned about the environmental problems of the nation's privately owned lands; to the contrary, they were among their generation's most creative thinkers on the subject. But the effective regulation of private land use in the United States has been particularly difficult to achieve. This, as much as our cultural affinity for compartmentalizing humans and nature, explains the lasting power of modern wilderness as the centerpiece of American preservationist politics.

A number of the other principles in the Wilderness Society platform focused specifically on the importance of seeing wilderness as one of a number of competing recreational ideals. One insisted that "scenery and solitude are intrinsically separate things, that the motorist is entitled to his full share of scenery, but that the motorway and solitude together constitute a contradiction." Too often, scenic preservation and tourist development had been one and the same process, the founders intimated. The job of wilderness preservation was to promote a deeper aesthetic experience in nature—to get people beyond scenic conventions and to disabuse them of the developmental imperatives that came with scenic tourism. "Our ability to perceive quality in nature," Leopold later wrote, "begins, as in art, with the pretty. It expands through successive stages of the beautiful to values as yet uncaptured by language."[5] Wilderness preservation was a push in this aesthetic direction.

This distinction between scenery and solitude also addressed the confounding tendency among contemporaries to see wilderness preservation as an elitist attempt to limit access. In response to these charges of elitism, the founders steadfastly maintained that wilderness recreation and motorized recreation were fundamentally different experiences. In the case of scenery, motorized access could be synonymous with democratic access since the automobile privileged the visual experience that was at the heart of scenic appreciation. But motorized access and the wilderness experience were at odds by definition, because the former undermined the latter. The founders of the Wilderness Society wanted to keep automobiles, not people, out of wilderness areas.

The founders were careful to support recreational provisioning for the mass of Americans who did not yet desire a wilderness experience. But they also insisted that people needed to understand the differences between such areas and wilderness. Thus, another principle stated that "outing areas in which people may enjoy the non-primitive forest are highly desirable for many pent-up city people who have no desire for solitude, but that such areas should not be confused in mental conception or administration with those reserved for wilderness." All of the founders recognized the importance of defining and working toward a varied and intelligently planned public recreational landscape. They did not advocate that wilderness preservation come at the expense of meeting the needs of other recreational constituencies, but they were concerned that overdevelopment for these constituencies would destroy wilderness.

Yet another principle suggested that "since the invasion of wilderness areas is generally boosted by powerful, countrywide organizations, it is essential that individuals and groups who desire to preserve the wilderness must unite in a country-wide movement in its defense." The founders, in other words, saw themselves as an organized minority operating in opposition to other organized minorities, all of whom vied to shape outdoor recreation and define the public utility of preserved areas. Moreover, in calling for a national organization to fight against well-organized recreational boosters, the founders recognized a growing frustration among activists who fought the same battles in different locales during this period. The Wilderness Society aimed to nationalize and coordinate previously localized wilderness politics.

A few other components of the Society's platform are worth mentioning. One, which bore MacKaye's distinct imprint, suggested that, given the "brutalizing pressure of metropolitan civilization," wilderness needed to be seen as a human necessity rather than a luxury. Another stressed the centrality of mechanization to the Wilderness Society critique by noting that existing wilderness areas were being "sacrificed to the mechanical invasion in its various killing forms." In their distaste for automotive access, the founders were in revolt against the "machine age" and much of what is stood for.[6] Another principle, bearing the imprint of the ESA's preservationist model, argued that Americans needed "to preserve, under scientific care, for the observation, study, and appreciation of generations to come, as many, as large, and as varied examples of the remaining primitive as possible." Leopold likely had pushed for the inclusion of this principle, though others certainly understood its importance. Finally, the founders insisted that the "means of achieving our objectives should be positive and creative as well as merely defensive." They understood, in other words, that by institutionalizing the wilderness idea, the Wilderness Society might slip into a defensive battle and lose the visionary thinking that underlay interwar wilderness advocacy.

After outlining the Society's general principles, the founders defined, in a section reminiscent of Marshall's contribution to the Copeland Report, the five "Types of Wilderness" that they sought to preserve. First and foremost, there were "Extensive Wilderness Areas," areas "which possess no means of mechanical conveyance and which are sufficiently spacious that a person may spend at least a week in travel in them without crossing his own tracks." Here was the familiar Leopold-Marshall definition of wilderness. Such areas had two attributes: "first, that visitors to them must depend largely upon their own efforts and their own competence for survival; and second, that they must be free from all mechanical disturbances." Second, they sought to preserve "Primeval Areas," which they defined as "virgin tracts in which human activities have never modified the normal processes of nature." Such areas, they suggested, were of both aesthetic and scientific value. They also sought to protect "Superlatively Scenic Areas." This inclusion of the scenic was likely Yard's doing, and it functioned largely as an expression of support for national park preservation, though there was a Forest Service variant on this theme dating from Arthur Carhart's protection of Trapper's Lake. But scenic preservation was not a top priority; indeed, the founders had important reservations about the implications of such preservation. The fourth category was "Restricted Wild Areas," preserves that were not large, scenically superlative, or "virgin," but that were "free from the sights and sounds of mechanization" and near areas of concentrated population. "Although one cannot obtain in them the adventure, the dependence on competence, and the emotional thrill of extensive wilderness areas," the statement read, echoing Marshall's advocacy, "they are the closest approximation to wilderness conditions available to millions of people." Fifth and finally, the founders, in a nod to Mac-Kaye's "wilderness way" concept, sought to preserve "Wilderness Zones," "strips along the backbone of mountain ranges [and] along rivers which, although they may be crossed here and there by railroads and highways, nevertheless maintain primitive travel conditions along their major axes." Protecting "extensive wilderness areas" was the core of the Wilderness Society program. These other four categories of wilderness reflected persistent individual variations on the theme and revealed how complicated a process defining wilderness was.

The statement of principles concluded with a list of "Common Types of Wilderness Invasion." Predictably, the first was "roads for motor vehicles." Also noted were mechanical invasions such as "[r]ailroads, cog roads, funiculars, cableways, etc." A sanction against the radio was included at the behest of MacKaye, who constantly fulminated against this particular "metropolitan influence."[7] The founders also mentioned motorboats and airplanes as threats, and they listed a series of infrastructural improvements that, though sometimes necessary for administrative purposes, were worthy of careful consideration. Graded trails, ski trails, footbridges, power lines, lookout cabins, fire and beacon towers, and shelters were to be kept to a minimum. Logging, they wrote, "should not be permitted in any sort of wilderness." This was a notable departure from previous contentions that timber harvests might continue on a small scale in wilderness areas as long as they did not require roads. The founders also suggested sanctions against water development. Where grazing existed, the Society would tolerate it (excepting in primeval and superlative scenic areas), though they did not want it to expand into preserved areas where it had not previously existed. Although this list provided stronger sanctions against resource development within wilderness areas, the founders targeted roads, mechanization, and infrastructural improvements as the primary threat to wilderness.

By defining and institutionalizing their advocacy, the founders of the Wilderness Society effectively transformed wilderness from an intellectual ideal into a political goal. What had been largely an intellectual process before 1935 became largely a political process afterwards. This is perhaps too neat a way to put it, for the founders had pursued political ends in the formative years of their advocacy, and they and others continued to rethink the intellectual content of wilderness in the coming years. Nonetheless, prior to World War II, wilderness politics were insular; after the war, they became increasingly popular, as changes in wilderness policy shifted the power to preserve wilderness from land managers to legislators and the public at large. These changes, along with a new set of threats to wilderness, made postwar wilderness advocacy different from interwar advocacy. Unfortunately, few of the founders would survive to witness these changes.

The First Generation Fades

In the years between 1935 and 1948, Aldo Leopold participated in the Wilderness Society from afar, but his participation was far from inconsequential. Still devoted to game—or increasingly "wildlife"—issues, Leopold gave greater attention to the importance of wilderness conditions to wildlife, foreshadowing some of the conclusions that his son, Starker, offered in a 1963 report on wildlife and the national parks. He also thought and wrote about wilderness in more scientific terms. This was partly a result of trips to Germany and Mexico, in 1935 and 1936 respectively, and partly because of the profound ecological problems America faced in the 1930s.[8]

Where Leopold had initially offered only a halfhearted call for the Wilderness Society to include scientific interests in their program, after 1935 he was a more vocal advocate for the importance of wilderness to ecological science. The essays that he wrote for *The Living Wilderness* during the last decade of his life revealed this expanded approach to wilderness. As early as the first issue, Leopold had argued for the need for wilderness preservation because of "the baffling obstacles" faced by land managers. But he made this scientific case most forcefully in a 1941 article entitled "Wilderness as a Land Laboratory." "The recreational value of wilderness has been often and ably presented," he wrote, "but its scientific value is as yet dimly understood." "The science of land health needs, first of all, a base-datum of normality," he continued, "a picture of how healthy land maintains itself as an organism." Wilderness areas, he suggested, could provide such a service.[9]

Yet Leopold's attention to the scientific role of wilderness never replaced his recreational critique. Rather, the insights of ecology—insights more a result of his keen powers of observation than the collective wisdom of ecologists—increasingly informed Leopold's aesthetic and ethical thought. "Good professional research in wilderness ecology is destined to become more and more a matter of perception," he predicted, and "good wilderness sports are destined to converge on the same point." The task was to preserve areas where this perceptual convergence could occur.[10] He was after much more than ecological protection when he advocated wilderness preservation. Yet, because ecological arguments became important to postwar wilderness advocacy, and because of a tendency to see these arguments as more sophisticated than recreational arguments, many have assumed that Leopold's central contribution to the American wilderness tradition was ecological. This assumption has obscured the centrality of his recreational critique. Ecology was but a servant to the advanced perception that he hoped wilderness recreation would foster.

A Sand County Almanac was Leopold's last word on the variety of conservation issues to which he had devoted his life. Throughout the volume's elegant essays, he sprinkled observations about the importance of wilderness, but wilderness was only one of many central themes. He spoke with equal facility about agricultural conservation, forestry, wildlife management, and environmental restoration. His dedication to wilderness, expanded as it had become, was but one body in a constellation of thought whose unifying principle was a new ethical and aesthetic relationship with the land. Aldo Leopold died of a heart attack while fighting a brush fire near his family's shack on April 21, 1948. *A Sand County Almanac* appeared a year later, and it soon became one of the canonical texts of the postwar environmental movement.

Robert Sterling Yard was the Wilderness Society's administrative workhorse during its first decade. He ran the Society from his home in Washington, D.C., his salary paid entirely by Bob Marshall. Yard's voluminous correspondence held the members of the fledgling society together, and his insistence that the cavalier collective establish a formal system of bylaws proved crucial to the organization's effectiveness. Yard single-handedly produced *The Living Wilderness* during the early years, though he could only muster one issue a year until 1942. He also brought to the Wilderness Society his old commitments to principled battle and no compromise. He continued to push for national park standards, blasting the Park Service when such standards seemed at risk. Finally, he initiated an important trend that carried over into the postwar era: cooperation with other major preservationist groups. He remained affiliated with the National Parks Association, and he acted as a mentor to Devereux Butcher, who became the NPA's Executive Secretary in the early 1940s. Butcher did much to smooth over Yard's rocky relationship with the NPA establishment, and he was an important ally in key postwar wilderness battles. And as early as 1936, Yard began crafting a relationship with the Sierra Club, an alliance that would prove even more important to the eventual passage of the Wilderness Act.[11]

Yard lived the final three decades of his life in austerity in order to fight for national park standards and, later, for wilderness. After a long bout with pneumonia, during which he ran the Society's affairs from his bed, Robert Sterling Yard died on May 17, 1945.[12] His passing signaled a major shift in Wilderness Society leadership, as it took three people to fill his shoes. Benton MacKaye replaced Yard as president, though that position quickly became a ceremonial one. The new Executive Secretary, Howard Zahniser, and Olaus Murie, who filled the newly created position of Director, ran the Society for most of the next two decades.[13] Zahniser followed most directly in Yard's footsteps. He took over the Society's magazine, making it into a quarterly, and he inherited Yard's mantle as the city-bound advocate who was more comfortable in the halls of Congress than in the wild. Fortunately for the cause of wilderness, Zahniser proved a better politician and diplomat than Yard.

Benton MacKaye died in 1975 at the age of ninety-six. He had lived through, and often been at the center of, a century of American environmental policy. Born in 1879, in the same decade that Congress created the first national park, he survived to see the flurry of landmark postwar environmental legislation in the 1960s and early 1970s. MacKaye had already lived a full and fascinating life when he helped to found the Wilderness Society, and he had another four decades ahead of him.

In 1936, MacKaye left the TVA and his circle of friends in Knoxville to return to Shirley Center, where he wrote, worked as a consultant, and peppered the other founders with letters about issues of philosophical and strategic importance. He also continued to be active in Appalachian Trail activities, despite the damaging split with Myron Avery and the Appalachian Trail Conference. In 1937, a CCC work crew finished the final few miles of the Appalachian Trail in Maine, just sixteen years after MacKaye first went public with his idea. The completed trail did not last long, however. The great hurricane of 1938 destroyed large stretches, and the construction of the Blue Ridge Parkway forced the rerouting and rebuilding of approximately 120 miles of the trail in Virginia. World War II slowed rebuilding efforts, and though Earl Shaffer completed the first official through-hike in 1948, the trail was not a continuous entity again until 1951.[14]

The hurricane of 1938 also prompted a career change for MacKaye. The storm caused considerable flooding in New England, and, in its wake, MacKaye went back to work for the Forest Service as a flood control and watershed management expert in the region. His boss during these years was Bernard Frank. MacKaye remained with the Forest Service through 1941, when he took a job with the Rural Electrification Administration in St. Louis. In 1945, he retired to Shirley Center for good, where he spent the last decades of his life working on a massive treatise that he called "The Geotechnics of North America." A combination of biography, geography, conservation and planning history, and personal philosophy, it was a hugely ambitious project that never made it into print.[15]

By the early 1950s, MacKaye was primarily a figurehead for the Wilderness Society, a philosopher in a group increasingly bent on legislative action. This was an all-too-familiar position for him. Ironically, it may have been MacKaye who first suggested the idea for a wilderness bill. In 1946, in keeping with a long tradition of working his ideas into legislative form, MacKaye, with Zahniser's help, drafted "a bill to establish a nation-wide system of wilderness belts."[16] This particular bill went nowhere, but MacKaye did hang around long enough to see the passage of the Wilderness Act as well as the National Trails System Act of 1968, which provided federal protection for the AT and made it the centerpiece of an expanding system of National Scenic Trails.[17]

Since his death, MacKaye's ideas about the architectonic role of preserved open space have experienced a renascence in various greenway and rail-to-trail efforts. Meanwhile, conservation biologists, island biogeographers, and a new generation of scientifically informed wilderness advocates have discovered the value in connecting fragmented habitats through the preservation of wilderness corridors. Although their vision is more biological than social, the architects of today's Wildlands Project seek to reweave wildness into the landscape and to control the tendency of modern infrastructures to fragment large areas of undeveloped nature in ways that are reminiscent of MacKaye's interwar vision. MacKaye's unique legacy is still being fleshed out in iterations that even he could not have imagined.[18]

Bob Marshall's tenure with the Wilderness Society was brief—a mere four years—but his contributions during those years were fundamental to the organization's institutional health. During 1935 and 1936, Marshall focused on building a wilderness wish list for the group; he and his friend Althea Dobbins catalogued and mapped the nation's remaining large-scale roadless areas, finding seventy-seven in all. Most were in the West, all but two were in public ownership, and quite a few were already provisionally protected by the Forest Service's Regulation L-20. But many more remained unprotected, including a number of "desert areas" withdrawn from public entry by the Taylor Grazing Act and controlled by the Grazing Service. While the founders tended to equate wilderness with forested land, such desert areas were among the largest remaining roadless areas in the nation, and they would be crucial to postwar wilderness battles. Marshall thus did a great service by putting these areas "on the map."[19]

Not long after releasing this inventory, Marshall assumed a powerful new position from which to coordinate the fight for wilderness. In 1937, Ferdinand Silcox named Marshall the first head of the Service's Division of Recreation and Lands. Two years later, and largely in response to Marshall's advocacy, the Forest Service instituted new regulations for wilderness preservation called the U Regulations. The U Regulations were the crowning achievement of Marshall's career with the agency.

Under 1929's Regulation L-20, Forest Service preservation of "primitive areas" had been makeshift. Decisions were made at the district level, and considerable latitude existed in terms of permissible development.

With the U Regulations, a number of things changed. There were three U Regulations issued by Agriculture Secretary Henry Wallace on September 19, 1939. Regulation U-1 read:

> Upon recommendations of the Chief, Forest Service, national forest lands in single tracts of not less than 100,000 acres may be designated by the Secretary as "wilderness areas," within which there shall be no roads or other provisions for motorized transportation, no commercial timber cutting, and no occupancy under special use permit for hotels, stores, resorts, summer homes, organization camps, hunting and fishing lodges, or similar uses.

This regulation made a number of important changes. First, there was yet another switch in terminology, from "primitive" back to "wilderness." More important, U-1 invested the power to recommend wilderness areas in the Chief of the Forest Service rather than in district foresters, and it put the ultimate power to preserve these areas in the hands of the Secretary of Agriculture. Unlike its predecessor, U-1 explicitly prohibited commercial timber harvests on designated wilderness areas, though it did allow grazing and water storage projects to exist if they did not necessitate roads. Finally, under U-1 only the Secretary of Agriculture could modify or eliminate these areas, and there had to be public notification and hearings on all changes in wilderness policy. Such mechanisms for public input democratized the designation process, which proved crucial to the postwar success of wilderness preservation.[20]

Regulation U-2 set up a similar protective formula for "wild areas," tracts of between 5,000 and 100,000 acres. The main difference between U-1 and U-2, aside from size, was that with U-2 all decisions rested with the Chief of the Forest Service alone. The same public procedures for establishing, altering, or abolishing these areas applied. Regulation U-3 provided for the creation of "roadless areas" of recreational value. These areas could be of any size; areas of less than 100,000 acres could be classified by the Chief of the Forest Service while areas over 100,000 acres had to receive the Secretary of Agriculture's approval. The feature that distinguished "roadless areas" from "wilderness" and "wild" areas was that they permitted timber cutting and other commodity uses.[21]

Together, the U Regulations created a stricter and more permanent system of preservation within the national forests. Fittingly, one of the first areas to be classified under the new regulations was the Bob Marshall Wilderness Area in Montana. It was cobbled together from three areas that had been preserved under Regulation L-20.[22] But the implementation of the U Regulations also presaged some of the difficulties that would arise after the passage of the Wilderness Act of 1964. Existing "primitive areas" had to go through a plodding reclassification process. Permanence and sanctions against timber extraction came at the expense of a more cumbersome bureaucratic process. Indeed, during the first two years under the U Regulations, only twelve of the seventy-six areas that had been set aside as "primitive areas" made the shift to either "wilderness" or "wild" status. When President Johnson signed the Wilderness Act into law in 1964, 5.5 million acres of Forest Service land, more than a third of the agency's protected acreage, remained under L-20's "primitive" classification.[23]

During his final years, Marshall also worked on two other important initiatives in his capacity as the Forest Service's chief recreational planner. First, in line with his commitment to civil liberties, he tried to eliminate racial and religious discrimination in the provision of recreation on the national forests. In western national forests, certain Term Permit hotels and resorts discriminated against Jews (and presumably against other minorities). Marshall fought a successful battle to overturn such restrictions and to prohibit future discrimination in Term Permit facilities. In the South, Forest Service campgrounds had become, through intimidation, de facto white-only facilities. To provide African Americans with recreational opportunities in the region, some foresters had proposed the creation of a system of segregated facilities. Marshall rejected the idea because, as he noted to his friend Roger Baldwin, head of the ACLU, it "would involve having the government openly recognize the principle of discrimination." Instead, he encouraged the Forest Service to state publicly that it would not discriminate based on race. Fearing political recriminations from Southern members of Congress, his superiors shot down this suggestion.[24]

Marshall also prioritized making Forest Service recreational areas accessible to more Americans. The Forest Service, at his urging, developed group camps to cater to low-income users. He also tried to outlaw the granting of term permits to resorts that were out of the financial reach of most, and he even sought to abolish the term permit system on the grounds that it was "monopolistic." Neither of these initiatives met with success, but Marshall clearly was concerned that recreation on the national forests not be elitist or exclusive.[25]

Marshall hiked incessantly during his final years. From 1935 through

1939, he spent each summer (and often spring and fall too) in the field, covering prodigious amounts of ground as a way of fact-checking the wilderness inventory he had assembled. He made it back to Alaska twice more, in the summers of 1938 and 1939, though he was distraught by some of the changes overtaking Wiseman. These included considerably more tourist traffic, an automobile that ferried men to nearby mining operations, and increased airplane service to the region.[26] Ironically, his earlier celebration of the area probably contributed to these developments.

Bob Marshall died of a heart attack in November 1939 at the age of thirty-eight, just months after the passage of the U Regulations. Initially, it was unclear how the Wilderness Society would maintain itself financially, as Marshall had covered almost all of its expenses in the early years. Fortunately, he provided for the Society, as well as for the other causes that were important to him, in his will. After leaving $3,000 to Herb Clark, the family's Adirondack guide, Marshall divided the remainder of his fortune into four parts. Two parts went into a trust whose purpose was the "promotion and advancement of an economic system in the United States based upon the theory of production for use and not for profit." That he chose to devote half of his wealth to the cause of socialism revealed just how important this effort was to him. Another part went into a trust that promoted "the cause of civil liberties in the United States of America." The last part went into a trust for the "preservation of the wilderness conditions in outdoor America." Over the next quarter century, the Robert Marshall Wilderness Trust provided considerable funding to the Wilderness Society in its fight for a Wilderness Act.[27]

The Postwar Lens

With the formation of the Wilderness Society in 1935, the battle over the nation's remaining wilderness was joined. During the next three decades, the Wilderness Society would be the central organizational force in the fight for and passage of the Wilderness Act of 1964. But the postwar success of wilderness advocacy occurred in a political climate that differed markedly from that of the interwar era. Although the trends that made modern wilderness an interwar invention continued into the postwar era, often in magnified form, there were also new threats to wilderness that remade the tenor and rhetoric of preservationist politics. These changes have obscured the historical importance and continuing relevance of interwar wilderness advocacy.

Before detailing these changes, it is important to stress the continuities between the interwar and postwar eras. Most obviously, road building continued to threaten wilderness after World War II, and roadlessness remained the defining characteristic of wilderness. Nationwide, road building escalated dramatically as the federal government pumped dollars into the nation's automotive infrastructure. The Federal Aid Highway Act of 1956, which created the Interstate system, made manifest a redoubled federal commitment to building highways and changing the face of the American landscape. Road building on the nation's wild lands also continued at an ambitious clip, in line with the Park Service's efforts to provide for rapid increases in visitation and the Forest Service's attempts to market its timber resources. Although millions of roadless acres had been protected as wilderness by the postwar era, millions more would disappear because of this onslaught.

Recreational interest in nature also grew by leaps and bounds, as postwar affluence translated into more automobiles and greater leisure time.[28] Visitation to the national parks and national forests grew in the decades after World War II at a pace that made the interwar years look tame by comparison. Between 1916 and 1941, annual visitation to the national parks had increased from about 360,000 people to more than 21 million. Although these figures dipped considerably during World War II, visitation was over 56 million by 1955. That year, Director Conrad Wirth announced an ambitious development initiative, Mission 66, which aimed at modernizing park facilities and accommodating anticipated increases in visitation by 1966—the fiftieth anniversary of the National Park Service. Mission 66, which provided $1 billion over ten years for the expansion of park roads and accommodations, was a fitting accompaniment to the delivery services provided by the burgeoning Interstate system as well as a logical extension of the Mather-Albright emphasis on tourist development. By 1966, more than 133 million people visited the national parks and other areas controlled by the Park Service. By 2000, that number was pushing 300 million.[29] The recreational use of the national forests increased at a similar pace after World War II. In 1950, more than 27 million people visited the national forests for recreation; by 1964, the total was almost 135 million, and by 1999 the Forest Service estimated that there were more than 330 million recreational visitor days.[30] Such increases were a mixed blessing; while they illustrated the tremendous interest among Americans in outdoor recreation and the preservation of public nature for such purposes, they also presented land managers with the challenge of keeping these spaces wild.

Technological developments continued to play an important role in shaping postwar outdoor recreation. The automobile, of course, remained the most important technology in the relationship between Americans and recreational nature, but a variety of other motorized forms of transport, from airplanes and off-road vehicles (ORVs) to snowmobiles and jet skis, emerged as new threats to the solitude sought by wilderness users. Importantly, these devices have provided the same sort of sport and mechanized intimacy with the landscape that early motorists had prized, and they have opened more of the landscape (and waterscape) to motorized travel.[31] These new forms of motorized transport have been to postwar wilderness politics what the automobile was during the years before World War II.

Supporting recreational technologies have played a more complicated postwar role. During the interwar years, most recreational gadgetry, both homespun and commercially produced, had been an extension of the mechanized mobility afforded by the automobile. Wilderness recreation remained technologically primitive, the way most adherents wanted it to be. Indeed, such technological primitivism was crucial to the interwar definition of wilderness. In the postwar years, however, a new generation of lightweight materials made wilderness recreation itself a technological experience. Nylon tents with aluminum poles superseded their heavy canvas predecessors while lightweight sleeping bags replaced bulkier bedrolls. Outdoor clothing became lighter, as synthetic materials challenged the traditional reliance on wool. Finally, aluminum frame backpacks allowed hikers to carry heavier loads with less discomfort, and nesting aluminum cookware replaced cast iron.[32]

These technological developments had a number of important implications for wilderness preservation. First, lightweight gear prompted more Americans to head for the backcountry. Although wilderness recreation remained a minority activity, the number of wilderness users grew substantially after World War II, providing an important base of political support for wilderness preservation. Moreover, new camping technologies went a long way toward replacing pack stock and hired guides, and they thus helped to change the image of wilderness users. Before World War II, the perception was that wilderness was a refuge for those who could afford to hire such an outfit. Wilderness advocates constantly had to fight this impression, and they did so largely by suggesting that one could (in fact, should) go into the wilderness simply. The postwar technological revolution helped to democratize wilderness recreation in the minds of Americans. Although obtaining this new gear may not have been that much cheaper than hiring a guide and stock, it nonetheless promoted a more egalitarian impression of wilderness recreation.

These technologies also shifted the relationship between wilderness recreation and consumerism. While some saw the new gear as a consumerist distraction and feared that wilderness recreation was increasingly about testing space-age technologies against the forces of nature, others lauded postwar equipment as a boon to the wilderness cause. In the end, these technologies have played both roles. By allowing people to buy their way into the wilderness (to paraphrase Leopold), and by blunting the reliance on competency and the thrill of facing nature alone (to paraphrase Marshall), such gear runs counter to the purpose of the wilderness experience as the interwar generation defined it. But these technological developments also have encouraged more Americans to visit preserved wilderness, and in ways that have not substantially undermined the roadless and unmechanized qualities that define such areas. Indeed, modern gear has been crucial to minimizing the impacts that come with increased recreational use of preserved wilderness, and it thus has helped to resolve a tension central to interwar advocacy: that the founders wanted wilderness to be a democratic ideal and yet they defined it in ways that were necessarily restrictive.

A final similarity worth noting between the interwar and postwar years was the federal role in studying outdoor recreation and sponsoring recreational development. In 1958, Congress created the Outdoor Recreation Resources Review Commission (ORRRC) to explore present trends in, and future needs for, outdoor recreation in America. The effort was reminiscent of the National Conference on Outdoor Recreation of the 1920s, though the ORRRC's work was more systematic and influential. In January 1962 the ORRRC published its report, *Outdoor Recreation for America,* along with twenty-seven study reports on various facets of outdoor recreation.

In a survey of more than 16,000 Americans, the ORRRC found that the most popular form of outdoor recreation was pleasure driving. The Commission also found that although there was substantial public acreage available for outdoor recreation, much of the land was distant from the centers of population. "The problem is not one of total acres," the report concluded, "but of *effective* acres." Despite this emphasis, the report did strongly recommend that Congress pass legislation creating a national system of wilderness areas. The ORRRC effort showed that the federal government remained an avid sponsor of recreational development on the public domain, though its final report called for the sort of recreational planning that would have heartened the founders of the Wilderness Society. Although the report did not produce the comprehensive national outdoor recreation policy that it proposed, the Johnson administration did use its findings to shape its environmental policy in significant ways.[33]

While many of these postwar trends built upon a foundation laid during the interwar years, they were matched, even overwhelmed, by a new series of threats to the nation's wild lands. During the first four decades of the Forest Service's existence, the national forests had seen only modest cutting; most commercial timber came from private timberlands. But by the end of World War II, as private timber supplies dwindled, the timber industry began to look amorously at the remaining old growth in the national forests. Military mobilization and the federally subsidized postwar housing boom heightened demand for public timber, and foresters, suddenly flush with a sense of usefulness that was shaped by the ideological demands of the Cold War, facilitated commercial harvests of public timber at every turn.[34]

Even as recreational use continued to rise on the national forests, foresters emphasized commodity production. Annual national forest timber harvests grew from about 3 billion board feet in 1945 to almost 12 billion in 1970, with the most dramatic growth occurring between 1950 and 1965.[35] Accompanying this profound increase was a massive road-building program designed to provide access to remote timber and to protect stands from fire and pests. Between 1945 and 1960, the system of forest development roads grew from approximately 100,000 miles to 160,000 miles.[36] By 1959, there was roughly one mile of road for every two square miles of national forest. That year, the Forest Service released a report, entitled "Operation Multiple Use," which called for an additional 392,600 miles of development roads by the year 2000. That figure was later revised downward, but there were still approximately 340,000 miles of forest development roads in existence by 1990.[37]

The Forest Service's enthusiasm for facilitating the liquidation of public timber transformed postwar wilderness politics.[38] Understandably, preservationists refocused their efforts on opposing timber harvests, the roads that facilitated them, and the production-oriented mindset of government foresters, and recreational users of all sorts were crucial allies in this effort. In the process, the recreational questions and concerns that had defined Forest Service wilderness policy during the interwar years were obscured.

Water development posed an equally compelling threat to wilderness after the war. This was particularly true in the West, a region transformed by federal spending during World War II and the Cold War.[39] During the New Deal, the federal government had sponsored a number of major dam projects, most of them in the name of river basin development and economic modernization for lesser-developed regions. Some conservationists and planners—among them wilderness advocates such as Benton MacKaye—had seen considerable promise in these government-sponsored projects. But during World War II, as the social goals of regional development yielded to the needs of war, the federal government shunted the waterpower from places such as the Tennessee and Columbia river valleys into military production and the creation of an atomic infrastructure. After the war, the politics of productivity and Cold War technological symbolism guided federal dam building; gone were the utopian visions of public hydropower as an engine for equitable and environmentally sound regional planning.[40]

For postwar wilderness activists, there was a more immediate problem posed by the federal enthusiasm for dams. Increasingly, the Bureau of Reclamation turned its eyes to the remote and undeveloped canyon country of the Southwest. The Colorado Plateau was a new type of wilderness, and the battle over the region sparked a new era of wilderness activism.[41] In the late 1940s, the Bureau of Reclamation announced plans to build a dam, just downstream from Echo Park, as part of the Colorado River Storage Project. In doing so, they would have invaded the boundaries of Dinosaur National Monument. After a bitter fight, activists saved Echo Park and Dinosaur National Monument from water development, though their victory came at the expense of some spectacular but unprotected canyon country downstream. The successful defense of Dinosaur galvanized the wilderness movement and introduced the nation to a new generation of wilderness advocates, including Howard Zahniser and Olaus Murie of the Wilderness Society and David Brower of the Sierra Club. A decade later, a successful campaign to keep water development out of the Grand Canyon strengthened the wilderness cause.[42]

Yet, with the exception of the outcomes, there was something eerily familiar about the Echo Park and Grand Canyon controversies. They seemed like Hetch Hetchy all over again. The politics of preservation again revolved around conflicts between those who would save nature for aesthetic and recreational enjoyment and those who sought to transform nature in the name of resource development.[43] The same battle lines that had defined Progressive conservation politics defined the legislative push for a Wilderness Act. Resource agencies, industrial interests, and the politicians who supported them provided the most spirited opposition to wilderness legislation, while support came not only from a group of committed activists but also from an expanding middle-class constituency of recreational users (including, importantly, organized labor).[44] To a great extent, the postwar fight for wilderness became a conflict between producers and consumers.

Once this political shift occurred, it was easy to read it back into the nation's environmental past. The behavior of postwar resource agencies, in their clash with a strengthened preservationist movement, provided a filter through which scholars and activists have viewed not only the history of Progressive conservation but also the entire trajectory of twentieth-century environmental thought and politics. As the postwar wilderness movement blossomed, a new generation of advocates found its identity in Progressive Era archetypes, and in the ideological opposition between the preservation and use of public nature. Activists were quick, for instance, to resurrect John Muir as a culture hero, and to see the conflict over Hetch Hetchy as the defining moment of American environmental history.[45] Historians generally have accepted this argument for continuity between the Progressive and postwar eras, fashioning the orthodoxy that American conservation has always had at its core an ideological conflict between preservation and use.

In emphasizing the similarities between these eras and in seeing the Progressive era through this postwar lens, activists and scholars have obscured and ignored the complex terrain of the interwar period. In doing so, they have not only failed to understand that modern wilderness advocacy grew out of questions about the automobile, roads, recreational development, and consumerism; they have also missed a period during which environmental thought and politics were defined less by an intractable debate between preservation and use than by debates *within* each of these categories. The historical lens ground by postwar wilderness politics created a blind spot that it is time to correct.

The New Problem of the Wilderness

Preservationist politics have changed yet again in the last decade or so, as dam building, forest liquidation, and other such threats to wilderness have slackened (though they have by no means disappeared) and as consumer challenges have again risen to the fore. In the process, a funny thing has happened: utilitarian conservation and preservation no longer seem like such antithetical agendas. Under pressure from the public and a new generation of insurgent foresters, the Forest Service has been flirting with ecosystem management, a model with the potential to rein in commodity production and refocus the attention of foresters on the important ecological functions performed by public forests. Moreover, environmental groups have initiated cooperative programs with farmers, ranchers, and landowners to figure out ways to preserve open space and counter the rising tide of subdivision. At the same time, the Park Service has had to shift its attention from industrial incursions to the impacts of visitation. In the process, they have begun a series of bold experiments to reform the relationship between Americans, their automobiles, and the national parks. Meanwhile, federal land managers from the Park Service, the Forest Service, the Bureau of Land Management, and other agencies are faced with the politics of motorized access in other forms—from deciding the fate of snowmobiles and jet skis in national parks and national recreation areas to managing ORV access on the national forests, wildlife refuges, and BLM lands. Indeed, in its 2000 report on the nation's most threatened wildlands, the Wilderness Society has argued that "ORVs are now the single fastest growing threat to the natural integrity of our public lands." Recreational challenges to wilderness have come to rival extractive threats, much as they did during the interwar era when modern wilderness was born.[46]

Unfortunately, some have concluded that under these circumstances the wilderness idea has ceased to be useful. The assumption behind this conclusion, of course, is that wilderness was a product of Progressive and postwar brands of preservation—that it was, and remains, merely a consumer ideal designed to counter the forces of production. Indeed, most recent critiques of wilderness have assumed this, and in so doing they have ignored the interwar generation of advocates and the context in which their advocacy was born. In the process, the wilderness idea has been caricatured. It is vital to rethink wilderness as a preservationist ideal, but such a reconsideration ought to begin with a more subtle understanding of the origins and content of the modern wilderness idea. For when we understand where modern wilderness came from and how it was made—when we seek a more multidimensional sense of its history—we begin to see that the wilderness idea can be a continuing and vital part of the search for environmental solutions.

One of the most pressing intellectual problems environmentalists face today is the affinity between environmental sentiment and consumerism. Not only do we need a much fuller understanding of the environmental impacts of consumption, but we also need to fathom the ways in which our roles and identities as consumers have shaped how we idealize and seek to protect nature. How, in other words, have we constructed nature as consumers? This question has been at the heart of recent critiques of wilderness, many of which are grounded in the assumption that wilderness itself is a consumer ideal. But for the interwar generation of wilderness advocates, the wilderness idea was built upon a critique of then-emergent consumer trends. The founders of the Wilderness Society offered wilderness as a new preservationist paradigm because they were concerned with how the automobile, roads, and a boom in outdoor recreation were changing both the natural world and Americans' relations with nature. As we rethink our preservationist ideals today, it is worth remembering that wilderness advocacy emerged during the interwar period as the product of a similarly critical endeavor.

NOTES

1. The Problem of the Wilderness

1. Robert Marshall, "The Problem of the Wilderness," *Scientific Monthly* 30, 2 (February 1930): 148.

2. Benton MacKaye, "An Appalachian Trail: A Project in Regional Planning," *Journal of the American Institute of Architects* 9 (October 1921): 325–30.

3. Harvey Broome, *Out Under the Sky of the Great Smokies: A Personal Journal* (1975), viii.

4. For details of this founding moment, see Broome, "Origins of the Wilderness Society," *The Living Wilderness* 5, 5 (July 1940): 13–15; Michael Nadel, "Genesis of the Wilderness Society," June 1973, TWSP, Box 11, Folder 19; Stephen Fox, "'We Want No Straddlers'," *Wilderness* 48 (Winter 1984): 5–19; Harvey Broome to Robert Sterling Yard, September 7, 1939, TWSP, Box 11, Folder 20. The Broome quote is from this letter. There is disagreement about who drafted the constitution at the center of this roadside discussion. Broome says it was MacKaye while Fox suggests it was Marshall. Neither documents their claim. A letter between Marshall and Anderson (October 24, 1934, TWSP, Box 11, Folder 15) suggests that it was MacKaye, perhaps with Broome's assistance. But MacKaye, in a letter to Yard (September 16, 1939, TWSP, Box 11, Folder 20), gives Marshall more of the credit.

5. Broome, Frank, MacKaye, and Marshall drafted an invitation dated October 19, 1934. Marshall then sent this invitation and a form letter to each of the other proposed founders. Most of those letters were dated October 25, 1934. RMP, Box 1, Folder 18.

6. Harold Anderson to Harvey Broome, December 11, 1939, TWSP, Box 11, Folder 20.

7. This quote is from a letter from Marshall to Leopold, February 21, 1930, ALP, 10-3, Box 4.

8. For a biographical sketch of Oberholtzer, see R. Newell Searle, *Saving Quetico-Superior: A Land Set Apart* (1977), 53–59.

9. See Nadel, "Genesis of the Wilderness Society," 4. In a letter to MacKaye (November 24, 1934, TWSP, Box 11, Folder 20), Marshall noted that, "Collier says he would be delighted to join with us, although I am not sure whether it would be wise from his standpoint."

10. See letter, John C. Merriam to Bob Marshall, November 3, 1934, TWSP, Box 11, Folder 20.

11. See "Minutes of the Wilderness Society," and the resulting statement, "The Wilderness Society," both in TWSP, Box 11, Folder 19.

12. The best example of this interpretation is Nash's *Wilderness and the American Mind* (1982). See also Alan Taylor, "'Wasty Ways': Stories of American Settlement," *Environmental History* 3, 3 (July 1998): 291–310.

13. Roderick Nash, *The Rights of Nature: A History of Environmental Ethics* (1989); Michael Frome, *The Battle for the Wilderness* (1997); J. Baird Callicott, *In Defense of the Land Ethic: Essays in Environmental Philosophy* (1989). On deep ecology, see Bill Devall and George Sessions, *Deep Ecology* (1985).

14. Samuel Hays's *Beauty, Health, and Permanence: Environmental Politics in the United States, 1955–1985* (1987) is the classic example here, though it deals mostly with the postwar era. For versions rooted in earlier eras, see Nash, *Wilderness and the American Mind;* Peter Schmitt, *Back to Nature: The Arcadian Myth in Urban America* (1969).

15. "A Summons to Save the Wilderness," *The Living Wilderness* 1, 1 (September 1935): 1.

16. On the new ecology, see Donald Worster, "The Ecology of Order and Chaos," in *The Wealth of Nature: Environmental History and the Ecological Imagination* (1993), 156–70; Worster, *Nature's Economy: A History of Ecological Ideas* (1994); Daniel Botkin, *Discordant Harmonies: A New Ecology for the Twenty-First Century* (1990); and Michael Barbour, "Ecological Fragmentation in the Fifties," in Cronon, ed., *Uncommon Ground: Rethinking the Human Place in Nature* (1996), 233–55. A polemical and less satisfying treatment is Stephen Budiansky, *Nature's Keepers: The New Science of Nature Management* (1995). Richard Sellars presents a nice overview of many of the management dilemmas faced by the National Park Service over the course of its history in *Preserving Nature in the National Parks: A History* (1997). Conservation biologists and island biogeographers have been a leading force in getting Americans to rethink the adequacy of traditional wilderness preservation. *Wild Earth,* the journal of the Wildlands Project, and the *Journal of Conservation Biology* are both excellent sources on how wilderness defenders have rethought the wilderness idea in recent years. On island biogeography, see David Quammen, *The Song of the Dodo: Island Biogeography in an Age of Extinctions* (1996). For other ecological critiques, see J. Baird Callicott and Michael P. Nelson, eds., *The Great New Wilderness Debate: An Expansive Collection of Writings Defining Wilderness from John Muir to Gary Snyder* (1998), particularly Part Four. On nature as a moving target in efforts to measure human-induced damage, see Jeffrey Wheelwright, *Degrees of Disaster: Prince William Sound: How Nature Reels and Rebounds* (1996). On the inappropriateness of wilderness as an ecological descriptor, see William Cronon, *Changes in the Land: Indians, Colonists, and the Ecology of New England* (1983), particularly 3-15. For a twist on the logic of human ubiquity, see Bill McKibben, *The End of Nature* (1989).

17. Francis Jennings made this argument decades ago in *The Invasion of America: Indians, Colonialism, and the Cant of Conquest* (1975). See also Cronon, *Changes in the Land.* Recent scholarship has argued that the wilderness ideal continued to dispossess even after American attitudes toward wilderness became favorable. See Mark Spence, *Dispossessing the Wilderness: Indian Removal*

and the Making of the National Parks (1999); Louis Warren, *The Hunter's Game: Poachers and Conservationists in Twentieth-Century America* (1997); Karl Jacoby, *Crimes Against Nature: Squatters, Poachers, Thieves, and the Hidden History of American Conservation* (2001). Theodore Catton, in his history of park preservation in Alaska, examines an instance in which wilderness and native subsistence have coexisted. See Catton, *Inhabited Wilderness: Indians, Eskimos, and National Parks in Alaska* (1997).

18. On the pristine myth, see William Denevan, "The Pristine Myth: The Landscape of the Americas in 1492," *Annals of the Association of American Geographers* 82, 3 (1992): 369–85. On the charge of ethnocentrism, see J. Baird Callicott, "The Wilderness Idea Revisited: The Sustainable Development Alternative," *The Environmental Professional* 13 (1991): 235–47. For a third-world critique, see Ramachandra Guha, "Radical American Environmentalism and Wilderness Preservation: A Third World Critique," *Environmental Ethics* 11 (Spring 1989): 71–83. All are reprinted in Callicott and Nelson, eds., *The Great New Wilderness Debate*. See also Roderick P. Neumann, *Imposing Wilderness: Struggles over Livelihood and Nature Preservation in Africa* (1998).

19. See Warren, *The Hunter's Game;* Karl Jacoby, "Class and Environmental History: Lessons from the 'War in the Adirondacks,'" *Environmental History* 2, 3 (July 1997): 324–42; Jacoby, *Crimes Against Nature;* Steven Hahn, "Hunting, Fishing, and Foraging: Common Rights and Class Relations in the Postbellum South," *Radical History Review* 26 (1982): 37–64; Benjamin Heber Johnson, "Conservation, Subsistence, and Class at the Birth of Superior National Forest," *Environmental History* 4, 1 (January 1999): 80–99. Two of the most famous cases of removing poor inhabitants to create national parks occurred in the Shenandoah and Great Smoky Mountains National Parks during the interwar era. On Shenandoah, see Charles M. Perdue, Jr. and Nancy Martin Perdue, "Appalachian Fables and Facts: A Case Study of the Shenandoah National Park Removals," *Appalachian Journal,* 7, 1–2 (Autumn/Winter 1979–80): 84–104, and "'To Build a Wall Around These Mountains': The Displaced People of Shenandoah," *The Magazine of Albemarle County History* 49 (1991): 48–71. On the Great Smokies, see Durwood Dunn, *Cade's Cove: The Life and Death of a Southern Appalachian Community* (1988). On the role of the state, see James Scott, *Seeing Like a State: How Certain Schemes to Improve the Human Condition Have Failed* (1998). On natural ideals masking class agendas, see Raymond Williams, *The Country and the City* (1973).

20. The most important recent critique of the wilderness idea, and one that largely sticks to cultural ground, is Cronon's "The Trouble with Wilderness; or, Getting Back to the Wrong Nature," in Cronon, ed., *Uncommon Ground,* 69–90. A number of other essays in *Uncommon Ground* tread on this territory. Particularly important is Richard White's "'Are You an Environmentalist or Do You Work for a Living?': Work and Nature," 171–85. There is a growing and vibrant literature on the connections between nature and tourism. See Hal Rothman, *Devil's Bargains: Tourism and the Twentieth-Century American West* (1998); Marguerite Shaffer, "'See America First': Re-Envisioning Nation and Region through Western Tourism," *Pacific Historical Review* 65, 4 (November 1996): 559–81;

Dona Brown, *Inventing New England: Regional Tourism in the Nineteenth Century* 1995); John Sears, *Sacred Places: American Tourist Attractions in the Nineteenth Century* (1989); John Jakle, *The Tourist: Travel in Twentieth-Century North America* (1985). On the connections between landscape design and the preservation of natural spaces, see Anne W. Spirn, "Constructing Nature: The Legacy of Frederick Law Olmsted," in *Uncommon Ground*, 91–113; Ethan Carr, *Wilderness By Design: Landscape Architecture and the National Park Service* (1998); Linda Flint McClelland, *Building the National Parks: Historic Landscape Design and Construction* (1998); Sellars, *Preserving Nature in the National Parks.*

21. See Worster, *Nature's Economy;* Ronald Tobey, *Saving the Prairies: The Life Cycle of the Founding School of American Plant Ecology, 1895–1955* (1981); Joel Hagen, *An Entangled Bank: The Origins of Ecosystem Ecology* (1992).

22. On federal highway policy, see Bruce E. Seely, *Building the American Highway System: Engineers as Policy Makers* (1987); Frederic L. Paxson, "The Highway Movement, 1916–1935," *American Historical Review* 51 (1946): 236–53.

23. On Ford, see James J. Flink, *The Automobile Age* (1988): 47–50; David Hounshell, *From the American System to Mass Production, 1800–1932: The Development of Manufacturing Technology in the United States* (1984), 217–61.

24. The best source on these years remains Nash, *Wilderness and the American Mind.*

25. On postwar wilderness politics, see Craig Allin, *The Politics of Wilderness Preservation* (1982); Michael Frome, *Battle for the Wilderness* (1997); Mark Harvey, *A Symbol of Wilderness: Echo Park and the American Conservation Movement* (1994); Michael Cohen, *The History of the Sierra Club, 1892–1970* (1988), 101–332; Nash, *Wilderness and the American Mind,* 209–37; and Stephen Fox, *The American Conservation Movement: John Muir and His Legacy* (1985), 250–90.

2. Knowing Nature through Leisure

1. Jesse Frederick Steiner, *Americans at Play: Recent Trends in Recreation and Leisure Time Activities* (1933), 34.

2. This phrase is adapted from Richard White's work on the connections between labor and nature. See White's chapter "Knowing Nature Through Labor" in *The Organic Machine: The Remaking of the Columbia River* (1995).

3. This is not to suggest that the process does not have an even deeper history. On that history, see John Sears, *Sacred Places;* Cindy S. Aron, *Working at Play: A History of Vacations in the United States* (1999); Anne Farrar Hyde, *An American Vision: Far Western Landscape and National Culture, 1820–1920* (1990); Earl Pomeroy, *In Search of the Golden West: The Tourist in Western America* (1957).

4. Information in this paragraph is drawn from John Higham, "The Reorientation of American Culture in the 1890s," in John Weiss, ed., *The Origins of Modern Consciousness* (1965), 25–48; Peter Schmitt, *Back to Nature;* and Nash, *Wilderness and the American Mind.* On the rise of sport hunting and its relationship to conservation, see John Reiger, *American Sportsmen and the Origins of Conservation* (1975), and Warren, *The Hunter's Game.* On the Country Life Movement, see William L. Bowers, *The Country Life Movement in America, 1900–1920* (1974).

5. T. J. Jackson Lears, "From Salvation to Self-Realization: Advertising and the Therapeutic Roots of Consumer Culture," in Lears and Richard Wightman Fox, eds., *The Culture of Consumption: Critical Essays in American History, 1880–1980* (1983). For a much fuller treatment of this shift, see Lears, *No Place of Grace: Antimodernism and the Transformation of American Culture, 1880–1920* (1981); and Warren Susman, *Culture as History: The Transformation of American Society in the Twentieth Century* (1984).

6. For another example of the use of this sort of sentiment, see Anne F. Hyde, "William Kent: The Puzzle of Progressive Conservationists," in William Deverell and Tom Sitton, eds., *California Progressivism Revisited* (1994), 34–56.

7. On G. Stanley Hall, see Lears, *No Place of Grace*, 247–51; Gail Bederman, *Manliness and Civilization: A Cultural History of Gender and Race in the United States, 1880–1917* (1995), 77–120.

8. Higham, "The Reorientation of American Culture in the 1890s," 29–30. Bederman, *Manliness and Civilization*.

9. On the issue of authenticity, see Miles Orvell, *The Real Thing: Imitation and Authenticity in American Culture, 1880–1940* (1989).

10. Daniel T. Rodgers, *The Work Ethic in Industrial America, 1850–1920* (1974), 108.

11. This notion of "spending" leisure time is borrowed from Robert and Helen Lynd, *Middletown: A Study in Contemporary American Culture* (1929), 225. On leisure and commercial amusements at the turn of the century, see John Kasson, *Amusing the Million: Coney Island at the Turn of the Century* (1978); Kathy Peiss, *Cheap Amusements: Working Women and Leisure in Turn-of-the-Century New York* (1986); David Nasaw, *Going Out: The Rise and Fall of Public Amusements* (1993).

12. Rodgers, 123. For another excellent discussion of the relationship between work and leisure, see Roy Rosenzweig, *Eight Hours for What We Will: Workers and Leisure in the Industrial City, 1870–1920* (1983).

13. In 1930, there were about 23,034,700 registered automobiles in the United States, and the population of the United States, according to the 1930 census, was 122,775,046. Those figures yield a ratio of 5.33 people for every automobile. Automobile statistics are from *Historical Statistics of the United States, Colonial Times to 1970,* Part 2 (1975), 716. Population data are from Clifford L. Lord and Elizabeth H. Lord, *Historical Atlas of the United States* (1944), 211.

14. *Middletown,* 251.

15. *Ibid.,* 253–55 (quote, 253). See also Richard Wightman Fox, "Epitaph for Middletown: Robert S. Lynd and the Analysis of Consumer Culture," in Fox and Lears, *The Culture of Consumption,* 101–42. Frank Stricker has shown that even by the late 1920s most working-class families did not own automobiles. See Stricker, "Affluence for Whom?—Another Look at Prosperity and the Working Classes in the 1920s," *Labor History* 24, 1 (Winter 1983): 5–33.

16. *Middletown,* 260–63.

17. See Samuel P. Hays, *Conservation and the Gospel of Efficiency* (1959).

18. On governmental sponsorship, see Andrew G. Truxal, *Outdoor Recreation Legislation and Its Effectiveness: A Summary of American Legislation for Public Out-*

door Recreation, 1915–1927 (1929). On the associative tenor of the era, see Ellis Hawley, "Herbert Hoover, the Commerce Secretariat, and the Vision of an 'Associative State,' 1921–1928," *Journal of American History* 61 (June 1974): 116–40.

19. Letter, A. S. Peck to The Forester, December 3, 1920, National Archives, Record Group 95, Entry 85, Box 68, Folder "U-Recreation, General, 1920–1921."

20. The notion of a "moral equivalent of war" comes from a famous William James essay by that title. See James, "The Moral Equivalent of War," *Popular Science Monthly* (October 1910), 400–410.

21. Fox, "Epitaph for Middletown," 103.

22. Jean-Christophe Agnew has referred to this consumer habit of seeing as "acquisitive cognition." See Agnew, "The Consuming Vision of Henry James," in Fox and Lears, *The Culture of Consumption,* 67.

23. Lynn Dumenil, *The Modern Temper: American Culture and Society in the 1920s* (1995), 77.

24. Dean MacCannell, *The Tourist: A New Theory of the Leisure Class* (1976), 41.

25. Jonathan Culler, "The Semiotics of Tourism," in *Framing the Sign: Criticism and Its Institutions* (1988), 153–67.

26. Ibid.

27. MacCannell, 1–3.

28. Warren Belasco has written the best study about the automobile during this period; he is particularly perceptive in his treatment of the ways the automobile affected the relationship between Americans and nature. See Warren James Belasco, *Americans on the Road: From Autocamp to Motel, 1910–1945* (1979). The following analysis is much indebted to Belasco's work.

29. The *New York Times* statistic is from J. C. Long and John D. Long, *Motor Camping* (1923), 1–2. The general figures are from Belasco, 73–74.

30. See Carey Bliss, *Autos Across America: A Bibliography of Transcontinental Automobile Travel, 1903–1940* (1982).

31. On these conflicts in the countryside, see Belasco; Reynold Wik, *Henry Ford and Grass Roots America* (1972); Michael Berger, *The Devil Wagon in God's Country: The Automobile and Social Change in Rural America* (1979); Joseph Interrante, "You Can't Go to Town in a Bath-tub: Automobile Movement and the Reorganization of Rural American Space, 1900–1930," *Radical History Review* 21 (Fall 1979): 151–68; Paul S. Sutter, "Paved with Good Intentions: Good Roads, the Automobile, and the Rhetoric of Rural Improvement in *Kansas Farmer,* 1890–1914," *Kansas History* 18, 4 (Winter 1995–1996): 284–99; Hal Barron, *Mixed Harvest: The Second Great Transformation in the Rural North, 1870–1930* (1997), 19–42.

32. Belasco, 74–76.

33. Ibid., 7–18.

34. John Burroughs, "A Strenuous Holiday," in *Under the Maples* (1921), 121.

35. Ibid.

36. "Harding in Camp with Noted Party; Chops Fire Wood," *New York Times,* July 24, 1921, 1–2.

37. See Kendrick Clements, "Herbert Hoover and Conservation, 1921–1933,"

American Historical Review 89, 1 (February 1984): 67–88; Clements, *Hoover, Conservation, and Consumerism: Engineering the Good Life* (2000); Carl E. Krog, "'Organizing the Production of Leisure': Herbert Hoover and the Conservation Movement in the 1920s," *Wisconsin Magazine of History* 67 (Spring 1984): 199–218; Fox, *The American Conservation Movement*, 183–217.

38. Allan Wallis, *Wheel Estate: The Rise and Decline of Mobile Homes* (1991), 32.

39. On the railroad experience, see Wolfgang Schivelbusch, *The Railway Journey: The Industrialization of Time and Space in the 19th Century* (1986).

40. Donna R. Braden, *Leisure and Entertainment in America* (1988), 349; see also the company history, "The First Fifty Years" (Wichita, Kans.: The Coleman Company, 1950), copy in the Kansas Collection, University of Kansas.

41. Braden, 323–52.

42. Wallis, 48. The information in this paragraph is drawn from my reading of *Popular Mechanics* issues from 1915 to 1940.

43. Braden, 330.

44. Michael Aaron Rockland, *Homes on Wheels* (1980), 44.

45. Wallis, 47.

46. "200,000 Trailers," *Fortune* (March 1937), 105. The "Fastest Growing U.S. Industry" quote is from one of the article's subtitles. See p. 106.

47. On the house trailer, see Carroll D. Clark and Cleo Wilcox, "The House Trailer Movement," *Sociology and Social Research* 22, 6 (July–August 1938): 503–19; and Donald Olen Cowgill, *Mobile Homes: A Study of Trailer Life* (1941). The best historical treatments are Wallis; Carlton M. Edwards, *Homes for Travel and Living: The History and Development of the Recreational Vehicle and Mobile Home Industries* (1977).

48. See F. E. Brimmer, *Auto-camping* (1923) and *Motor Campcraft* (1923); Elon Jessup, *The Motor Camping Book* (1921) and *Roughing It Smoothly* (1923). For other examples of this literature, see J. C. Long and John D. Long, *Motor Camping* (1923); Melville F. Ferguson, *Motor Camping on Western Trails* (1925); and Porter Varney, *Motor Camping* (1935). The literature on trailer life coincided with the late-1930s boom in trailer production. See Blackburn Sims, *The Trailer Home: With Practical Advice on Trailer Life and Travel* (1937); Charles Edgar Nash, *Trailer Ahoy!* (1937). The most famous guide to woodcraft before the automobile was Horace Kephart's *The Book of Camping and Woodcraft* (1906).

49. Quotes from Long and Long, *Motor Camping*, 15–16.

50. Brimmer, *Motor Campcraft*. Brimmer was the "Motor Camping Editor" of *Outers' Recreation*. He also wrote a book for the Coleman entitled *Coleman Motor Campers' Manual* (1926).

51. Brimmer, *Motor Campcraft*, 4.

52. This notion of companionate leisure is from Belasco, 64.

53. See Belasco, 55–58; Aron, *Working at Play*.

54. Rinehart quoted in Belasco, 85. Virginia Scharff also quotes Rinehart in *Taking the Wheel: Women and the Coming of the Motor Age* (1991), 138.

55. Belasco, 84.

56. On household modernization, see Ruth Schwartz Cowan, *More Work for Mother: The Ironies of Household Technology from Open Hearth to the Microwave*

(1983); Ronald Tobey, *Technology and Freedom: The New Deal and Electrical Modernization of the American Home* (1996). Tobey begins his book by invoking the Sinclair Lewis character, Samuel Dodsworth, and his dream of a high-tech trailer. For Tobey, "Dodsworth's Dream" was the consummate example of the modernized domestic ideal. Tobey, 1–5.

57. This phrase appeared throughout touring memoirs and prescriptive literature of the era. Irving Cobb used it as a title to his book, *Roughing It De Luxe* (1914).

58. See "The Great American Roadside," *Fortune* (September 1934), 53–63, 172, 174, 177 (quote, 56).

59. Belasco, 71–72.

60. See Sellars, 69; Thomas Cox, *The Park Builders: A History of State Parks in the Pacific Northwest* (1988), 3–13; James L. Greenleaf, "The Study and Selection of Sites for State Parks," *Landscape Architecture* 15, 4 (July 1925): 227–34. Greenleaf, president of the American Society of Landscape Architects, objected to this recreational logic. Instead, he suggested the motto, "a state park wherever nature smiles, a motor camp every hundred miles."

61. In 1921, only 20 states had state parks; by 1927, there were more than 500 state parks in 43 states. See Truxal, 99–106 (quote, 106).

62. Steiner, *Americans at Play,* 34–37.

63. On the Lincoln Highway, see Drake Hokanson, *The Lincoln Highway: Main Street Across America* (1988). For a list of the transcontinental marked routes, see F. E. Brimmer, *Motor Campcraft* (1923), 209–20.

64. Belasco, 89.

65. Ibid., 90–92.

66. Ibid., 75, 89.

67. E. P. Meinecke, *Camp Planning and Camp Reconstruction* (n.d. [circa 1934]), 3. Emphasis is mine.

68. Linda Flint McClelland, *Presenting Nature: The Historic Landscape Design of the National Park Service, 1916–1942* (1993), 161–66 (quote, 166). McClelland points out that "the term 'meineckizing' campgrounds became a common term among landscape designers and CCC supervisors in the 1930s" (163). McClelland published a revised version of this government report as *Building the National Parks: Historic Landscape Design and Construction* (1998).

69. Steiner, *Americans at Play,* 49–50.

70. On outboard motors, see David Backes, *Canoe Country: An Embattled Wilderness* (1991), 82–83; Steiner, 49–54.

71. Steiner, 42–44.

72. The first generation of hiking clubs, which included groups like the Appalachian Mountain Club and the Sierra Club, had been formed in the late 1800s. By the 1920s, in part as a response to being driven off the country roads by automobiles, a second generation of clubs came into being. On the history of hiking in the Northeast, see Laura and Guy Waterman, *Forest and Crag: A History of Hiking, Trail Blazing, and Adventure in the Northeast Mountains* (1989).

73. William Cronon has an anecdotal discussion of this shift in his "Epilogue" to *Nature's Metropolis: Chicago and the Great West* (1991), 379–83.

74. Brimmer, *Motor Campcraft*, 90.

75. Hays, *Conservation and the Gospel of Efficiency,* 127–30, 139–41; Fox, *The American Conservation Movement,* 130.

76. According to Kendrick Clements, 128 different organizations were represented at the NCOR. See Clements, "Herbert Hoover and Conservation, 1921–1933," 67.

77. Coolidge's opening remarks can be found in the "Proceedings," National Conference on Outdoor Recreation (1924), 13–14.

78. "Proceedings of the Meeting of the Advisory Council," National Conference on Outdoor Recreation, December 11–12, 1924 (1925), 13.

79. "Proceedings" (1924), 71.

80. See, for instance, the testimony of Dr. Joseph Hyde Pratt, "Proceedings" (1924), 86.

81. For connections between the American landscape and national identity, see Hyde, *An American Vision;* Shaffer, "See America First"; Sears, *Sacred Places.*

82. This recreational safety valve thesis replaced the older frontier safety valve thesis. On that more venerable thesis, see Henry Nash Smith, *Virgin Land: The American West as Symbol and Myth* (1950), particularly 201–10; William Deverell, "To Loosen the Safety Valve: Eastern Workers and Western Lands," *Western Historical Quarterly* 19, 3 (August 1988): 269–85.

83. "Proceedings" (1924), 33–34.

84. See *Recreation Resources of Federal Lands* (1928).

85. "Proceedings" (1924), 116–18; "Proceedings of the Second National Conference on Outdoor Recreation," January 20–21, 1926 (1926), 141.

86. "Proceedings of the Meeting of the Advisory Council" (1925), 8–11; "Proceedings" (1926), 111–12.

87. "Proceedings" (1926), 147.

88. For this debate, see "Proceedings" (1926), 147–54.

89. "Proceedings" (1926), 13–17 (lengthy quote, 16).

90. For a good example of how this discourse on Americanism played itself out in the NCOR, see the "Report of the Committee on Citizenship Values of Outdoor Recreation," appended to the "Proceedings" (1924), 203–5.

91. "Proceedings" (1926), 15.

92. "Proceedings" (1926), 64.

93. Steiner, *Americans at Play,* 12.

94. Jesse Steiner focuses on the NRA codes and the shorter workweek in his article "Challenge of the New Leisure," *New York Times Magazine* (September 24, 1933), 1–2, 16.

95. *Annual Report of the Secretary of the Interior* (1933), 186–87. In 1933, the Park Service had an annual appropriation of $4.5 million for park roads.

96. John C. Paige, *The Civilian Conservation Corps and the National Park Service, 1933–1942: An Administrative History* (1985), 48–49; *Annual Report of the Secretary of the Interior* (1933), 156–57.

97. Paige, 132.

98. Phoebe Cutler, *The Public Landscape of the New Deal* (1985), 8–28.

99. Most of these RDAs eventually became state parks. See Cutler, 70–75. See

also Richard F. Knapp and Charles E. Hartsoe, *Play for America: The National Recreation Association* (1979), 119.

100. Swain, "The National Park Service and the New Deal," 317–19. Swain cites an increase from 3.5 million visitors in 1933 to 16.75 million in 1940. Much of this increase came with the addition of military and capitol parks to the system, but even without these additions, there was still a marked increase during the Depression. In 1935, Arno Cammerer noted that, excluding these additional areas, visitation had risen 22 percent that year. See *Annual Report of the Secretary of the Interior* (1935), 179–80.

101. *Annual Report of the Secretary of the Interior* (1935), 179.

102. On the rise of landscape architecture as a profession, see Cutler, 83–89. For an overview of the impact of New Deal programs on the National Park Service and the parks themselves, see Sellars, 132–42.

103. On the development of roads and truck trails on the national forests during the New Deal years, see *Report[s] of the Chief of the Forest Service* (1933, 1934, 1935).

104. William C. Tweed, *Recreation Site Planning and Improvement in the National Forests, 1891–1942* (1981), 16.

105. Ibid., 16–26.

106. This is Phoebe Cutler's main point in *The Public Landscape of the New Deal.*

107. Although Grazing Service lands were not administered to serve recreational ends, the reversion to permanent public ownership did open up the possibility for the recreational use of such lands by the American public. See, for instance, *A Study of the Park and Recreation Problem of the United States* (1941).

3. A Blank Spot on the Map: Aldo Leopold

1. James P. Gilligan, "The Development of Policy and Administration of Forest Service Primitive and Wilderness Areas in the Western United States" (1953), 82; Donald Baldwin, *The Quiet Revolution: The Grass Roots of Today's Wilderness Preservation Movement* (1972).

2. Craig Allin, "The Leopold Legacy and American Wilderness," in Thomas Tanner, ed., *Aldo Leopold: The Man and His Legacy* (1987), 25; Susan Flader, "'Let the Fire Devil Have His Due': Aldo Leopold and the Conundrum of Wilderness Management," in David W. Lime, ed., *Managing America's Enduring Wilderness Resource* (1990), 89; R. Nash, *Wilderness and the American Mind,* 187.

3. Curt Meine, *Aldo Leopold: His Life and Work* (1988), 3–83.

4. Ibid., 87–123.

5. *Yearbook of the Department of Agriculture* (1912), 58.

6. These concerns are particularly evident in "Forest Service" sections of the "Report[s] of the Secretary," *Yearbook of the Department of Agriculture* (1910–15).

7. This picture of the national forests is drawn from a number of sources. See David Clary, *Timber and the Forest Service* (1986), particularly chapters 1–3; Meine, chapters 6–7; "Forest Service" sections of the "Report[s] of the Secretary," *Yearbook of the Department of Agriculture* (1910–115); and the "Annual Report[s]

of the Forester" (1910–15). Both Steen and Clary point out that timber famine was unlikely. As Clary notes (p. 66), per capita timber consumption was declining at the very moment that Pinchot and others painted a future of scarcity. Another excellent source on the Forest Service during this period, a study whose date of publication makes it essentially a primary source, is John Ise, *The United States Forest Policy* (1920).

8. Ise, 254–98.

9. Meine, 87–123.

10. Ibid., 125. Good sources on Muir and the Hetch Hetchy episode include Nash, *Wilderness and the American Mind,* 122–40, 161–81; Fox, *The American Conservation Movement,* 3–99; and Smith, *Pacific Visions: California Scientists and the Environment, 1850–1915* (1987), 143–85.

11. Nash, *Wilderness and the American Mind,* 180–81; Meine, 144–45; Alfred Runte, *National Parks: The American Experience* (1987), 83–85; John Miles, *Guardians of the Parks: A History of the National Parks and Conservation Association* (1995); Shaffer, "See America First: Tourism and National Identity, 1905–1930."

12. Meine, 135–37; Warren, *The Hunter's Game,* 71–105.

13. Meine, 144–46; Aldo Leopold and Don Johnston, "Grand Canyon Working Plan: Uses, Information, Recreational Development," December, 1916, ALP, 10–11, Folio 1.

14. Paul Schullery, ed., *The Grand Canyon: Early Impressions* (1981); Hal K. Rothman, *Devil's Bargains: Tourism in the Twentieth-Century American West* (1998), 50–80.

15. Smith, *Pacific Visions,* 187; Shaffer, "See America First," 168–70; Robert W. Rydell, *All the World's a Fair: Visions of Empire at American International Expositions, 1876–1916* (1984), 208–33.

16. The most egregious offender was Ralph Cameron, later a Republican senator from Arizona, who managed to cobble together numerous claims totaling thousands of acres. For Cameron's sordid history, see Robert Shankland, *Steve Mather of the National Parks* (1970), 227–42.

17. Information on the act appears in *The Use Book: A Manual of Information about the National Forests* (1918), 145–47. See also William Tweed's very useful study, *Recreation Site Planning and Improvement in the National Forests, 1891–1942* (1981), 3. Many of the permits predating the Term Permit Act were for the development of spas near mineral springs. See, for example, the "Report of the Secretary," in *Yearbook of the Department of Agriculture* (1913), 50.

18. In 1916, there were 2,118 permits for residence sites out of a total of 19,289 in effect that year. See "Report of the Forester," in *Annual Reports of the Department of Agriculture* (1917). By 1924, there were 8,349 cabins and residences and 724 camps, resorts, and hotels (the latter up from 359 in 1917) according to a report by E. A. Sherman, "Outdoor Recreation on the National Forests," copy in SAFC, Box 72. See also Tweed, 3–4. On recreation in the Angeles National Forest, see Abraham Hoffman, "Angeles Crest: The Creation of a Forest Highway System in the San Gabriel Mountains," *Southern California Quarterly* 50, 3 (September 1968): 309–45; and Hoffman, "Mountain Resorts and Trail

Camps in Southern California's Great Hiking Era, 1884–1938," *Southern California Quarterly* 58, 3 (Fall 1976): 381–406.

19. Leopold, "Notes on the Lake Mary Public Use Area," ALP, 10–11, Box 2, Folder 4.

20. For an example of this interpretation, see Hal K. Rothman, "'A Regular Ding-Dong Fight': Agency Culture and Evolution in the NPS-USFS Dispute, 1916–1937," *Western Historical Quarterly* 20, 2 (May 1989): 141–62.

21. "Report of the Forester," 1925, 41–44; Gilligan, 73–74. Visitation increased from three million in 1917 to about eleven million in 1924. Gilligan, 95. Counting visitors was still an inexact science. For instance, these statistics probably included motorists simply passing through the national forests. Later, the Forest Service would attempt to distinguish between "transient motorists" and recreational users.

22. Craig W. Allin makes this argument in *The Politics of Wilderness Preservation* (1982), 62, 64.

23. O. C. Merrill stated that the Forest Service had built and/or improved only 2,000 miles of road between its inception in 1905 and 1916. See O. C. Merrill, "Opening Up the National Forests by Road Building," *Yearbook of the Department of Agriculture* (1917), 521–29. James Gilligan maintains that, as of 1907, the national forests contained less than 5,000 miles of roads (presumably the 2,000 Merrill referred to brought the total up to somewhere around 7,000 miles by 1916). The annual "Report[s] of the Forester" provide breakdowns of yearly road appropriations as well as mileage of roads constructed and maintained. The "Report of the Forester" for 1935 lists a planned system of 20,924 miles of forest highways and 100,024 miles of forest development roads for a total of 120,948 miles. As of 1935, 31,796 miles of this system remained to be built. This 1935 figure reflects much of the frenetic activity in the national forests that resulted from New Deal conservation and work initiatives.

24. Tweed, 2–7.

25. Frank Waugh, *Recreation Uses on the National Forests* (1918); Tweed, 6–7.

26. Tweed, 7–10. Donald Baldwin's book, *The Quiet Revolution,* is a detailed treatment of Carhart's activities during the late teens and early twenties.

27. Background information from Tweed, 7–10; for the specifics of his plans, see Carhart, "General Working Plan: Recreational Development of the San Isabel National Forest, Colorado," December 1919, and "Recreation Plan, San Isabel National Forest, Colorado," ACP. The quote is from the "General Working Plan," 59. For Carhart's accomplishments, see Robert W. Cermak, "In the Beginning: The First National Forest Recreation Plan," *Parks and Recreation* 9, 11 (November 1974). Carhart provided similar plans for the Pike and Uncompahgre National Forests in Colorado and the Superior National Forest in Minnesota.

28. Baldwin, 29–42.

29. Carhart, "Memorandum to Mr. Leopold, District 3," December 10, 1919, ACP, Box 1.

30. David Backes explores the tensions between Carhart's preservationist vi-

sions and his prodevelopment plan in "Wilderness Visions: Arthur Carhart's 1922 Proposal for the Quetico-Superior Wilderness," *Forest and Conservation History* 35 (July 1991): 128–37.

31. L. F. Kniepp to "District Forester," October 23, 1920, and A. S. Peck to "The Forester," December 3, 1920, NARA, RG 95, Entry 85, Box 68, Folder entitled "U-Recreation-General—1920–21." Although this is speculation, Carhart may well have ghostwritten Peck's report.

32. For responses, see: Frank Pooler (District 3) to "The Forester," December 17, 1920; R. H. Rutledge (District 4) to "The Forester," November 8, 1920; Paul Reddington and L. A. Barrett (District 5) to "The Forester," November 17, 1920; George Cecil (District 6) to "The Forester," November 10, 1920; and F. W. Reed (District 7) to "The Forester," December 3, 1920. All letters from NARA, RG 95, Entry 85, Box 68, "U-Recreation-General—1920–21."

33. L. F. Kniepp to "District Forester," March 28, 1921. For other participants in this discussion, see the following letters: George Cecil to "The Forester," December 27, 1920; L. F. Kniepp to Cecil, January 19, 1921; Fred Morrell to Kniepp, February 2, 1921; and Kniepp to Morrell, February 10, 1921. All are in NARA, RG 95, Entry 85, Box 68, "U-Recreation-General—1920–21."

34. See Tweed, 11; Baldwin, 62–71.

35. Tweed, 10–11.

36. These frustrations are clear in a series of letters between Carhart and Sherman that can be found in the Carhart Papers. Donald Baldwin reprinted them as an appendix to *The Quiet Revolution,* 273–83.

37. Tweed, 10–12; Baldwin, 125–28.

38. Meine, 175.

39. Warren, 90–93.

40. Susan Flader, *Thinking Like a Mountain* (1974), 17–18. Flader and others cite Leopold's 1923 essay, "Some Fundamentals of Conservation in the Southwest," as his ecological coming out. The essay is reprinted in Flader and Callicott, eds., *The River of the Mother of God and Other Essays by Aldo Leopold* (1991), 86–97. In the same volume, see Leopold's essay, "Grass, Brush, Timber, and Fire in Southern Arizona," 114–22.

41. For examples of these connections, see Leopold's inspection reports of the Tajique District, Manzano National Forest (1920), the Gila National Forest (1922), and the Manzano National Forest (1923), ALP, 10–11, Box 3.

42. Leopold, "The Popular Wilderness Fallacy: An Idea That Is Fast Exploding," *Outer's Book–Recreation* 58, 1 (January 1918): 43–46. Reprinted in Flader and Callicott, 49–52.

43. Carhart, "Memorandum to Mr. Leopold, District 3."

44. Leopold, "The Wilderness and Its Place in Forest Recreational Policy," *Journal of Forestry* 19, 7 (November 1921): 718–21; reprinted in Flader and Callicott, 78–81 (quote, 78–79). Italics are mine.

45. Ibid., 79. Italics are mine.

46. Ibid., 79–80.

47. Ibid., 81.

48. Leopold, "General Inspection Report on the Gila National Forest, May

21–June 27, 1922," and appended to this report, "Report on the Proposed Wilderness Area," October 2, 1922, ALP, 10–11, Box 3, Folder 9.

49. See Leopold, "A Plea for Wilderness Hunting Grounds," *Outdoor Life* 56, 5 (November 1925): 348–50.

50. "Report on the Proposed Wilderness Area," 1.

51. Leopold, "A Wilderness Area Program," no date, though archivists have written in "about 1922?" ALP, 10–6, Box 16, Folder 4.

52. Ibid., 2–3.

53. "Notes and Comment: The Organization of the Ecological Society of America, 1914–1919," *Ecology* 19, 1 (January 1938): 164–66 (quote, 166). Victor E. Shelford, ed., *Naturalist's Guide to the Americas* (1926). See also the ESA's brief report, *Preservation of Natural Conditions* (1922); James Pritchard, *Preserving Yellowstone's Natural Conditions: Science and the Perception of Nature* (1999), 43–44.

54. See Joel Hagen, *An Entangled Bank: The Origins of Ecosystem Ecology* (1992), 29.

55. "The Wilderness and Its Place in Forest Recreational Policy," in Flader and Callicott, 79.

56. Victor E. Shelford, "Preserves of Natural Conditions," *Transactions of the Illinois Academy of Science* 13 (1920): 57–58.

57. Shelford, in offering a "model description" for a preserved area, describes a 20-acre riverside flood plain and savanna forest. Shelford, "Preserves," 55. Pearson suggests areas of from 80 to 640 acres as ideal. The *Naturalist's Guide* affirms this point on scale.

58. G. A. Pearson, "Preservation of Natural Areas in the National Forests," *Ecology* 3, 4 (October 1922): 284–87.

59. On the ESA, the Ecologists' Union, and the evolution of The Nature Conservancy, see Robert Croker, *Pioneer Ecologist: The Life and Work of Victor Ernest Shelford, 1877–1968* (1991), 120–46; Sara Tjossem, "Preservation of Nature and Academic Respectability: Tensions in the Ecological Society of America, 1915–1979" (1994).

60. Leopold, "A Criticism of the Booster Spirit," in Flader and Callicott, 98–105 (quotes, 99, 103).

61. Ibid., 103.

62. See Warren, *Hunter's Game.*

63. Leopold, "The River of the Mother of God," in Flader and Callicott, 123–24.

64. Ibid., 124.

65. Ibid., 126–27.

66. Gopher Prairie was the fictional town in Sinclair Lewis's 1920 novel, *Main Street.* For other Leopold allusions to Babbitt, see "Game and Wildlife Conservation," "Conservation Ethic," "The Arboretum and the University," and "The Ecological Conscience," in Flader and Callicott.

67. The best example of this argument is Leopold's "The Pig in the Parlor" (1925), in Flader and Callicott, 133.

68. Leopold, "Conserving the Covered Wagon," *Sunset Magazine* 54, 3 (March 1925): 21, 56; reprinted in Flader and Callicott, 128–32.

69. Ibid., 128. On Leopold and Turner as neighbors, see Meine, 233. On the connections between frontier nostalgia and tourism in the Southwest, see Kerwin L. Klein, "Frontier Products: Tourism, Consumerism, and the Southwestern Public Lands, 1890–1990," *Pacific Historical Review* 62, 1 (February 1993): 38–71.

70. Leopold, "Conserving the Covered Wagon," 130.

71. Leopold, "A Plea for Wilderness Hunting Grounds."

72. Leopold, "Wilderness as a Form of Land Use," *Journal of Land and Public Utility Economics* 1, 4 (October 1925): 398–404; reprinted in Flader and Callicott, 134–42.

73. Leopold, "The Last Stand of the Wilderness," *American Forests and Forest Life* 31 (October 1925): 599–604. Tellingly, this article is subtitled "A Plea for Preserving a Few Primitive Forests, Untouched by Motor Cars and Tourist Camps, Where Those Who Enjoy Canoe or Pack Trips May Fulfill Their Dreams."

74. Howard R. Flint, "Wasted Wilderness," *American Forests and Forest Life* 32 (1926): 407–10, and "Comment" by Aldo Leopold, 410–11 (quotes, 408, 410).

75. Flint, 410.

76. Leopold, "Comment," 410–11.

77. Ibid., 411. Italics are Leopold's.

78. Gilligan, 100.

79. *Recreation Resources of Federal Lands* (1928), 86–103. (Leopold quoted, 86–88).

80. Gilligan, 91–92; W. B. Greeley, "Wilderness Recreation Areas," *Service Bulletin* (October 18, 1926), 1–3.

81. Greeley, 3.

82. Manly Thompson, "A Call from the Wilds," *Service Bulletin* 12, 20 (May 14, 1928): 2.

83. Robert Marshall, "The Wilderness as Minority Right," *Service Bulletin* (August 27, 1928), 5–6.

84. Leopold and Marshall corresponded as early as 1926. See Leopold to Marshall, September 29, 1926, RMP, Box 7, Folder 19.

85. Meine, 246–48. Leon Kniepp drafted the regulation in the wake of a survey that he had conducted of the remaining roadless areas in the national forests. For details on Kniepp's survey, see Gilligan, 125–26; Kniepp, "These Tame National Forests," *Service Bulletin* 11, 10 (March 7, 1927): 2–3. Gilligan includes a portion of Regulation L-20 as Appendix A of his dissertation (quote, A-1).

86. Gilligan, 126. According to Gilligan, in 1930 the Forest Service further divided "research reserves" into "experimental forests," "experimental ranges," and "natural areas." The first two acted as natural standards against which foresters could measure conditions, while "natural areas" were reserved for ecological research.

87. L. F. Kniepp, "What Shall We Call Protected Recreation Areas in the National Forests?" *American Planning and Civic Annual* 1 (1929): 35. Kniepp's change in terms bore an inverse relation to terms being used within the Park Service and the National Parks Association. Robert Sterling Yard, for instance, used "primitive" to refer to pristine areas and "wilderness" to refer to roadless areas.

88. Meine, 254–56; Thomas Dunlap, *Saving America's Wildlife* (1988), 65–70; Flader, *Thinking Like a Mountain,* 84–87.

89. Letter from Leopold to Pooler quoted in Meine, 270. For some of Leopold's thoughts on these problems, see Leopold, "Game Management in the National Forests," *American Forests and Forest Life* 36 (July 1930): 412, 414.

90. Meine, 271.

91. Leopold, "Foreword" (originally written as a foreword to *Great Possessions,* an earlier title for what became *A Sand County Almanac;* it did not appear in the final version), reprinted in J. Baird Callicott, ed., *Companion to A Sand County Almanac: Interpretive and Critical Essays* (1987), 277–88 (quote, 284–85).

92. Meine, 307.

93. Research done by Leopold's own students helped confirm that habitat was the vital element. See Dunlap, *Saving America's Wildlife,* chapter 5.

94. On Leopold's concerns about the intensification of agriculture, see Curt Meine, "The Farmer as Conservationist: Leopold on Agriculture," in Tanner, ed., *Aldo Leopold: The Man and His Legacy,* 41–42. On the intensification and mechanization of agriculture during the period, see Donald Worster, *Dust Bowl: The Southern Plains in the 1930s* (1979), 87–94.

95. Aldo Leopold, *Game Management* (1933), 3.

96. Ibid., 394.

97. The phrase "controlled wild culture" is from Leopold, "The Conservation Ethic," Flader and Callicott, 190–91.

98. *Game Management,* 134.

99. Leopold, "The Conservation Ethic," *Journal of Forestry* 31, 6 (October 1933): 634–43; reprinted in Flader and Callicott, 181–92 (quote, 188).

100. For Leopold's efforts to define a democratic game policy, see "Report to the American Game Conference on an American Game Policy" (1930), "Game Methods: The American Way" (1931), and "Game and Wildlife Conservation" (1932), all in Flader and Callicott, 150–68.

101. See Meine, 304–5.

102. Stephen Fox makes a similar point about *A Sand County Almanac* in *The American Conservation Movement,* 248–49.

103. "The Conservation Ethic," 184.

104. Leopold, *A Sand County Almanac,* 214.

105. Leopold, "Conservation Ethic," 187–88. Italics are Leopold's.

106. Leopold, "Conservation Economics," in Flader and Callicott, 196.

107. Ibid. Curt Meine suggests that Leopold's strong feelings about heavy reliance on public ownership may have come in reaction to arguments made by people like Bob Marshall, whose *The People's Forests* posited the need for public ownership of all forest lands. See Meine, 322.

108. Leopold, "Conservation Economics," 191.

109. Leopold, "Land Pathology," in Flader and Callicott, 212–17. "Conservation Economics" discusses similar themes.

110. Leopold, "Conservation Economics," in Flader and Callicott, 193–202. See also Meine, 321–24.

111. On Forest Service activities during the New Deal, see Tweed, *Recreation Site Planning*, 16–25.

112. Aldo Leopold, "Why the Wilderness Society?" *The Living Wilderness* 1, 1 (September 1935): 6.

113. See Flader, 28–30; Meine, 351.

114. Leopold, "Wilderness," in Flader and Callicott, 227. See also his articles, "*Naturschutz* in Germany," *Bird-Lore* 38, 2 (March–April 1936): 102–11; "Deer and *Dauerwald* in Germany: I. History," *Journal of Forestry* 34, 4 (April 1936): 366–75; "Deer and *Dauerwald* in Germany: II. Ecology and Policy," *Journal of Forestry* 34, 5 (May 1936): 460–66. On Leopold's trip, see Meine, 351–60; Flader, *Thinking Like a Mountain*, 139–44.

115. Susan Flader suggested that Leopold's participation in founding the Society "marked a reorientation in his thinking from a historical and recreational to a predominantly ecological and ethical justification for wilderness." Flader, *Thinking Like a Mountain*, 29. See also Worster, *Nature's Economy*, 285.

116. See, for instance, Meine, 343.

117. Leopold to Marshall, October 29, 1934, TWSP, Box 11, Folder 20. The split with ecologists lingered on. In a 1940 letter to Robert Cushman Murphy of the American Museum of Natural History (June 8, 1940, ALP, 10–2, Box 9, Folder 1), Leopold wrote: "the [Wilderness] Society as now constituted is interested mainly in wilderness *recreation*. Another group, the Ecological Society, is interested mainly in wilderness *study*."

118. Leopold, "Why the Wilderness Society?" 6.

119. That spring brought tremendous dust storms to the southern plains, confirming for Leopold the pathology that had crept into the human-land relationship. Leopold also came in contact with the ideas of ecologists like Paul Sears, whose *Deserts on the March* appeared in 1935. Meine, 348–49; Worster, *Dust Bowl*, 198–209.

120. Leopold, "Conservationist in Mexico," in Flader and Callicott, 239.

121. Leopold, "Foreword," in Callicott, ed., *Companion to A Sand County Almanac*, 285–86.

122. See Leopold, "Wilderness as Land Laboratory," *The Living Wilderness* 6 (July 1941). Part of the problem with questions about whether Leopold was thinking ecologically is the term "ecological." Do we mean to imply that Leopold was thinking more like an ecological scientist? Or that he was anticipating an ecological consciousness that emerged with the postwar environmental movement, a consciousness that valued biocentrism, stability, and interconnectedness? Susan Flader noted the complexity of the question: "Leopold may be said to have been thinking ecologically, in a functionalist or holistic sense before ecological science had evolved a conceptual framework capable of supporting such thought." *Thinking Like a Mountain*, 17. In a sense, Flader argued that Leopold was thinking ecologically before ecologists were, which is a strange thing to say. What she meant is that Leopold was thinking about wilderness holistically and functionally before most professional ecologists were. The term "ecological" has served, in describing the evolution of Leopold's thought, as a synonym for these more specific changes. My point is that we need to say what we mean.

123. Leopold, "Conservation Esthetic," *Bird-Lore* 40, 2 (March–April 1938): 101–9. Reprinted in *A Sand County Almanac* (1949), 165–77 (quote, 165–66).

124. Ibid., 176–77.

4. Advertising the Wild: Robert Sterling Yard

1. Biographical information is from *The Living Wilderness* 10, 14–15 (December 1945): 1–17; John C. Miles, *Guardians of the Parks: A History of the National Parks and Conservation Association* (1995); *The National Cyclopaedia of American Biography*, 43 (1961): 132.

2. Shankland, *Steve Mather of the National Parks* (1970), 7. See also Horace Albright and Marian Albright Schenck, *Creating the National Park Service: The Missing Years* (1999).

3. Donald C. Swain, "The Passage of the National Park Service Act of 1916," *Wisconsin Magazine of History* (Autumn 1966), 7; Yard, "Historical Basis of National Park Standards," *National Parks Bulletin* 57 (November 1929); Shaffer, "See America First."

4. Miles, 13; Swain, *Wilderness Defender: Horace M. Albright and Conservation* (1970), 42–43; Yard, "Historical Basis of National Park Standards."

5. Yard, "Historical Basis of National Park Standards."

6. Shankland, 95.

7. See Christopher Wilson, "The Rhetoric of Consumption: Mass Market Magazines and the Demise of the Gentle Reader," in Fox and Lears, *The Culture of Consumption*, 39–64.

8. See Swain, *Wilderness Defender*, 46–52.

9. Jackson Lears, *Fables of Abundance: A Cultural History of Advertising in America* (1994), 219. On interwar advertising, see Roland Marchand, *Advertising the American Dream: Making Way for Modernity, 1920–1940* (1985).

10. David Kennedy, *Over Here: The First World War and American Society* (1980), 90. For a discussion of Lippmann and the concept of public opinion, see Warren Susman, "Culture and Civilization: The Nineteen-Twenties," in *Culture as History: The Transformation of American Society in the Twentieth Century* (1984), 105–21.

11. The quote is from Lears, *Fables of Abundance*, 201.

12. U.S. Department of the Interior, *National Parks Portfolio* (1916). See also Marguerite S. Shaffer, "Negotiating National Identity: Western Tourism and 'See America First,'" in Hal Rothman, ed., *Reopening the American West* (1998), 139–42.

13. On these parks see John Ise, *Our National Park Policy* (1961), 136–42.

14. U.S. Department of the Interior, *Glimpses of Our National Parks* (1916). On the funding arrangement, see Runte, *National Parks*, 109–10; Shaffer, "See America First," 186.

15. Swain, "The Passage of the National Park Service Act of 1916," 4–17; Ise, *Our National Park Policy*, 185–93 (quote, 191–92); Sellars, 28–46.

16. Swain, *Wilderness Defender*, 61–67; Horace Albright and Robert Cahn, *The Birth of the National Park Service: The Founding Years, 1913–1933* (1985), 54; Miles, 15–16.

17. Information for this paragraph was taken from Swain, *Wilderness Defender,* 66–67; Miles, 14–16.

18. Miles, 16–18; Albright and Schenck, 211–12. A number of scholars, Richard Sellars most recently, have pointed to the importance of a policy statement that Albright drafted, and that Franklin Lane issued in 1918. Known as the "Lane Letter," the statement provided a strong prioritization of tourist development. See Sellars, 57. The letter is reprinted in Albright and Cahn, 69–73.

19. Miles, 16–21.

20. Yard recounted the first meeting of the NPA in the pages of its publication, *Bulletin* 1 (June 6, 1919). Information on NPA's founding is from Miles, 16–21. In a reminiscence of this episode that has only recently been published, Albright suggested that relations between Yard and Mather were quite tense during this transition, and that Mather may have been less supportive of the NPA than his public statements suggest. See Albright and Schenck, 297–98.

21. *Report of the Director* (1924), 82.

22. As Marguerite Shaffer and Anne Hyde have shown, national parks and national identity were tightly intertwined during this period. Shaffer, "See America First"; Anne Farrar Hyde, *An American Vision: Far Western Landscape and National Culture, 1820–1920* (1990).

23. Mather, *Report of the Director of the National Park Service* (1920), 14.

24. See *Use of Automobiles in National Parks,* 62nd Cong., 2d sess., March 13, 1912, S. Doc. 433.

25. Ibid.

26. Shankland, 66; Stephen Mather, "Progress in the Development of the National Parks" (1916), 10.

27. *Report of the Director* (1916), 83.

28. Runte, *National Parks,* 156.

29. Mather, "Progress in the Development of the National Parks," 10; *Report of the Director* (1917), 39–40. The latter report was written by Albright as Acting Director.

30. Belasco, 3–4.

31. Mather, *Report of the Director* (1919), 16–17.

32. Mather, *Report of the Director* (1920), 37–39.

33. Mather, *Report of the Director* (1919), 17–18.

34. Mather, *Report of the Director* (1919), 20–21.

35. Mather, *Report of the Director* (1919), 17–23.

36. Mather, *Report of the Director* (1919), 25–28; Ethan Carr, *Wilderness By Design: Landscape Architecture and the National Park Service* (1998), 88–89. For a detailed look at the Punchard era, see Linda Flint McClelland, *Building the National Parks: Historic Landscape Design and Construction* (1998), 136–57.

37. Carr, 1.

38. Albright, *Report of the Director* (1917), 10–13.

39. See Shaffer, "See America First," for a thorough discussion of these groups.

40. Yard, "Historical Basis of National Park Standards."

41. *Proceedings of the National Park Conference, Berkeley, California on March 13, 14, 15, 1915* (1915), 151.

42. See Roy Rosenzweig and Elizabeth Blackmar, *The Park and the People: A History of Central Park* (1992).

43. *Bulletin* (NPA) 1 (June 6, 1919): 8.

44. Yard's concept was similar to what Alfred Runte has referred to as "monumentalism." See Runte, *National Parks*. Yard also voiced what Marjorie Hope Nicolson called the "aesthetics of the infinite." Nicolson traced the development of this aesthetic in her classic study, *Mountain Gloom, Mountain Glory: The Development of an Aesthetics of the Infinite* (1959).

45. Yard, *The Book of the National Parks* (1919), 3. Also see Yard, "The People and the National Parks," 547–53, 583. It is important to note that it was a short journey from the standards with which Yard worked—particularly the biological standards—and the contemporary intellectual interest in eugenics. Susan Schrepfer has shown that many of the founders of the Save the Redwoods League and some important figures in early national parks educational efforts were sympathetic with the eugenics movement and were quick to see the big trees as a sort of master race. See Schrepfer, *The Fight to Save the Redwoods: A History of Environmental Reform* (1983), 43–45. Stephen Fox cites rhetoric from Yard's days with *Century Magazine* that suggests an undercurrent of racism in his thought, but I found little evidence of this in my extensive review of his writings after 1915. See Fox, *The American Conservation Movement,* 348.

46. Yard, "The People and the National Parks," *The Survey* 48; 13 (August 1, 1922): 547.

47. Ibid., 550.

48. See *NPA Bulletins* for 1920 and 1921; *Reports of the Director* (1920, 1921); Donald Swain, *Federal Conservation Policy, 1921–1933* (1963), 123–43.

49. Miles, 43–48. See also the NPA's *Bulletins* 22–24 (November 8, 1921; December 14, 1921; January 30, 1922).

50. For visitation figures, see *Report of the Director* (1924), Appendix B, 82–83.

51. See Yard, "The People and the National Parks," 583; Yard, "Economic Aspects of Our National Parks Policy," *Scientific Monthly* 16 (April 1923): 380–88. Yard also noted the economic argument in *The Book of the National Parks,* 19. See also Cammerer, "Selling the National Parks," *Western Advertising* (October 1920), 20–22. In the midst of the passage of the National Park Service Act, Yard published a piece that made an early case for the economic importance of scenic tourism. Yard, "Making a Business of Scenery," *The Nation's Business* 4 (June 1916): 10–11.

52. Yard, "Statement of the National Parks Association before the Public Lands Committee of the House on the Slemp Bill to create the Appalachia National Park," January 3, 1923; reprinted in the NPA's *Bulletin* 33 (March 8, 1923).

53. Ibid.

54. On the spate of national park bills, see Yard, "Thirteen New National Parks Asked of Congress," *National Parks Bulletin* 37 (January 21, 1924).

55. Yard, "Statement of the National Parks Association before the Public Lands Committee."

56. Ibid.

57. "Difference Between Publicity and Propaganda," no author (probably Yard), *National Parks Bulletin* 49 (March 1926).

58. Yard, "Statement of the National Parks Association before the Public Lands Committee."

59. Yard, "Scenic Resources of the United States," *Proceedings of the National Conference on Outdoor Recreation* (1924), 45–49. Miles, 65–66. A few years later the NPA and AFA, presumably on the strength of their collaboration, gave serious consideration to a merger.

60. Shankland, 221–23.

61. See John Jakle, *The Tourist* (1985), 146–48. Jakle called this the "impulse towards constant driving."

62. See Rothman, *Devil's Bargains,* 143–67.

63. *Report of the Director* (1922). Existing park roads, designed for horse and stage traffic and built by the Army in the 1800s, were proving inadequate in the face of increased automobile traffic, and as late as 1923 the Park Service was working with annual road appropriations of only about $200,000.

64. Mather, *Report of the Director* (1924), 14.

65. This letter is quoted in Sellars, 63. Italics are mine.

66. Yard to Hubert Work, October 13, 1925, reprinted in *National Parks Bulletin* 45 (October 24, 1925): 1–2.

67. See John Campbell Merriam, *The Garment of God: Influence of Nature on Human Experience* (1943); Merriam, *Published Papers and Addresses of John Campbell Merriam* (1938).

68. Yard, "Five Minutes with the News," *National Parks Bulletin* 49 (March, 1926). In the same edition of the *National Parks Bulletin,* Yard reprinted Merriam's address to the second meeting of the NCOR.

69. Yard, "Five Minutes with the News."

70. Ibid.

71. Alfred Runte, *Yosemite: The Embattled Wilderness* (1990), 116–17.

72. See Yard's "Memo on Adult Educational Projects in Connection with National Parks," which was attached to a letter to Merriam, March 17, 1926, JCMP, Box 186.

73. Yard to Merriam, January 12 (no year, but probably 1926 or 1927), JCMP, Box 186.

74. Yard, "Memo on Adult Educational Projects in Connection with National Parks."

75. Miles, 67.

76. "Extract from the Annual Report of the Secretary of the Interior, Fiscal Year 1928, Relating to the National Park Service" (1928), 3–4.

77. The Division was renamed the Branch for Research and Education. See Shankland, 262.

78. See *Report of the Director* for 1929 and 1930; Shankland, 257–63.

79. See Swain, *Wilderness Defender;* Albright and Cahn; Joseph W. Ernst, ed., *Worthwhile Places: Correspondence of John D. Rockefeller, Jr. and Horace Albright* (1991).

80. See *Report of the Director* throughout the 1920s; Ralph H. Lewis, *Museum Curatorship in the National Park Service, 1904–1982* (1993).

81. On his knowledge of this policy and its emphases, see Yard to Merriam, February 24, 1926, JCMP, Box 186.

82. While I do not have direct evidence of this Leopold-Yard correspondence, Leopold refers to receiving a letter from Yard in a letter to Arthur Ringland. See Leopold to Ringland, December 29, 1925, RMP, Box 7, Folder 19.

83. Yard, "But We Must Hold Our Heritage," *National Parks Bulletin* 47 (January 1926): 2.

84. Ibid.

85. Yard to Grinnell, September 18, 1926, JCMP, Box 186.

86. Yard, "Needless Road Project Endangers Yosemite: Only Way to Save a Few Wildernesses In as well as Out of our National Parks is to Keep Motor Roads Out of Them," *National Parks Bulletin* 52 (February 1927): 16.

87. Ibid.

88. Ibid.

89. Yard, "The Motor Tourist and the National Parks, II," *National Parks Bulletin* 53 (July 1927): 17–19.

90. Horace M. Albright, "The Everlasting Wilderness," *Saturday Evening Post* 201 (September 29, 1928): 28, 63, 66, 68.

91. On the perceptual experience imparted by modern roads, see Christopher Tunnard and Boris Pushkarev, *Man-Made America: Chaos or Control?* (1963), 157–275.

92. Alfred Runte makes a similar argument. See Runte, *National Parks,* 159.

93. Yard, *Our Federal Lands: A Romance of American Development* (1928), 13.

94. *Recreation Resources on Federal Lands.* The discussion of wilderness policy came primarily in the context of the section on national forests, and Aldo Leopold's ideas had a clear influence on the section.

95. A good example of Yard's thought on these threats is his draft of an article entitled "National Parks in New Peril," which he attached to a letter to Merriam, January 11, 1927, JCMP, Box 186.

96. Yard, "Serious Attack on National Forest and National Park Policies," February 2, 1928, attached to a letter to Merriam, February 4, 1928, JCMP, Box 187.

97. See, for instance, Yard to George Shiras, January 13, 1926, JCMP, Box 186.

98. In a 1930 letter, Yard wrote of Mammoth, and of rumors of a Carlsbad Caverns National Park, that, "I can't get used to calling a cave a park, for it's exactly what it isn't." Yard to Merriam, March 22, 1930, JCMP, Box 187.

99. Yard, "Politics in Our National Parks: Shall Standards Be Lowered to Serve Political Expediency?" *American Forests* 32, 392 (August 1926): 485. Mammoth Cave had a long history as a tourist attraction. See Sears, 31–48.

100. Yard, "Important Information for Use of Our Allies," March 1927, attached to letter to Mark Squires, March 14, 1927, JCMP, Box 186.

101. Runte, *National Parks,* 117. For good discussions of Rockefeller's role in purchasing land for national parks during this era, see Fox, *The American Conservation Movement,* 219–23.

102. Yard to Ray Lyman Wilbur, February 25, 1930, attached to copy of letter

Yard wrote to Merriam, February 25, 1930, JCMP, Box 187. On Merriam's suggestion that the Everglades be kept as a biological preserve, see Schrepfer, 59. On the proposal, see Runte, *National Parks,* 128–31.

103. Yard to Merriam, October 24, 1930, JCMP, Box 187.

104. Yard to Merriam, December 12, 1930, JCMP, Box 187.

105. For an expression of these concerns, see Yard to "Mr. Colton," attached to letter to Merriam, received January 5, 1931, JCMP, Box 187.

106. Yard to Wallace Atwood, January 22, 1931, JCMP, Box 187.

107. Ward to Yard, March 14, 1931, JCMP, Box 187.

108. Yard to Ward, March 20, 1931, JCMP, Box 187.

109. Yard to Merriam, August 5, 1932, JCMP, Box 188.

110. Frederick Law Olmsted, Jr., and William P. Wharton, "The Florida Everglades: Where the Mangrove Forests Meet the Storm Waves of a Thousand Miles of Water," *American Forests* 38 (March 1932): 142–47, 192.

111. Alfred Runte refers to the Everglades as the first wilderness national park. See Runte, *National Parks,* 128–36.

112. Yard, "Notes on the History of the Primitive in National Parks," manuscript, 1932, JCMP, Box 188.

113. Yard quoted this section from the National Park Service Act of 1916 in "Notes on the History of the Primitive."

114. See *Report of the Director* (1929, 1930).

115. *Annual Report of the Secretary of the Interior.* "National Parks" (1935), 180.

116. "Speech by Roosevelt, Two Medicine Chalet, Glacier National Park, August 5, 1934," reprinted in Herbert B. Nixon, ed., *Franklin D. Roosevelt and Conservation, 1911–1945,* 1 (1957): 321–24.

117. On the inclusion of these new areas, see Donald Swain, "The National Park Service and the New Deal, 1933–1940," *Pacific Historical Review* 41, 3 (August 1972): 313–14.

118. Miles, 100–104.

119. Yard to Merriam, August 15, 1932, JCMP, Box 188.

120. Yard to "My dear Harry," February 2, 1935, TWSP, Box 11, Folder 3.

121. Yard to Merriam, October 2, 1932, JCMP, Box 188. Marshall, "The Problem of the Wilderness."

122. Yard to Merriam, March 29, 1934, JCMP, Box 188. Letter is also cited in Miles, 103.

123. Miles, 93–95; Swain, *Wilderness Defender,* 247–53 (quote, 247).

5. Wilderness as Regional Plan: Benton MacKaye

1. MacKaye was not alone among interwar thinkers in linking technological modernism with environmental reform. See Richard White, *The Organic Machine* (1995); James W. Carey and John J. Quirk, "The Mythos of the Electronic Revolution," *American Scholar* 39 (Spring 1970): 219–41; David Nye, *Electrifying America: Social Meanings of a New Technology* (1990).

2. Robert Gottlieb, one of the few historians of American environmental thought who has looked at MacKaye, makes this point in his book, *Forcing the*

Spring: The Transformation of the American Environmental Movement (1993), 71–75.

3. There are several good sources on MacKaye's life. See Paul T. Bryant, "The Quality of the Day: The Achievement of Benton MacKaye" (1965); John R. Ross, "Benton MacKaye: The Appalachian Trail," in Donald A. Krueckeberg, ed., *The American Planner: Biographies and Reflections* (1983), 196–207; and John L. Thomas, "Lewis Mumford, Benton MacKaye, and the Regional Vision," in Thomas P. and Agatha C. Hughes, eds., *Lewis Mumford, Public Intellectual* (1990), 66–99. On Steele MacKaye, see his son Percy's memoir, *Epoch: The Life of Steele MacKaye* (1927).

4. "Work Record of Benton MacKaye," undated manuscript, BMKP, Box 185; Bryant, 82–84.

5. James MacKaye, *The Economy of Happiness* (1906). James later wrote *Americanized Socialism: A Yankee View of Capitalism* (1918). On the importance of James's influence, see Bryant, "The Quality of the Day," 73–77. Evidence of Lippmann's influence is found in a letter from him to MacKaye, December 3, 1910, BMKP, Box 165. On Lippmann's life and thought, see Ronald Steel, *Walter Lippmann and the American Century* (1980).

6. See MacKaye, "The Forest Cover on the Watersheds Examined by the Geological Survey in the White Mountains, New Hampshire," in B. MacKaye, M. O. Leighton, and A. C. Spencer, *The Relation of Forests to Stream Flow* (1913), BMKP, Box 181; Robert McCullough, *The Landscape of Community: A History of Communal Forests in New England* (1995), 209–10. On the stream-flow controversy, see Harold K. Steen, *The U.S. Forest Service* (1976), 122–29; Donald Pisani, "Forests and Reclamation, 1891–1911," in Pisani, *Water, Land, and Law in the West: The Limits of Public Policy, 1850–1920* (1996), 141–58; Gordon B. Dodds, "The Stream-Flow Controversy: A Conservation Turning Point," *Journal of American History* 56, 1 (June 1969): 59–69.

7. Bryant, "The Quality of the Day," 90–93; "Work Record of Benton MacKaye, 1905–1943."

8. On the region's slash fires see Stephen Pyne, *Fire in America: A Cultural History of Wildland and Rural Fire* (1982), 198–218. On agricultural settlement in the "cutover," see Robert Gough, *Farming the Cutover: A Social History of Northern Wisconsin, 1900–1940* (1997).

9. MacKaye, "Settling the Timber Lands," June 1916. MacKaye also completed a much larger report titled "Colonization of Timber Lands" (1917) that focused on Minnesota and Wisconsin. BMKP, Box 181.

10. See *Report of the Secretary of Labor* (1915), 43–45.

11. Ross, "Benton MacKaye: The Appalachian Trail," 197; John L. Thomas, *Alternative America: Henry George, Edward Bellamy, Henry Demarest Lloyd, and the Adversary Tradition* (1983), 225–32.

12. MacKaye to Louis Post, December 17, 1915, BMKP, Box 165.

13. MacKaye, "Notes on the IWW/Forest Camps—November 1916," BMKP, Box 181; Carlos Schwantes, *The Pacific Northwest: An Interpretive History* (1989), 260–65; Richard White, *"It's Your Misfortune and None of My Own": A New History of the American West* (1991), 351. See also Norman H. Clark, *Mill Town: A Social*

History of Everett, Washington (1970); Robert L. Tyler, *Rebels of the Woods: The I.W.W. and the Pacific Northwest* (1967); and William J. Williams, "Bloody Sunday Revisited," *Pacific Northwest Quarterly* 71 (1980): 50–62. A year later, in Ludlow, Colorado, a group of striking coal miners and their families were attacked in the tent colony they had erected after being evicted from company housing for striking. The attack left 39 dead. That same year also, Wobbly organizer Frank Little was lynched in Butte, Montana. See White, 349, 351.

14. MacKaye, "Recreational Possibilities of Public Forests," *Journal of the New York State Forestry Association* 3 (October 1916): 4–10, 29–30.

15. See MacKaye, "Memorandum for Mr. Zon," December 21, 1917, and Raphael Zon, "Memorandum for Miss Judson," December 31, 1917, BMKP, Box 165.

16. Bill G. Reid, "Franklin K. Lane's Idea for Veterans' Colonization, 1918–1921," *Pacific Historical Review* 33, 4 (November 1964): 447–61.

17. MacKaye, "Some Social Aspects of Forest Management," *Journal of Forestry,* 16, 2 (February 1918): 210–14 (quotes, 210, 211).

18. Ibid., 212–13.

19. MacKaye, "The First Soldier Colony—Kapuskasing, Canada," *The Public* 22, 1122 (November 15, 1919): 1066–68. Reprinted in Paul T. Bryant, ed., *From Geography to Geotechnics* (1968), 115–20.

20. MacKaye, *Employment and Natural Resources: Possibilities for Making New Opportunities for Employment through the Settlement and Development of Agricultural and Forest Lands and Other Resources* (1919), 10–12.

21. On Powell's critique, see his *Report on the Lands of the Arid Region* (1878); Donald Worster, "The Legacy of John Wesley Powell," in *An Unsettled Country: Changing Landscapes of the American West* (1994), 1–30. In 1891, when MacKaye was a boy, he saw Powell speak at the National Geographic Society. Later, his brother James worked with Powell at the U.S. Geological Survey, adding a personal connection to the intellectual influence. See Bryant, 32, 42.

22. *Employment and Natural Resources,* 18.

23. See Hays, *Conservation and the Gospel of Efficiency* (1959); Robert L. Dorman, *Revolt of the Provinces: The Regionalist Movement in America, 1920–1945* (1993), 316. On the distinction between social and business efficiency, see Robert Westbrook, "Tribune of the Technostructure: The Popular Economics of Stuart Chase,"*American Quarterly* 32, 4 (Fall 1980): 387–408.

24. Two bills came out of MacKaye's efforts: the National Colonization Bill (H.R. 11329, 64th Cong., 1st sess., 1916) sponsored by Rep. Robert Crosser of Ohio, and the Public Construction Bill (H.R. 15672, 65th Cong., 2d sess., 1919) sponsored by Rep. M. Clyde Kelly of Pennsylvania. Hearings were held on each bill, but neither passed. See MacKaye, "From Homesteads to Valley Authorities," *The Survey* 86, 11 (November 1950): 496–98; reprinted in Bryant, *From Geography to Geotechnics,* 33–43.

25. MacKaye to Rep. M. Clyde Kelly, October 2, 1919, BMKP, Box 165. See also MacKaye, "Lessons of Alaska," *The Public* 22 (August 30, 1919): 930–32, and "The First Soldier Colony—Kapuskasing, Canada," *The Public* 22 (November 15, 1919): 1066–68.

26. Bryant, 90–93. I am indebted to Larry Anderson for much of my understanding of MacKaye during this period. Information in this section is gleaned from his essay, "'A Retreat from Profit': MacKaye's Path to the Appalachian Trail, 1919–1921," presented at the symposium entitled "Benton MacKaye and the Appalachian Trail," University at Albany, State University of New York, November 22, 1996.

27. MacKaye, Herbert Brougham, and Charles Harris Whitaker to Ludwig C.A.K. Martens, March 23, 1920. Those on the list included Benton and Betty MacKaye, Brougham, Whitaker, Stuart Chase, C. H. Chase, Paul Wallace Hanna, Jacob Kotinsky, Aaron Kravitz, Leland Olds, and Horace Warner Truesdell. See also Martens to MacKaye, April 4, 1920, BMKP, Box 165. On Martens and the Palmer Raids, see David M. Kennedy, *Over Here: The First World War and American Society* (1980), 287–92. One of the best treatments of the attractiveness of the Soviet experiment to Americans during this period is Richard Pells, *Radical Visions and American Dreams: Culture and Social Thought in the Depression Years* (1998 [1973]).

28. Kennedy, 25–26, 75–78, 289–90; Berger to MacKaye, November 22, 1922, BMKP, Box 166; Sally Miller, *Victor Berger and the Promise of Constructive Socialism, 1910–1920* (1973).

29. See folders of clippings from the *Leader*, BMKP, Box 182. On MacKaye's departure, see MacKaye to Elizabeth Thomas, December 12, 1920, BMKP, Box 165.

30. Veblen got his idea from a movement among engineers in the late teens, led by Morris Cooke and Henry Gantt, to reject their profession's subservience to business and chart an independent course for the profession. Most scholars have concluded that Veblen was quite mistaken to see revolutionary potential among a group that turned out to be conservative. See William Akin, *Technocracy and the American Dream: The Technocratic Movement, 1900–1940* (1977), 14–26; Edwin Layton, *The Revolt of the Engineers: Social Responsibility and the Engineering Profession* (1971); John M. Jordan, *Machine-Age Ideology: Social Engineering and American Liberalism, 1911–1939* (1994).

31. MacKaye recounted this episode in letters he wrote telling friends of Betty's death and responding to letters of condolence. See, for instance, MacKaye to Anna Brunnacker, and MacKaye to Mary Clayton, both dated May 7, 1921, BMKP, Box 165.

32. MacKaye, "From Homesteads to Valley Authorities," in Bryant, *From Geography to Geotechnics*, 36.

33. Thomas, "Lewis Mumford, Benton MacKaye, and the Regional Vision," 72.

34. MacKaye, "Memorandum on Regional Planning" (1921), BMKP, Box 182.

35. Ibid. Anderson, "'A Retreat from Profit,'" 14–15.

36. MacKaye, "Regional Planning and Social Readjustment," BMKP, Box 182.

37. Ibid. Ronald Foresta has argued that MacKaye's social vision for the AT faded because he and others failed to anticipate the trail's appropriation by the middle class and its transformation into a solely recreational facility. But MacKaye clearly understood this danger. See Foresta, "Transformation of the Appalachian Trail," *Geographical Review* 77, 1 (January 1987): 76–85.

38. MacKaye, "An Appalachian Trail: A Project in Regional Planning," *Journal of the American Institute of Architects* 9 (October 1921): 325.

39. Ibid., 326.

40. See Dorman, *Revolt of the Provinces.*

41. "An Appalachian Trail," 327. Italics are mine.

42. Thomas, "Lewis Mumford, Benton MacKaye, and the Regional Vision," 67–69.

43. See Aron, *Working at Play* (1999).

44. MacKaye, "An Appalachian Trail," 327. Italics are MacKaye's.

45. Ibid., 327.

46. Ibid., 329.

47. Ibid., 327–29.

48. For treatments that pick up MacKaye at this point in his career, see Nash, *Wilderness and the American Mind,* 189–90; Fox, *The American Conservation Movement,* 210; Gottlieb, *Forcing the Spring;* Fox, "'We Want No Straddlers.'"

49. Clarence Stein, "Introduction," to reprints of MacKaye, "An Appalachian Trail," BMKP, Box 183.

50. MacKaye to Stein, November 9, 1921, BMKP, Box 165.

51. MacKaye to Stein, December 11, 1921, BMKP, Box 165.

52. MacKaye, "Progress Toward the Appalachian Trail," *Appalachia* 15, 3 (December 1922): 244–52.

53. Ibid., 244.

54. Kermit C. Parsons, "Collaborative Genius: The Regional Planning Association of America," *Journal of the American Planning Association* 60, 4 (Autumn 1994): 462–82; Lubove, *Community Planning in the 1920s* (1963).

55. Evidence of this financial relationship is gleaned from the MacKaye-Stein correspondence, BMKP, Box 166.

56. See, for instance, Mumford to MacKaye, December 18, 1924, BMKP, Box 166.

57. There are several good historical treatments of the group and its work. See Thomas, "Lewis Mumford, Benton MacKaye, and the Regional Vision"; Spann, *Designing Modern America* (1996); Lubove; Parsons; Peter Hall, *Cities of Tomorrow* (1988), particularly his chapter "The City in the Region," 136–73; and Daniel Schaffer, *Garden Cities for America: The Radburn Experience* (1982). Carl Sussman brought together many of the most important writings of the group in his collection, *Planning the Fourth Migration: The Neglected Vision of the Regional Planning Association of America* (1976). In particular, see Sussman's "Introduction."

58. Sussman, "Introduction," 7–9; Lubove, 17–29.

59. On bioregionalism, see Kirkpatrick Sale, *Dwellers in the Land: The Bioregional Vision* (1985).

60. Stuart Chase, *The Tragedy of Waste* (1927); Robert Westbrook, "Tribune of the Technostructure."

61. On the Progressive influence, see Westbrook, "Tribune of the Technostructure"; Spann.

62. Before MacKaye, Chase, and Whitaker had made their Soviet entreaties in 1920, they had paid careful attention to the economic coordination that the

federal government had achieved during the war. Only when wartime planning collapsed did they contact Martens.

63. On such transatlantic borrowings in urban planning, see Daniel Rodgers, *Atlantic Crossings: Social Politics in a Progressive Age* (1998), 160–208.

64. Sussman, "Introduction," 11. For Howard's vision, see his *Garden Cities of Tomorrow* (1902). For a thorough discussion of the garden city ideal, see Schaffer, *Garden Cities for America*, 15–22.

65. The classic expression of Geddes's ideas is his *Cities in Evolution* (1915). On the promise of electricity, see Carey and Quirk; Nye. Geddes's imprint was particularly strong on MacKaye, who later adopted another term from the Geddes lexicon—"geotechnics"—to describe his regional planning methodology.

66. On the regionalist sentiment of the era, see Dorman. On the Nashville Agrarians, see Twelve Southerners, *I'll Take My Stand: The South and the Agrarian Tradition* (1930). For Jane Jacobs's critique of modernist planning, see *The Death and Life of Great American Cities* (1961).

67. Both Mumford and MacKaye were cautious about transforming traditions, rooted in participation and a close connection with place, into facades and fetishes. When, for instance, MacKaye began referring to his ideal middle landscape of Shirley Center as the "Colonial environment," Mumford jumped on him: "What we look for, as an alternative to metropolitanism, is not a revivalism of the old; it is a fresh growth of something new." "[T]he indigenous," MacKaye responded, "is that which is *permanent* rather than that which is *past*." See Mumford to MacKaye, July 25, 1927, and MacKaye to Mumford, July 30, 1927, BMKP, Box 166.

68. Mumford, "Introduction," to *The New Exploration* (1962), xv. For the issue itself, see *Survey Graphic* 7 (May 1925). All of the articles were reprinted in Sussman's *Planning the Fourth Migration*.

69. Mumford, "The Fourth Migration," in Sussman, 59.

70. Ibid., 56.

71. Mumford, "Regions—To Live In," in Sussman, 92.

72. MacKaye, "The New Exploration: Charting the Industrial Wilderness," in Sussman, 109–10.

73. I borrow the term "legibility" from James Scott's *Seeing Like a State: How Certain Schemes to Improve the Human Condition Have Failed* (1998).

74. Lewis Mumford, *The Golden Day: A Study in American Literature and Culture* (1934 [1926]), 79–81.

75. See Dorman, 84–93; Warren Susman, "The Frontier Thesis and the American Intellectual," in *Culture as History*, 27–38; Casey Nelson Blake, *Beloved Community: The Cultural Criticism of Randolph Bourne, Van Wyck Brooks, Waldo Frank, and Lewis Mumford* (1990); David M. Wrobel, *The End of American Exceptionalism: Frontier Anxiety from the Old West to the New Deal* (1993). Bernard DeVoto mounted the most spirited response; his *Mark Twain's America* (1932) was a pointed reply to this criticism of pioneering.

76. See MacKaye's "Preface" to *The New Exploration*, xxiv–xxv.

77. MacKaye to Mumford, December 3, 1926, BMKP, Box 166.

78. MacKaye to Mumford, May 21, 1927, BMKP, Box 166. Italics are mine.

79. Ronald Foresta makes this argument in "Transformation of the Appalachian Trail."

80. Leopold to MacKaye, February 3, 1926, BMKP, Box 166. Leopold noted that his letter came in response to one MacKaye had written, but I could not locate that letter.

81. MacKaye to Walter Pritchard Eaton, August 11, 1926, BMKP, Box 166.

82. Runte, *National Parks,* 115–17.

83. MacKaye to Clarence Stein, March 8, 1925, BMKP, Box 166.

84. Anderson to MacKaye, November 22, 1927, BMKP, Box 166.

85. Chase to MacKaye, December 17, 1928; Oberholtzer to MacKaye, January, 1929 (no day); and MacKaye to Oberholtzer, February 4, 1929, BMKP, Box 166.

86. MacKaye, "Outdoor Culture—The Philosophy of Through Trails," *Landscape Architecture* 17, 3 (April 1927): 163–71; reprinted in Bryant, *From Geography to Geotechnics,* 169–79 (quote, 169).

87. John Kasson's discussion of Progressive and radical responses to Coney Island in *Amusing the Million* (pp. 104–9) provides an excellent look at these tensions.

88. "Outdoor Culture," in Bryant, 174.

89. Ibid., 176–77. On the influence of Spengler's *Decline of the West,* see Spann, 91.

90. MacKaye, *The New Exploration.* MacKaye spent the last third of his life working on his magnum opus, "The Geotechnics of North America," which exists in a variety of manuscript forms. Copies in BMKP, Box 196.

91. MacKaye, *The New Exploration,* 214.

92. Parsons, "Collaborative Genius," 466.

93. MacKaye, *The New Exploration,* 179.

94. Ibid. Italics are mine.

95. Mumford, "Introduction" to *The New Exploration,* xx.

96. MacKaye, "Wilderness Ways," *Landscape Architecture* 19, 4 (July 1929): 237–38.

97. Ibid., 249.

98. MacKaye's "Super By-Pass for Boston: A New Proposal for the Bay Circuit" appeared in the *Boston Globe* on October 31, 1930, copy in BMKP, Box 184. In 1937, MacKaye's Bay Circuit proposal was published in pamphlet form; copy in BMKP, Box 185.

99. On "highwayless towns," see Schaffer, *Garden Cities for America,* especially 152–53.

100. MacKaye, "The Townless Highway," *New Republic* (March 12, 1930), 94.

101. MacKaye to Crosser, January 1931, BMKP, Box 167.

102. Anderson to MacKaye, April 3, 1930, BMKP, Box 167. Anderson's piece, "The Recreational Value of Hiking," appeared in *American Motorist* in May 1930.

103. Anderson to MacKaye, April 3, 1930, BMKP, Box 167; Marshall, "The Problem of the Wilderness."

104. Broome to MacKaye, July 16, 1931, BMKP, Box 167.

105. See Broome to MacKaye, February 28, 1921, and MacKaye to Broome, March 15, 1931, BMKP, Box 167.

106. MacKaye, "The Appalachian Trail: A Guide to the Study of Nature," *Scientific Monthly* 34 (April 1932): 330.

107. MacKaye produced a number of other short manifestoes on the subject of nature study. See "The Appalachian Trail: A Guide to the Study of Nature"; "Memorandum re: A Nature Guide Service for Appalachia"; and "The Trail as a Dramatizer of Nature," all in BMKP, Box 184.

108. Anderson to MacKaye, June 10, 1932, BMKP, Box 167.

109. Anderson to MacKaye, October 14, 1932, BMKP, Box 167.

110. MacKaye to Anderson, October 19, 1932, BMKP, Box 167.

111. The University of Virginia Round-Table Conference on Regionalism was held in January 1931. See Spann, 127–29.

112. See Stein to MacKaye, March 6 and March 25, 1931, BMKP, Box 167. MacKaye's reference here was to a strange and fascinating utopian vision, offered two decades earlier by Edgar Chambless, of cities built out into the countryside along roadways rather than vertically in the form of skyscrapers. He described this vision in *Roadtown* (1910). MacKaye's comment about minimal taxes is somewhat cryptic, but presumably he referred to reforming tax policies that encouraged timber owners to liquidate timber and abandon land rather than pay taxes on a sitting resource.

113. Despite interest in Roosevelt, MacKaye ended his letter by hinting that he would vote for Socialist candidate Norman Thomas. See MacKaye to Stein, September 24, 1932; MacKaye to Stein, November 4, 1932; both in BMKP, Box 167.

114. Ellis Hawley, "Herbert Hoover, the Commerce Secretariat, and the Vision of an 'Associative State,' 1921–1928," *Journal of American History* 61, 1 (June 1974): 116–40.

115. MacKaye, "Memorandum to the Secretary of Labor: Re Forest Communities," BMKP, Box 184; MacKaye to Clarence Stein, April 7, 1933, BMKP, Box 167.

116. MacKaye, "Tennessee—Seed of a National Plan," *Survey Graphic* 22, 5 (May 1933): 251–54, 293–94. Reprinted in Bryant, *From Geography to Geotechnics,* 132–48. See also MacKaye, "The Tennessee River Project: First Step in a National Plan," *New York Times* (April 16, 1933); MacKaye, "The Challenge of Muscle Shoals," *The Nation* 136 (April 19, 1933): 445–46.

117. MacKaye, "Memorandum to the Commissioner of Indian Affairs," May 24, 1933, BMKP, Box 184. In the memo, MacKaye argued strongly for preserving Indian villages as "indigenous" environments. "The basic need of this balance of settings," he said in this memo, "has been revealed by the invention of the motor car (with the subsequent motorway). This has invaded each environment in turn—especially in upsetting village life and strewing the rural wayside with an endless motorslum."

118. MacKaye recounted meeting Leopold in a letter to Harvey Broome, July 5, 1933, BMKP, Box 167. See MacKaye and Leopold, "Sandia Cooperative Flood Control Project," June 1933, BMKP, Box 184.

119. MacKaye to Fritz Gutheim, August 20, 1933, BMKP, Box 167. In "Work Record of Benton MacKaye," he referred to a "report on plans for carrying out

the policy of Commissioner John Collier for developing the Navajo Reservations as a sphere of native Navajo culture."

120. MacKaye, "Memorandum to the Commissioner of Indian Affairs."

121. Collier to MacKaye, August 31, 1933; MacKaye to Broome, September 3, 1933; Anderson to MacKaye, September 11, 1933; all in BMKP, Box 167.

122. Mumford to MacKaye, September 14, 1933, BMKP, Box 167.

123. During MacKaye's illness, Broome and Stein corresponded. See folder entitled "Harvey Broome/Clarence Stein re Benton MacKaye's Illness," BMKP, Box 167.

124. See Anderson to MacKaye, September 13, September 27, and September 28, 1933; Anderson to Harold Ickes, September 27, 1933, BMKP, Box 167.

125. Myron Avery to Carlos Campbell, December 1, 1933, BMKP, Box 167.

126. Cammerer to Avery, December 2, 1933, BMKP, Box 167.

127. MacKaye to Cammerer, December 30, 1933, BMKP, Box 167.

128. Anderson to MacKaye, January 7, 1934, BMKP, Box 167.

129. MacKaye to Ned Richards, December 13, 1933, BMKP, Box 186.

130. See the scrapbook, "The Philosopher's Club—The Two Damnedest Years," prepared for MacKaye by the club's members, ATCA.

131. MacKaye to Stein, April 11, 1934; MacKaye to Stein, June 17, 1934; both in BMKP, Box 167.

132. Daniel Schaffer, "Benton MacKaye: The TVA Years," *Planning Perspectives* 5 (1990): 8. Also see Schaffer, "Ideal and Reality in 1930s Regional Planning: The Case of the Tennessee Valley Authority," *Planning Perspectives* 1 (1986): 27–44.

133. These memos can be found in BMKP, Box 186.

134. See Schaffer, "Benton MacKaye: The TVA Years," 11–14.

135. Ibid., 13–14.

136. MacKaye, "Memo #1" in folder titled "Recreational Development of the Upper Tennessee Basin," October 1935, BMKP, Box 186.

137. MacKaye to Judson King, May 2, 1936, BMKP, Box 168. Daniel Schaffer argues that it was MacKaye's influence on these younger members of the TVA that constituted his greatest intellectual influence on the agency. See Schaffer, "Benton MacKaye: The TVA Years," 17–18; Robert M. Howes, "The Knoxville Years," *The Living Wilderness* 39, 132 (January/March 1976): 23–26.

138. On the Green Mountain Parkway, see Hannah Silverstein, "No Parking: Vermont Rejects the Green Mountain Parkway," and Hal Goldman, "James Taylor's Progressive Vision: The Green Mountain Parkway," in *Vermont History* 63 (1995): 133–79.

139. MacKaye, "Frankline vs. Skyline," *Appalachia* 20 (1934): 104–5.

140. Ibid., 106.

141. Ibid., 106–8.

142. Avery to MacKaye, September 10, 1934, BMKP, Box 167.

143. On this schism, see George J. Ellis, "The Path Not Taken," M.A. thesis, Shippensburg University, 1993.

144. Cammerer to MacKaye, September 14, 1934, BMKP, Box 167.

145. MacKaye to Cammerer, September 21, 1934, BMKP, Box 167.

146. Cammerer to MacKaye, October 3, 1934, BMKP, Box 167; Justin B. Reich,

"Re-Creating the Wilderness: Shaping Narratives and Landscapes in Shenandoah National Park," *Environmental History*, 6, 1 (January 2001), 95–117.

147. Harvey Broome to Harold Anderson, August 22, 1934, TWSP. See also, Broome, "Origins of the Wilderness Society," *The Living Wilderness* (July 1940), 13–15.

148. Anderson to MacKaye, September 12, 1934; MacKaye to Anderson, September 20, 1934; both in TWSP, Box 11.

149. MacKaye to Chase, July 20, 1935, BMKP, Box 168.

150. Schaffer, "Benton MacKaye: The TVA Years," 18.

151. Benton MacKaye, "Why the Appalachian Trail?" *The Living Wilderness* 1, 1 (September 1935): 7.

152. This critique is inspired in part by Jane Jacobs. See Jacobs, 374–75.

6. The Freedom of the Wilderness: Bob Marshall

1. Roderick Nash discusses Marshall's career and accomplishments in "The Strenuous Life of Bob Marshall," *Forest History* 10 (1966): 18–25, and in *Wilderness and the American Mind*, 200–208. The most thorough treatment of Marshall's life is James M. Glover, *A Wilderness Original: The Life of Bob Marshall* (1986). See also Glover, "Romance, Recreation, and Wilderness: Influences on the Life and Work of Bob Marshall," *Environmental History Review* 14, 4 (Winter 1990): 22–39, and James and Regina Glover, "Robert Marshall: Portrait of a Liberal Forester," *Journal of Forest History* 30, 3 (July 1986): 112–19. Another treatment is David Bernstein, "Bob Marshall: Wilderness Advocate," *Western States Jewish Historical Quarterly* 13 (1980): 26–37.

2. On the connection between wilderness and physicality, see Christopher Sellers, "Thoreau's Body: Towards an Embodied Environmental History," *Environmental History* 4, 4 (October 1999): 503–4.

3. Nash does not really mention these commitments in any depth in *Wilderness and the American Mind*. In *A Wilderness Original*, Glover goes into more depth, but he tends to be apologetic about what he sees as Marshall's naiveté. In *The American Conservation Movement*, Stephen Fox is explicit about this separateness: "[Marshall's] interests were neatly compartmentalized into socialism, civil liberties, and conservation." Fox, 209. Robert Gottlieb's *Forcing the Spring* is the exception. See Gottlieb, 15–19.

4. Glover, *A Wilderness Original*, 244–45.

5. Again, Gottlieb's *Forcing the Spring* is an exception.

6. Catton, *Inhabited Wilderness*, especially 131–56.

7. Glover, *A Wilderness Original*, 15.

8. Aron, *Working at Play*, 183–84.

9. Glover, *A Wilderness Original*, 20–24. On the Adirondack camps, especially Knollwood, see Harvey H. Kaiser, *Great Camps of the Adirondacks* (1982), 140–44.

10. Philip Terrie claimed that Murray's book initiated a "recreational revolution" in the region. See Terrie, *Forever Wild: Environmental Aesthetics and the Adirondack Forest Preserve* (1985), 71. See also David Strauss, "Toward a Con-

sumer Culture: 'Adirondack Murray' and the Wilderness Vacation," *American Quarterly* 39, 2 (Summer 1987): 270–86. On Headley and Murray, see Terrie, *Forever Wild,* 45, 69; Nash, *Wilderness and the American Mind,* 61–62, 116.

11. Terrie, *Forever Wild,* 92–93.

12. Terrie, *Forever Wild,* 93–95. For a national treatment of the rise of these concerns, see Donald Pisani, "Forests and Conservation, 1865–1890," *Journal of American History* 72, 2 (September 1985): 340–59. Both stress the importance of George Perkins Marsh's *Man and Nature; or, Physical Geography as Modified by Human Action* (1864). For the international context, see Ian Tyrrell, *True Gardens of the Gods: Californian-Australian Environmental Reform, 1860–1930* (1999).

13. Terrie, *Forever Wild,* 95–97; Jacoby, "Class and Environmental History," 326.

14. This act of preservation was itself a very complicated one. See Terrie, *Contested Terrain: A New History of Nature and People in the Adirondacks* (1997), 90–105.

15. Terrie, *Forever Wild,* 108.

16. Reading this political episode as an example of evolving preservationist sentiment can be confusing business. Was this or was this not an explicit preservation of wilderness? Ultimately, that depends on what one means by wilderness. Scholars such as Roderick Nash argue that it was. See Nash, *Wilderness and the American Mind.* By contrast, Philip Terrie has argued that wilderness was preserved with the passage of the article, but that such preservation was largely accidental—by which he means that it was not motivated by what he recognizes as a wilderness aesthetic. See Terrie, *Forever Wild,* 104–6; Terrie, "The Adirondack Forest Preserve: The Irony of Forever Wild," *New York History* 62, 3 (July 1981): 260–88.

17. Terrie, *Contested Terrain,* 118–24; Jacoby, "Class and Environmental History."

18. Terrie, *Contested Terrain,* 107–14.

19. Terrie, *Forever Wild,* 109–35.

20. Glover, 24.

21. Frank Graham, Jr., *The Adirondack Park: A Political History* (1978), 170–71. On the shift to recreational development, see Terrie, *Forever Wild,* 109–35.

22. Glover, 29–30; Terrie, *Forever Wild,* 90–91.

23. Glover, *A Wilderness Original,* 15–54, 66; Nash, *Wilderness and the American Mind,* 201.

24. The exceptions here were the few trails to the top of Adirondack peaks and portage trails built for those who traveled the region by canoe. Terrie, *Forever Wild,* 46.

25. Bob Marshall, "Circling Lake Walk—July 22, 1918," in *Adirondack Notebook II, 1919–1921.* See also "Scarface Mountain—July 14, 1919," *Adirondack Notebook II, 1919–1921,* both in RMP, Carton 1, Folder 3; Graham, 146–47.

26. Although Bob, George, and Herb Clark did make it to some of the more remote sections of the park, even they showed signs of significant human habitation and impact. See Glover, *A Wilderness Original,* 32–36.

27. Terrie, *Contested Terrain,* 137–39.

28. Glover, *A Wilderness Original,* 37–38. On Louis Marshall's role in creating the College of Forestry, see Graham, 147, 167.

29. Glover, 42–50. Marshall recorded his summer explorations in a journal entitled "Weekend Trips in the Cranberry Lakes Region," RMP, Carton 1.

30. Robert Marshall, *The High Peaks of the Adirondacks* (1922); Glover, *A Wilderness Original,* 50–51.

31. Russell Carson, *Peaks and Peoples of the Adirondacks* (1927).

32. Marshall, "Recreational Limitations to Silviculture in the Adirondacks," *Journal of Forestry* 23 (1925): 173.

33. Ibid., 173–74.

34. Ibid., 177–78.

35. On Forest Service plans for Alaska in the early 1920s, see Clary, *Timber and the Forest Service,* 72.

36. Glover, 55–60. Marshall's efforts to get a position in Alaska are documented in a series of letters from Forest Service officials, RMP, Box 11, Folder 21. See also Richard E. McArdle, with Elwood Maunder, "Wilderness Politics: Legislation and Forest Service Policy,"*Journal of Forest History* 10 (1975): 166.

37. Glover, *A Wilderness Original,* 60–61, 64. Marshall's master's thesis was published as *The Growth of Hemlock Before and After Release from Suppression,* Harvard Forest Bulletin No. 11 (1927). On the history of the Harvard Forest and the region, see Hugh Raup, "The View from John Sanderson's Farm: A Perspective for the Use of the Land," *Forest History* 10, 1 (April 1966): 2–11; Hugh M. Raup and Reynold E. Carlson, *The History of Land Use in the Harvard Forest,* Harvard Forest Bulletin No. 20 (1941).

38. Glover, *A Wilderness Original,* 66–68.

39. The most representative policy changes of the Greeley era came packaged in the Clarke-McNary Act of 1924, which provided for the cooperation of the federal government, state governments, and private industry on a number of issues—particularly fire prevention. See Clary, 67–74; Steen, *The U.S. Forest Service,* 173–95; Swain, *Federal Conservation Policy, 1921–1933,* 9–29. These attitudes are also quite evident in Greeley's annual *Report[s] of the Forester,* 1920 to 1928.

40. See Glover, *A Wilderness Original,* 68–69; William Greeley, *Report of the Forester* (1926), 21. Greeley provides an extensive description of the fire in his report.

41. Marshall, "Mountain Ablaze," *Nature Magazine* 46, 6 (June–July 1953): 290.

42. Ibid., 291.

43. Ibid.; George Marshall, "The Growth of a Forester: Letters from Robert Marshall from Montana and Idaho—June 19, 1925–August 1928," RMP, Carton 2, Folder 24; Glover, 70–73.

44. For an example of Marshall's postfire research, see Robert Marshall and Clarence Averill, "Soil Alkalinity on Recent Burns," *Ecology* 9, 4 (October 1928): 533.

45. See "Broadmore Camp 1, Kaniksu National Forest, Idaho—10/23/25," in "Growth of a Forester"; Glover, 75.

46. Marshall, "Growth of a Forester," letter excerpt from August 31, 1925.

47. Richard McArdle did not recall Marshall mentioning wilderness preservation during the summer they worked together, though Marshall was always going off for hikes in the area. See McArdle and Maunder, 178.

48. Gilligan, 85–86. In his 1925 letter about his Clearwater trip, Marshall had referred to the area as a "wilderness" in a way that might be construed as recognizing an official status. Both Gilligan and Glover (citing Gilligan) say that the other preserved wilderness areas were in the Grand Tetons (now Grand Teton National Park), the Two-Ocean Pass area (now part of the Teton Wilderness), a portion of the Absaroka National Forest (now the Absaroka Wilderness), the country surrounding the Middle Fork of the Salmon River (now the Frank Church—River of No Return Wilderness), and "parts of the Clearwater country in Montana, now in the Bob Marshall Wilderness Area" (quote from Glover, 94). Reference to this last area seems mistaken, since the Clearwater country is mostly in Idaho in what is now the Selway-Bitterroot Wilderness, and the Bob Marshall Wilderness Area is in the Flathead River country east of Missoula and south of Glacier National Park. Gilligan probably meant the Clearwater area. It is thus likely that in Marshall's August 31, 1925, letter, when he referred to "the 28,000 square mile wilderness," he was using the term "wilderness" officially rather than generically.

49. Leopold to Arthur Ringland, December 29, 1925, RMP, Box 7, Folder 19.

50. Leopold to Marshall, September 29, 1926, RMP, Box 7, Folder 19.

51. On Forest Service inventory efforts, see L. F. Kniepp, "These Tame National Forests," *Service Bulletin* 11, 10 (March 7, 1927): 2–3.

52. Marshall, "The Growth of a Forester," quoted in Glover, *A Wilderness Original,* 82–83.

53. The quotes are from Marshall, "Impressions from the Wilderness," *Nature Magazine* 44 (November 1951): 481–84.

54. Glover, *A Wilderness Original* 86–87; Graham, 185–86.

55. Carson to Marshall, October 23, 1928, RMP, Box 4, Folder 5.

56. Marshall to L. F. Kniepp, September 8, 1927, RMP, Box 11, Folder 21.

57. See Robert Marshall and Althea Dobbins, "Largest Roadless Areas in the United States," *The Living Wilderness* 2, 2 (November 1936): 11–13.

58. Manly Thompson, "A Call from the Wilds," *Service Bulletin* 12, 20 (May 4, 1928): 2–3; Aldo Leopold, "Mr. Thompson's Wilderness," *Service Bulletin* 12, 26 (June 25, 1928): 1–2; Robert Marshall, "The Wilderness as Minority Right," *Service Bulletin* (August 27, 1928), 5–6.

59. As it turns out, though Thompson's figure was not supported in his article, it may have been accurate. Richard McArdle's recollection was that, as of 1962, only about 0.5 percent of recreational use of the national forests occurred in designated wilderness areas. See McArdle and Maunder, 170.

60. Marshall, "The Wilderness as Minority Right," 5–6.

61. Glover, *Wilderness Original* 83–84, 97.

62. Marshall to Meyer Wolff, February 1, 1928, RMP, Box 1, Folder 2. Wolff later played a key role in the creation of the Bob Marshall Wilderness Area. See Lawrence Merriam, "The Irony of the Bob Marshall Wilderness," *Journal of Forest History* 33, 2 (1989): 83.

63. See Marshall's letter home from Missoula, February 20, 1928, excerpted in

"The Growth of a Forester." Marshall mentioned Elers Koch as one of his supporters. On Koch's wilderness sentiments, see Koch, "The Passing of the Lolo Trail," *Journal of Forestry* 33, 2 (1935): 98–104; Koch, *Forty Years a Forester, 1903–1943*, ed. Peter Koch (1998).

64. Marshall, "An Experimental Study of the Water Relations of Seedling Conifers with Special Reference to Wilting," *Ecological Monographs* 1, 1 (January 1931): 37–98.

65. Glover, *A Wilderness Original*, 99–102. For a history of LID, see Bernard K. Johnpoll and Mark R. Yerburgh, eds., *The League for Industrial Democracy: A Documentary History*, 3 vols. (1980), and Robert Westbrook's review of these volumes in *International Labor and Working Class History* 20 (1981): 73–78.

66. Robert Westbrook, *John Dewey and American Democracy* (1991), 431.

67. John Dewey, *Individualism Old and New* (1929), 17–18.

68. On the interwar history of the ACLU and its antimajoritarian sentiment, see Edward J. Larson, *Summer for the Gods: The Scopes Trial and America's Continuing Debate over Science and Religion* (1997), 60–83.

69. Bob Marshall, "The North Fork of the Koyukuk," in George Marshall, ed., *Arctic Wilderness* (1956), 4. *Arctic Wilderness* is a collection of Marshall's letters and memoirs from his trips to Alaska, which his brother assembled after his death. It was reprinted as *Alaska Wilderness: Exploring the Central Brooks Range* (1970).

70. "The North Fork of the Koyukuk," 14.

71. See Richard White, "'Are You an Environmentalist or Do You Work for a Living?': Work and Nature," in Cronon, ed., *Uncommon Ground*, 176–78.

72. See, for instance, Marshall to a "Dr. Wise," February 15, 1932, RMP, Box 1, Folder 8.

73. Glover, *A Wilderness Original*, 103. George Ahern, *Deforested America* (1929).

74. In 1919 and 1920, Pinchot had leveled similar charges at the forestry profession. On this earlier controversy, see Steen, 173–85; Clary, 70–71.

75. Marshall, "Forest Devastation Must Stop," *The Nation* 129 (August 28, 1929): 218–19.

76. Ibid., 219.

77. Ibid.

78. Glover, *A Wilderness Original*, 112–13.

79. Marshall to Gerry and Lilly Kempff, March 3, 1930, RMP, Box 1, Folder 3.

80. "A Letter to Foresters," *Journal of Forestry* 28 (April 1930): 456–58 (quote, 456). The signatories of the letter were Marshall, Pinchot, George Ahern, E. N. Munns, Ward Shepard, W. N. Sparhawk, and Raphael Zon.

81. R. S. Kellogg, "As I See It," *Journal of Forestry* 28 (April 1930): 461.

82. Marshall, "A Proposed Remedy for Our Forestry Illness," *Journal of Forestry* 28, 3 (March 1930): 273–80.

83. Ibid., 274–75.

84. Marshall, *The Social Management of American Forests* (1930), 22.

85. Robert Marshall, "The Problem of the Wilderness," *Scientific Monthly* 30, 2 (February 1930): 141–48.

86. William James, "The Moral Equivalent of War" (1910).

87. "The Problem of the Wilderness," 143.

88. See Lears, *No Place of Grace,* particularly his section "The Psychological Uses of the Martial Ideal: The Cult of Experience and the Quest for Authentic Selfhood," 117–24.

89. "The Problem of the Wilderness," 141. Marshall quoted Leopold's "The Last Stand of the Wilderness," which made a strong argument against modern recreational developments.

90. See David E. Shi, *The Simple Life: Plain Living and High Thinking in American Culture* (1985), 125–53.

91. "The Problem of the Wilderness," 144.

92. Ibid., 146.

93. Ibid., 148.

94. Leopold to Marshall, February 27, 1930, RMP, Box 7, Folder 19; Anderson to Marshall, April 3, 1930, RMP, Box 7, Folder 27.

95. Marshall to Lincoln Ellison, April 3, 1930, RMP, Box 1, Folder 3.

96. Marshall to Al Cline, July 15, 1930, RMP, Box 1, Folder 4.

97. Marshall, *Arctic Village* (1933).

98. Daines Carrington, "An Arctic Middletown," *Saturday Review of Literature* 9 (May 13, 1933): 589. See also H. L. Mencken, "Utopia in Little," *American Mercury* 29 (May 1933): 124–26.

99. See Robert and Helen Lynd, *Middletown: A Study in Contemporary American Culture* (1929); Richard Wightman Fox, "Epitaph for Middletown: Robert S. Lynd and the Analysis of Consumer Culture," in Fox and Lears, *The Culture of Consumption,* 101–41.

100. Marshall to Margaret Mead, January 28, 1933, RMP, Box 1, Folder 11; Ruth Benedict, "The Happiest People," *The Nation* 136 (June 7, 1933): 647.

101. Marshall administered Stanford-Binet (IQ) tests to the residents of Wiseman and used the results to argue that Wiseman's Eskimo residents were of equal, if not superior, intelligence compared with a standard American control group. Marshall, *Arctic Village,* 53–54, 78–79. On the history of statistical intelligence testing, see Olivier Zunz, *Why the American Century?* (Chicago: University of Chicago Press, 1998), 50–57. On Boasian anthropology, see Richard Handler, "Boasian Anthropology and the Critique of American Culture," *American Quarterly* 42, 2 (June, 1990): 252–73; George Stocking, Jr., "Ideas and Institutions in American Anthropology: Toward a History of the Interwar Period," in *The Ethnographer's Magic and Other Essays in the History of Anthropology* (1992), 114–77.

102. Benedict, "The Happiest People."

103. Marshall to Pinchot, January 11, 1932, RMP, Box 1, Folder 8.

104. Marshall to Earl Clap [*sic*], August 17, 1932, RMP, Box 1, Folder 10; Glover, 144–45. See *A National Plan for American Forestry* (1933), and the summary of the report "Major Problems and the Next Big Step in American Forestry" (1933). Marshall wrote three sections of the larger report. The most substantial and important was "The Forest for Recreation," volume 1, 463–87. See also his shorter entries, "The National Parks and the National Monuments," volume 1, 633–36, and "A Program for Forest Recreation," volume 2, 1543–46.

105. General information from the report summary, "Major Problems and the Next Big Step in American Forestry." Quotes from the attached "Letter of Transmittal," v–vii.

106. "The Forest for Recreation," 464–65.

107. Ibid., 470.

108. Conversations that he had been having with John C. Merriam during the course of the report's preparation pushed Marshall's thought on this subject. On Merriam's influence, see Marshall to Ralph Hosmer, April 28, 1932, RMP, Box 1, Folder 9; "The Forest for Recreation," 471–73.

109. "The Forest for Recreation," 473–74.

110. Ibid., 474–76.

111. Ibid., 476–78, 480.

112. Recall Howard Flint's "Wasted Wilderness," *American Forests and Forest Life* 32 (1926): 407–10.

113. "The Forest for Recreation," 482–83.

114. Ibid., 484–85.

115. On the circumstances of the meeting between Marshall and Pinchot, see Marshall to Raphael Zon, January 23, 1933, RMP, Box 1, Folder 11. For the actual letter, written by Marshall but submitted by Pinchot, see Pinchot to Roosevelt, January 20, 1933, Personal Papers File 289, FDRP. See Glover, *A Wilderness Original*, 150–51.

116. Marshall to Norman Thomas, January 23, 1933, RMP, Box 1, Folder 11.

117. Marshall, "A Letter on Economics," RMP, Carton 2, Folder 38.

118. Robert Marshall, *The People's Forests* (1933), 219.

119. Franklin Reed, "Review of the *The People's Forests*," *Journal of Forestry* 32 (1934): 104–7.

120. See L. F. Kniepp to Marshall, June 29, 1933, RMP, Box 11, Folder 21; Marshall to L. F. Kniepp, July 12, 1933, RMP, Box 1, Folder 15.

121. Marshall to Ickes, March 28, 1933, RMP, Box 1, Folder 12; T. H. Watkins, *Righteous Pilgrim: The Life and Times of Harold Ickes, 1874–1952* (1990), 469.

122. Glover, *A Wilderness Original*, 157–61; Lawrence C. Kelly, "The Indian Reorganization Act: The Dream and the Reality," *Pacific Historical Review* 44, 3 (August 1975): 291–312.

123. See file folder, "Wilderness: Indian Reservations," RMP, Carton 5, Folder 24; Glover, *A Wilderness Original*, 203–5, 209–13.

124. See Jennings, *The Invasion of America*; Cronon, *Changes in the Land*; Spence, *Dispossessing the Wilderness*; Cronon, "The Trouble with Wilderness"; Kenneth Olwig, "Reinventing Common Nature: Yosemite and Mount Rushmore—A Meandering Tale of a Double Nature," in *Uncommon Ground*, 379–408.

125. Marshall, "A Policy for Navajo Roads, Truck Trails, Etc.," March 21, 1934, RMP, Carton 5, Folder 24.

126. Press Release, Department of the Interior, 1937, RMP, Carton 5, Folder 24. Italics are mine. See also "Wilderness Now on Indian Lands," *Living Wilderness* 3, 3 (December 1937): 3–4; William Zimmerman, Jr., "Wilderness Areas on Indian Lands," *Living Wilderness* 5, 5 (July 1940): 10–11.

127. Marshall, "Comments on the Report on Alaska's Recreational Resources

and Facilities," Appendix B, *Alaska and Its Resources* (1938), 213. For the link between Marshall's Alaskan suggestion and the ideal of "inhabited wilderness," see Catton, *Inhabited Wilderness*, 136–37.

128. Marshall was aware of the ecological limitations imposed by marginal reservation lands. See Marshall, "Ecology and the Indians," *Ecology* 18, 1 (January 1937): 159–61.

129. The best example of how Marshall grafted his critique of American society onto Indian policy is an article he co-wrote with John Collier and Ward Shephard titled "The Indians and Their Lands," *Journal of Forestry* 31, 8 (December 1933): 905–10.

130. The first significant challenge to Marshall's policy came in a 1957 Navajo petition that asked the Interior Department to reconsider the wilderness policy in light of desires to develop uranium and copper mines, and to get in on the benefits of federal water development. Glover, *A Wilderness Original,* 210–12. The policy of termination, adopted in the 1950s, contributed to the abrogation of these administratively created wilderness areas. See Kenneth R. Philp, "Termination: A Legacy of the Indian New Deal," *Western Historical Quarterly* 14, 2 (April 1983): 165–80; Clayton Koppes, "From New Deal to Termination: Liberalism and Indian Policy, 1933–1953," *Pacific Historical Review* 46 (November 1977): 543–66. By 1962, only two of the remaining "roadless" and "wild" areas on Indian reservations remained. See "Wilderness and Recreation—A Report on Resources, Values, and Problems" (Outdoor Recreation Resources Review Commission Study Report 3, 1962), 23. For an interesting take on the policy, see John Collier, "Wilderness and Modern Man," in his *From Every Zenith: A Memoir and Some Essays on Life and Thought* (1963), 269–84.

131. See R. Newell Searle, *Saving Quetico-Superior: A Land Set Apart* (1977); David Backes, *Canoe Country: An Embattled Wilderness* (1991). For Carhart's plan, see Backes, "Wilderness Visions: Arthur Carhart's 1922 Proposal for the Quetico-Superior Wilderness," *Forest and Conservation History* 35 (July 1991): 128–37.

132. See "Report to the President of the United States on the Quetico-Superior Area by the Quetico-Superior Committee," 1938, FDRP, Official Papers, OF 1119.

133. Marshall, "Should the Journal of Forestry Stand for Forestry?" *Journal of Forestry* 32 (1934): 904–8.

134. Marshall to R. Y. Stuart, March 17, 1933, RMP, Box 1, Folder 12. A letter three months later proved Marshall's concerns well founded. Kniepp told Marshall that, though the Forest Service had received a tremendous amount of emergency conservation work money, the Service's own budget had been reduced by 17.5 percent for that fiscal year. See L. F. Kniepp to Marshall, June 29, 1933, RMP, Box 11, Folder 21.

135. Marshall to Willard Van Name, July 31, 1933, RMP, Box 1, Folder 16.

136. Anderson to Marshall, August 9, 1933, RMP, Box 3, Folder 27.

137. Marshall to Van Name, July 31, 1933, RMP, Box 1, Folder 16.

138. Marshall to Yard, June 4, 1934, RMP, Box 12, Folder 13.

139. Ickes to Marshall, August 26, 1934, RMP, Box 6, Folder 35.

140. Information and quotes in this paragraph are from Harvey Broome, "Origins of the Wilderness Society," *Living Wilderness* 5, 5 (July 1940): 13–14.

141. Glover, *A Wilderness Original,* 175–76.

142. Marshall, "Memorandum for the Secretary," June 9, 1935, TWSP, Box 11, Folder 15. In March 1935, Marshall submitted another memo to Ickes, with similar recommendations, on the proposed Green Mountain Parkway in Vermont. See Marshall, "Memorandum for the Secretary: The Proposed Green Mountain Parkway," March 29, 1935, TWSP, Box 11, Folder 15.

143. Ickes to Cammerer, February 5, 1937, RMP, Box 6, Folder 35.

144. MacKaye to Harry Slattery (Assistant Secretary, Department of the Interior), October 22, 1934, BMKP, Box 167 (quote is from this letter). A month later, Marshall wrote to Ickes from the Klamath reservation in Oregon, apologizing that his speech on "forest aesthetics" had caused such an uproar. Marshall to Ickes, November 22, 1934, RMP, Box 1, Folder 18.

145. "Resolution Passed at the 59th Annual Meeting of the American Forestry Association, Held at the Andrew Johnson Hotel, Knoxville, Tennessee, October 7–20, 1934," AFAC.

146. Broome, "Origins of the Wilderness Society"; Stephen Fox, "'We Want No Straddlers.'"

147. Marshall to Ickes, February 14, 1935, RMP, Box 1, Folder 20.

148. Marshall to Broome, February 28, 1935, RMP, Box 1, Folder 20.

149. Marshall to Irving Clark, January 26, 1935, RMP, Box 1, Folder 19. See also Marshall to Irving Clark, March 19, 1935, RMP, Box 1, Folder 21.

150. Again, Harold Ickes was a key supporter of this policy shift. On the "wilderness" or "primeval" parks, and Ickes's support for them, see Donald Swain, "The National Park Service and the New Deal, 1933–1940," *Pacific Historical Review* 41, 3 (August 1972): 312–32; Watkins, *Righteous Pilgrim,* 549–79. Other parks created in this mold were Kings Canyon, the Everglades, and the Great Smoky Mountains. None were completely roadless, but all were marked by limited road development.

151. Marshall to Ferdinand Silcox, September 6, 1935, RMP, Box 1, Folder 26.

152. Glover, 200; Marshall, "Interview with Secretary Ickes," December 7, 1936, TWSP, Box 11, Folder 15.

153. Catherine Bauer to Marshall, June 1, 1937, BMKP, Box 168.

154. Yard to Marshall, July 9, 1937, RMP, Box 12, Folder 21.

155. Broome to Marshall, January 1, 1938, RMP, Box 3, Folder 58.

156. Marshall to Yard, August 24, 1937, RMP, Box 2, Folder 23.

157. On the gospel of efficiency, see Hays, *Conservation and the Gospel of Efficiency.* Hays argued against those scholars who saw Progressive conservation as a democratic and anticorporate movement. For that interpretation, see J. Leonard Bates, "Fulfilling American Democracy: The Conservation Movement, 1907 to 1921," *Mississippi Valley Historical Review* 44 (June 1957): 29–57; Roy M. Robbins, *Our Landed Heritage: The Public Domain, 1776–1936* (1942).

158. The term "island communities" is from Robert Wiebe's classic study, *The Search for Order, 1877–1920* (1967).

159. Donald Worster, "Wild, Tame, and Free: Comparing Canadian and American Views of Nature," in Ken Coates and John Findlay, eds., *On Brotherly Terms: Canadian-American Relations West of the Rockies* (2002).

Epilogue: A Living Wilderness

1. Yard to MacKaye, October 26, 1934, TWSP, Box 11, Folder 20.

2. Yard to Merriam, March 29, 1934 and April 6, 1934, JCMP, Box 188.

3. "The Wilderness Society," January 21, 1935, TWSP, Box 11, Folder 14.

4. Edward Abbey, "Polemic: Industrial Tourism and the National Parks," in *Desert Solitaire: A Season in the Wilderness* (1968), 45–67.

5. Leopold, "Marshland Elegy," in *A Sand County Almanac*, 96.

6. On the "machine age," see Richard Guy Wilson, Dianne H. Pilgrim, and Dickran Tashjian, *The Machine Age in America, 1918–1941* (1986).

7. In a letter to Harold Anderson discussing an early draft of principles, MacKaye implored him, "DON'T forget the RADIO. That is the one curse that has no redeeming feature whatever." MacKaye to Anderson, September 23, 1934, TWSP, Box 11, Folder 20.

8. On the connections between wilderness and wildlife, see Leopold's 1936 article, "Threatened Species," in Flader and Callicott, 230–34. For a discussion of the Leopold Report, see Sellars, 204–66.

9. Leopold, "Why the Wilderness Society?" *The Living Wilderness* 1, 1 (September 1935): 6; "Wilderness as Land Laboratory," *The Living Wilderness* 6, 6 (July 1941): 3.

10. Leopold, "Wilderness Values," *The Living Wilderness* 7, 7 (March 1942): 24.

11. On Butcher, see Miles, *Guardians of the Parks*. Yard to Marshall, July 16, 1936, RMP, Box 12, Folder 18.

12. "A Great Loss Is a Great Challenge," *The Living Wilderness* 13, 10 (July 1945): 1.

13. For Yard's passing and subsequent changes in leadership, see *The Living Wilderness* 10, 14–15 (December 1945).

14. *Appalachian Trail Conference Member's Handbook*, 13th edition (1988), 35–36.

15. MacKaye did manage to work portions of the manuscript into publishable articles that appeared in serial form in *The Survey* during the early 1950s; and, in 1968, Paul Bryant included these essays in an edited collection of MacKaye's writings titled *From Geography to Geotechnics*.

16. George Marshall cites this 1946 effort in his reminiscence, "Benton as Wilderness Philosopher," *The Living Wilderness* 39, 132 (January/March 1976): 13. See also Ed Zahniser, "Walk Softly and Carry a Big Map: Historical Roots of Wildlands Network Planning," *Wild Earth* 10, 2 (Summer 2000): 33–38.

17. *Appalachian Trail Conference Members Handbook*, 49.

18. On contemporary trail and greenway efforts, see Charles E. Little, *Greenways for America* (1990). On efforts to use natural corridors to connect island preserves, see the work of the Wildlands Project and the journal *Wild Earth*. On

the connections between MacKaye's thought and the Wildlands Project, see Zahniser, "Walk Softly and Carry a Big Map."

19. Robert Marshall and Althea Dobbins, "Largest Roadless Areas in the United States," *The Living Wilderness* 2, 2 (November 1936): 11–13. Marshall and Dobbins identified forested areas of at least 300,000 acres and desert areas of at least 500,000 acres. The Tayor Grazing Act of 1934 led to the withdrawal of most of the remaining public domain, which was then turned over to the Grazing Service. See Roy Robbins, *Our Landed Heritage*, 421–23; Hal K. Rothman, *The Greening of a Nation? Environmentalism in the United States since 1945* (1998), 69–73.

20. Regulation U-1 quoted in "Wilderness and Recreation—A Report on Resources, Values, and Problems: Report to the Outdoor Recreation Resources Review Commission by the Wildland Research Center, University of California" (ORRRC Study Report 3, 1962), 21.

21. Ibid., 21–22. The U-3 or "roadless" category was crafted primarily to include the preserved portions of the Superior National Forest. See Searle, *Saving Quetico-Superior*, 71–89; Allin, 77–79. Ironically, this "roadless" designation allowed roads, though it's not clear what sorts of roads were allowed. Donald Cate suggested that the roads were allowed on a temporary basis for resource extraction but were closed to the public. See Cate, "Recreation and the U.S. Forest Service: A Study of Organizational Response to Changing Demands" (1963), 145–46.

22. "Wilderness and Recreation," 22; Lawrence C. Merriam, "The Irony of the Bob Marshall Wilderness," *Journal of Forest History* 33, 2 (April 1989): 80–87.

23. "Wilderness and Recreation," 22; Gilligan, 198–99; Allin, 84–85.

24. Marshall to Roger Baldwin, February 11, 1938, RMP, Box 2, Folder 28; Marshall to Raphael Zon, March 3, 1938, RMP, Box 2, Folder 29; Glover, *A Wilderness Original*, 229–30.

25. On these efforts, see *Report of the Chief of the Forest Service* (1938), 51–52, which included for the first time a section on "Facilities for Lower Income Groups"; Marshall to Zon, March 3, 1938, RMP, Box 2, Folder 29. See also sections that Marshall wrote in Russell Lord, ed., *Forest Outings by Thirty Foresters* (1940), particularly the section "The Ill-To-Do," 259–62, 293.

26. On his return trips, see his *Arctic Wilderness*, ed. George Marshall, 110–65; Donald Worster, "Alaska: The Underworld Erupts," in *Under Western Skies: Nature and History in the American West* (1992), 168–70; Glover, *A Wilderness Original*, 241–42.

27. Quotes are from Bob Marshall's will, a copy of which can be found in the Gardner Jackson Papers, FDRP, Box 79. See also Fox, "'We Want No Straddlers,'" 14.

28. See *Outdoor Recreation for America: A Report to the President and to the Congress by the Outdoor Recreation Resources Review Commission* (1962). The ORRRC also produced twenty-seven study reports on a variety of different issues, including "Wilderness and Recreation—A Report of Resources, Values, and Problems." See also Allin, 125–26; Hays, *Beauty, Health, and Permanence*.

29. On the Interstate, see Mark H. Rose, *Interstate: Express Highway Politics,*

1941–1956 (1979); Tom Lewis, *Divided Highways: Building the Interstate High-ways, Transforming American Life* (1997). On Mission 66, see Runte, *National Parks*, 173; Sellars, 201–12. The National Park Service website has historical vis-itation figures since 1904. See "Recreation Visits: Decade Reports, 1904–1999," at: www2.nature.nps.gov/stats/decademain.htm.

30. Statistics are from the *Report[s] of the Chief of the Forest Service* for 1950 and 1964. The 1999 figure is from U.S. Forest Service, "National Recreation Use Pilot Study: Results, Analysis, and Recommendations" (Washington, D.C.: GPO, 1999). Forest Service visitation numbers have been generated various ways over the course of the agency's history, and it is difficult to compare vis-itor numbers as a result. For instance, beginning in 1965, the Forest Service cal-culated recreational use in terms of visitor days, each one equivalent to twelve hours of recreational activity. Use continued to climb, from 151 million visitor days in 1966 to 233.5 million visitor days in 1980.

31. On the history of the snowmobile, see Leonard S. Reich, "Ski-Dogs, Pol-Cats, and the Mechanization of Winter," *Technology and Culture* 40, 3 (1999): 484–516.

32. Louis Bignami, "Past and Present Tents," *Westways* 73 (1981): 34–37.

33. *Outdoor Recreation for America* (quote, 4). On Johnson's use of the report, see Hays, *Beauty, Health, and Permanence*, 57.

34. See Paul Hirt, *A Conspiracy of Optimism: Management of the National Forests Since World War Two* (1994).

35. Ibid., xliv.

36. Ibid., 114. These figures are for forest development roads only, and they do not include the more developed system of forest highways—a system that grew at a much slower pace during this period.

37. Ibid., 198–99. Hirt says that the Forest Service had, by 1990, revised their projected road mileage figures down to 403,000, of which 85 percent had been built. My figure of approximately 340,000 miles as of 1990 is extrapolated from these figures.

38. The Forest Service continued to protect the various areas established under the L-20 and U regulations, but the total wilderness acreage in the na-tional forests remained static between 1940 and 1960. Ibid., 163.

39. On the importance of World War II to the American West, see White, *It's Your Misfortune*, 496–533.

40. See William Chandler, *The Myth of the TVA: Conservation and Development in the Tennessee Valley, 1933–1983* (1984); White, *The Organic Machine;* Thomas P. Hughes, *American Genesis: A Century of Innovation and Technological Enthusi-asm, 1870–1970* (1989). On western water development, see Marc Reisner, *Cadillac Desert: The American West and Its Disappearing Water* (1986); Donald Worster, *Rivers of Empire: Water, Aridity, and the Growth of the American West* (1985). On the "politics of productivity," see Charles Maier, "The Politics of Pro-ductivity: Foundations of American International Economic Policy after World War II," *International Organization* 31 (Autumn 1977): 607–33.

41. Mark Harvey, *A Symbol of Wilderness: Echo Park and the American Conser-vation Movement* (1994), xvi.

42. On this new generation of advocates, see Harvey; Miles; Michael Cohen, *The History of the Sierra Club, 1892–1970* (1988); Fox, "'We Want No Straddlers'"; Fox, *The American Conservation Movement*, 250–90.

43. Stephen Fox has aptly referred to the two decades after World War II as "Muir Redux." See Fox, *The American Conservation Movement*, 250–90.

44. See Hays, "From Conservation to Environment" (1982); Scott Dewey, "Working for the Environment: Organized Labor and the Origins of Environmentalism in the United States, 1948–1970," *Environmental History* 3, 1 (January 1998): 45–63.

45. Fox, *The American Conservation Movement*, 250.

46. The Wilderness Society, *Protecting America's Wildlands: The 15 Most Endangered Wildlands 2000 Report* (2000).

SOURCES

Manuscript Collections

ACP   Arthur Carhart Papers, Denver Public Library, Denver, Colorado.

AFAC American Forestry Association Collection, Forest History Society, Durham, North Carolina.

ALP   Aldo Leopold Papers, University of Wisconsin Archives, Madison, Wisconsin.

ATCA Appalachian Trail Conference Archives, Appalachian Trail Conference, Harpers Ferry, West Virginia.

BMKP Benton MacKaye Papers, Special Collections, Dartmouth College Library, Hanover, New Hampshire.

FDRP Franklin D. Roosevelt Papers, Franklin D. Roosevelt Library, Hyde Park, New York.

GJP   Gardner Jackson Papers, Franklin D. Roosevelt Library, Hyde Park, New York.

HBP   Harvey Broome Papers, McClung Collection, Knox County Public Library, Knoxville, Tennessee.

JCMP John C. Merriam Papers, Library of Congress, Washington, D. C.

KC   Kansas Collection, Spencer Research Library, University of Kansas, Lawrence, Kansas.

NARA Records of the National Park Service and United States Forest Service, National Archives and Records Administration, Washington, D.C.

RMP   Robert Marshall Papers, Bancroft Library, University of California, Berkeley, California.

SAFC Society of American Foresters Collection, Forest History Society, Durham, North Carolina.

TWSP Wilderness Society Papers, Denver Public Library, Denver, Colorado.

Selected Bibliography

Abbey, Edward. *Desert Solitaire: A Season in the Wilderness.* New York: Ballantine Books, 1968.

Agnew, Jean-Christophe. "The Consuming Vision of Henry James." In *The Culture of Consumption: Critical Essays in American History,* edited by Richard Wightman Fox and T. J. Jackson Lears, 65–100. New York: Pantheon, 1983.

308

Ahern, George. *Deforested America.* 70th Cong., 2d sess. S. Doc. 216. Washington, D.C.: GPO, 1929.

Akin, William. *Technocracy and the American Dream: The Technocratic Movement, 1900–1940.* Berkeley: University of California Press, 1977.

Albright, Horace. "The Everlasting Wilderness." *Saturday Evening Post* 201 (September 29, 1928): 28, 63, 66, 68.

Albright, Horace and Robert Cahn. *The Birth of the National Park Service: The Founding Years, 1913–1933.* Salt Lake City: Howe Brothers, 1985.

Albright, Horace and Marian Albright Schenck. *Creating the National Park Service: The Missing Years.* Norman: University of Oklahoma Press, 1999.

Allin, Craig. *The Politics of Wilderness Preservation.* Westport, Conn.: Greenwood Press, 1982.

———. "The Leopold Legacy and American Wilderness." In *Aldo Leopold: The Man and His Legacy,* edited by Thomas Tanner, 25–38. Ankeny, Iowa: Soil Conservation Society of America, 1987.

Anderson, Larry. "'A Retreat from Profit': MacKaye's Path to the Appalachian Trail, 1919–1921." Paper presented at symposium "Benton MacKaye and the Appalachian Trail," University at Albany, State University of New York, November 1996.

Aron, Cindy S. *Working at Play: A History of Vacations in the United States.* New York: Oxford University Press, 1999.

Avery, Myron. "Trends in Materials and Designs in Light-Weight Camping Equipment." *Appalachia* (June 1937).

Backes, David. *Canoe Country: An Embattled Wilderness.* Minocqua, Wisc.: North Word Press, 1991.

———. "Wilderness Visions: Arthur Carhart's 1922 Proposal for the Quetico-Superior Wilderness." *Forest and Conservation History* 35 (July 1991): 128–37.

———. *A Wilderness Within: The Life of Sigurd F. Olson.* Minneapolis: University of Minnesota Press, 1997.

Baldwin, Donald. *The Quiet Revolution: The Grass Roots of Today's Wilderness Preservation Movement.* Boulder, Colo.: Pruett Publishing Company, 1972.

Barbour, Michael. "Ecological Fragmentation in the Fifties." In *Uncommon Ground: Rethinking the Human Place in Nature,* edited by William Cronon, 233–55. New York: Norton, 1996.

Barron, Hal S. *Mixed Harvest: The Second Great Transformation in the Rural North, 1870–1930.* Chapel Hill: University of North Carolina Press, 1997.

Bates, J. Leonard. "Fulfilling American Democracy: The Conservation Movement, 1907 to 1921." *Mississippi Valley Historical Review* 44 (1957): 29–57.

Bederman, Gail. *Manliness and Civilization: A Cultural History of Gender and Race in the United States, 1880–1917.* Chicago: University of Chicago Press, 1995.

Belasco, Warren James. *Americans on the Road: From Autocamp to Motel, 1910–1945.* Cambridge: MIT Press, 1979.

Benedict, Ruth. "The Happiest People." *The Nation* 136 (June 7, 1933): 647.

Berger, Michael. *The Devil Wagon in God's Country: The Automobile and Social Change in Rural America.* Hamden, Conn.: Anchor Books, 1979.

Bernstein, David. "Bob Marshall: Wilderness Advocate." *Western States Jewish Historical Quarterly* 13 (1980): 26–37.

Bess, Michael. "Ecology and Artifice: Shifting Perceptions of Nature and High Technology in Postwar France." *Technology and Culture* 36, 4 (October 1995): 830–62.

Bignami, Louis. "Past and Present Tents." *Westways* 73 (1981): 34–37.

Birch, Thomas H. "The Incarceration of Wilderness: Wilderness Areas as Prisons." *Environmental Ethics* 12 (Spring 1990): 3–26.

Blake, Casey Nelson. *Beloved Community: The Cultural Criticism of Randolph Bourne, Van Wyck Brooks, Waldo Frank, and Lewis Mumford.* Chapel Hill: University of North Carolina Press, 1990.

Bliss, Carey. *Autos Across America: A Bibliography of Transcontinental Automobile Travel, 1903–1940.* New Haven: Jekins and Reese Co., 1982.

Botkin, Daniel. *Discordant Harmonies: A New Ecology for the Twenty-First Century.* New York: Oxford University Press, 1990.

Bowden, M. J. "The Invention of American Tradition." *Journal of Historical Geography* 18, 1 (1992): 3–26.

Bowers, William L. *The Country Life Movement in America, 1900–1920.* Port Washington, N.Y.: Kennikat Press, 1974.

Boyer, Paul. *The Urban Masses and Moral Order in America, 1820–1920.* Cambridge: Harvard University Press, 1978.

Braden, Donna R. *Leisure and Entertainment in America.* Dearborn, Mich.: Henry Ford Museum and Greenfield Village, 1988.

Brimmer, F. E. *Motor Campcraft.* New York: Macmillan, 1923.

———. *Auto-camping.* Cincinnati: Stewart Kidd, 1923.

———. *Coleman Motor Campers' Manual.* Wichita: The Coleman Company, 1926.

Broome, Harvey. "Origins of the Wilderness Society." *The Living Wilderness* 5, 5 (July 1940): 13–15.

———. "The Last Decade, 1935–1945." *The Living Wilderness* 10, 14/15 (December 1945): 16.

———. *Out Under the Sky of the Great Smokies: A Personal Journal.* Knoxville, Tenn.: The Greenbrier Press, 1975.

Brown, Dona. *Inventing New England: Regional Tourism in the Nineteenth Century.* Washington, D.C.: Smithsonian Institution Press, 1995.

Bryant, Paul T. "The Quality of the Day: The Achievement of Benton MacKaye." Ph.D. dissertation, University of Illinois, 1965.

Budiansky, Stephen. *Nature's Keepers: The New Science of Nature Management.* New York: The Free Press, 1995.

Burroughs, John. "A Strenuous Holiday." In *Under the Maples,* 109–26. New York: Houghton Mifflin Company, 1921.

Callicott, J. Baird. *In Defense of the Land Ethic: Essays in Environmental Philosophy.* Albany: State University of New York Press, 1989.

———. "The Wilderness Idea Revisited: The Sustainable Development Alternative." *The Environmental Professional* 13 (1991): 236–47.

Callicott, J. Baird, editor. *Companion to A Sand County Almanac: Interpretive and Critical Essays*. Madison: University of Wisconsin Press, 1987.

Callicott, J. Baird, and Michael P. Nelson, eds. *The Great New Wilderness Debate: An Expansive Collection of Writings Defining Wilderness from John Muir to Gary Snyder*. Athens: University of Georgia Press, 1998.

Cammerer, Arno. "Selling the National Parks." *Western Advertising* (October 1920), 20–22.

Carey, James W. and John J. Quirk. "The Mythos of the Electronic Revolution." *American Scholar* 39 (Spring 1970): 219–41.

Carr, Ethan. *Wilderness By Design: Landscape Architecture and the National Park Service*. Lincoln: University of Nebraska Press, 1998.

Carrington, Daines. "An Arctic Middletown." *Saturday Review of Literature* 9 (May 13, 1933): 589.

Cate, Donald. "Recreation and the U.S. Forest Service: A Study of Organizational Response to Changing Demands." Ph.D. dissertation, Stanford University, 1963.

Catton, Theodore. *Inhabited Wilderness: Indians, Eskimos, and the National Parks in Alaska*. Albuquerque: University of New Mexico Press, 1997.

Cermak, Robert W. "In the Beginning: The First National Forest Recreation Plan." *Parks and Recreation* 9, 11 (November 1974): 20–24, 29–33.

Chambless, Edgar. *Roadtown*. New York: Roadtown Press, 1910.

Chandler, William. *The Myth of the TVA: Conservation and Development in the Tennessee Valley, 1933–1983*. Cambridge, Mass.: Ballinger, 1984.

Chase, Stuart. *The Tragedy of Waste*. New York: Macmillan, 1927.

Clark, Carroll D. and Cleo Wilcox. "The House Trailer Movement." *Society and Social Research* 22, 6 (July–August 1938): 503–19.

Clark, Norman H. *Mill Town: A Social History of Everett, Washington*. Seattle: University of Washington Press, 1970.

Clary, David. *Timber and the Forest Service*. Lawrence: University Press of Kansas, 1986.

Clements, Kendrick. "Herbert Hoover and Conservation, 1921–1933." *American Historical Review* 89, 1 (February 1984): 67–88.

———. *Hoover, Conservation, and Consumerism: Engineering the Good Life*. Lawrence: University Press of Kansas, 2000.

Cobb, Irving. *Roughing It De Luxe*. New York: Doran, 1914.

Cohen, Lizabeth. *Making a New Deal: Industrial Workers in Chicago, 1919–1939*. New York: Cambridge University Press, 1990.

Cohen, Michael. *The History of the Sierra Club, 1892–1970*. San Francisco: Sierra Club Books, 1988.

———. "The Bob: Confessions of a Fast-Talking Urban Wilderness Advocate." In *Wilderness Tapestry: An Eclectic Approach to Preservation*, edited by Samuel Zeveloff, L. Mikel Vause, and William McVaugh, 135–52. Reno: University of Nevada Press, 1992.

Collier, John. *From Every Zenith: A Memoir and Some Essays on Life and Thought*. Denver: Sage Books, 1963.

Conkin, Paul K. *The Southern Agrarians.* Knoxville: University of Tennessee Press, 1988.

Cowan, Ruth Schwartz. *More Work for Mother: The Ironies of Household Technology from Open Hearth to the Microwave.* New York: Basic Books, 1983.

Cowgill, Donald Olen. *Mobile Homes: A Study of Trailer Life.* Washington, D.C.: American Council on Public Affairs, 1941.

Cox, Thomas. *The Park Builders: A History of State Parks in the Pacific Northwest.* Seattle: University of Washington Press, 1988.

Cranz, Galen. *The Politics of Park Design: A History of Urban Parks in America.* Cambridge: MIT Press, 1982.

Croker, Robert. *Pioneer Ecologist: The Life and Work of Victor Ernest Shelford, 1877–1968.* Washington, D.C.: Smithsonian Institution Press, 1991.

Cronon, William. *Changes in the Land: Indians, Colonists, and the Ecology of New England.* New York: Hill and Wang, 1983.

———. *Nature's Metropolis: Chicago and the Great West.* New York: Norton, 1991.

———. "The Trouble with Wilderness; or, Getting Back to the Wrong Nature." In *Uncommon Ground: Rethinking the Human Place in Nature,* edited by William Cronon, 69–90. New York: Norton, 1996.

Cronon, William, ed. *Uncommon Ground: Rethinking the Human Place in Nature.* Paperback edition. New York: Norton, 1996.

Culler, Jonathan. "The Semiotics of Tourism." In *Framing the Sign: Criticism and Its Institutions,* 153–67. Norman: University of Oklahoma Press, 1988.

Cutler, Phoebe. *The Public Landscape of the New Deal.* New Haven: Yale University Press, 1985.

Denevan, William. "The Pristine Myth: The Landscape of the Americas in 1492." *Annals of the Association of American Geographers* 82, 3 (September 1992): 369–85.

Devall, Bill and George Sessions. *Deep Ecology.* Layton, Utah: Gibbs Smith, 1985.

Deverell, William. "To Loosen the Safety Valve: Eastern Workers and Western Lands." *Western Historical Quarterly* 19, 3 (August 1988): 269–85.

DeVoto, Bernard. *Mark Twain's America.* Moscow: University of Idaho Press, 1932.

Dewey, John. *Individualism Old and New.* New York: Minton, Balch, and Co., 1929.

Dodds, Gordon B. "The Stream-Flow Controversy: A Conservation Turning Point." *Journal of American History* 56, 1 (June 1969): 59–69.

Dorman, Robert. *Revolt of the Provinces: The Regionalist Movement in America, 1920–1945.* Chapel Hill: University of North Carolina Press, 1993.

Dumenil, Lynn. *The Modern Temper: American Culture and Society in the 1920s.* New York: Hill and Wang, 1995.

Dunlap, Thomas R. *Saving America's Wildlife: Ecology and the American Mind, 1850–1990.* Princeton: Princeton University Press, 1988.

Dunn, Durwood. *Cade's Cove: The Life and Death of a Southern Appalachian Community.* Knoxville: University of Tennessee Press, 1988.

Ecological Society of America, *Preservation of Natural Conditions.* Springfield, Ill.: Schnepp and Barnes, 1922.

Edwards, Carlton M. *Homes for Travel and Living: The History and Development of the Recreational Vehicle and Mobile Home Industries.* East Lansing, Mich.: Carl Edwards and Associates,1977.

Ellis, George J. "The Path Not Taken." M.A. thesis, Shippensburg University, 1993.

Ernst, Joseph W., ed. *Worthwhile Places: Correspondence of John D. Rockefeller, Jr. and Horace Albright.* New York: Fordham University Press, 1991.

Ferguson, Melville F. *Motor Camping on Western Trails.* New York: The Century Company, 1925.

Fishman, Robert. "The Mumford-Jacobs Debate." *Planning History Studies* 10, 1–2 (1996): 3–11.

Flader, Susan. *Thinking Like a Mountain: Aldo Leopold and the Evolution of an Ecological Attitude toward Deer, Wolves, and Forests.* Madison: University of Wisconsin Press, 1974.

———. "'Let the Fire Devil Have His Due': Aldo Leoopold and the Conundrum of Wilderness Management." In *Managing America's Enduring Wilderness Resource,* edited by David W. Lime, 88–95. St. Paul: University of Minnesota Press, 1990.

Flink, James J. *American Adopts the Automobile, 1895–1910.* Cambridge: MIT Press, 1970.

———. *The Car Culture.* Cambridge: MIT Press, 1975.

———. *The Automobile Age.* Cambridge: MIT Press, 1988.

Flint, Howard R. "Wasted Wilderness." *American Forests and Forest Life* 32 (1926): 407–10.

Flores, Dan. "Place: An Argument for Bioregional History." *Environmental History Review* 18, 4 (Winter 1994): 1–18.

Foresta, Ronald. "Transformation of the Appalachian Trail." *Geographical Review* 77, 1 (January 1987): 76–85.

Fox, Richard Wightman. "Epitaph for Middletown: Robert S. Lynd and the Analysis of Consumer Culture." In *The Culture of Consumption,* edited by Richard Wightman Fox and T. J. Jackson Lears, 100–141. New York: Pantheon, 1983.

Fox, Richard Wightman and T. J. Jackson Lears, eds. *The Culture of Consumption: Critical Essays in American History, 1880–1980.* New York: Pantheon, 1983.

———. *The Power of Culture: Critical Essays in American History.* Chicago: University of Chicago Press, 1993.

Fox, Stephen. "'We Want No Straddlers.'" *Wilderness* 48 (Winter 1984): 5–19.

———. *The American Conservation Movement: John Muir and His Legacy.* Madison: University of Wisconsin Press, 1985.

Frome, Michael. *Battle for the Wilderness.* Revised edition. Salt Lake City: University of Utah Press, 1997.

Galambos, Louis. "The Emerging Organizational Synthesis in Modern American History." *Business History Review* 44 (Autumn 1970): 279–90.

———. "Technology, Political Economy, and Professionalization: Central Themes of the Organizational Synthesis." *Business History Review* 57 (Winter 1983): 471–93.

Gates, Paul Wallace. *History of Public Land Law Development.* Washington, D.C.: GPO, 1968.

Geddes, Patrick. *Cities in Evolution.* 1915. Reprint, New York: Oxford University Press, 1950.

Gilligan, James P. "The Development of Policy and Administration of Forest Service Primitive and Wilderness Areas in the Western United States." Ph.D. dissertation, University of Michigan, 1953.

Glover, James M. *A Wilderness Original: The Life of Bob Marshall.* Seattle: The Mountaineers, 1986.

———. "Romance, Recreation, and Wilderness: Influences on the Life and Work of Bob Marshall." *Environmental History Review* 14, 4 (Winter 1990): 22–39.

Glover, James M. and Regina Glover. "Robert Marshall: Portrait of a Liberal Forester." *Journal of Forest History* 30, 3 (July 1986): 112–19.

Goldman, Hal. "James Taylor's Progressive Vision: The Green Mountain Parkway." *Vermont History* 63 (1995): 158–79.

Gottlieb, Robert. *Forcing the Spring: The Transformation of the American Environmental Movement.* Washington, D.C.: Island Press, 1993.

Gough, Robert. *Farming the Cutover: A Social History of Northern Wisconsin, 1900–1940.* Lawrence: University of Kansas Press, 1997.

Graham, Frank, Jr. *The Adirondack Park: A Political History.* New York: Alfred A. Knopf, 1978.

"The Great American Roadside." *Fortune* (September 1934), 53–63, 172, 174, 177.

Greeley, W. B. "Wilderness Recreation Areas." *Service Bulletin* 10, 41 (October 18, 1926): 1–3.

Greenleaf, James L. "The Study and Selection of Sites for State Parks." *Landscape Architecture* 15, 4 (July 1925): 227–34.

Guha, Ramachandra. "Radical American Environmentalism and Wilderness Preservation: A Third World Critique." *Environmental Ethics* 11 (Spring 1989): 71–83.

Hagen, Joel. *An Entangled Bank: The Origins of Ecosystem Ecology.* New Brunswick: Rutgers University Press, 1992.

Hahn, Stephen. "Hunting, Fishing, and Foraging: Common Rights and Class Relations in the Postbellum South." *Radical History Review* 26 (1982): 37–64.

Hall, Peter. *Cities of Tomorrow: An Intellectual History of Urban Planning and Design in the Twentieth Century.* London: Basil Blackwell, 1988.

Handler, Richard. "Boasian Anthropology and the Critique of American Culture." *American Quarterly* 42, 2 (June 1990): 252–73.

Harvey, Mark. *A Symbol of Wilderness: Echo Park and the American Conservation Movement.* Albuquerque: University of New Mexico Press, 1994.

Hawley, Ellis. "Herbert Hoover, the Commerce Secretariat, and the Vision of an 'Associative State,' 1921–1928." *Journal of American History* 61 (June 1974): 116–40.

Hays, Samuel P. *Conservation and the Gospel of Efficiency: The Progressive Conservation Movement, 1890–1920.* Cambridge: Harvard University Press, 1959.

———. "From Conservation to Environment: Environmental Politics in the

United States Since World War Two." *Environmental Review* 6, 2 (Fall 1982): 14–41.

———. *Beauty, Health, and Permanence: Environmental Politics in the United States, 1955–1985.* New York: Cambridge University Press, 1987.

Higham, John. "The Reorientation of American Culture in the 1890s." In *The Origins of Modern Consciousness,* edited by John Weiss, 25–48. Detroit: Wayne State University Press, 1965.

Hirt, Paul W. *A Conspiracy of Optimism: Management of the National Forests Since World War Two.* Lincoln: University of Nebraska Press, 1994.

Historical Statistics of the United States, Colonial Times to 1970, Part 2. Washington, D.C.: GPO, 1975.

Hoffman, Abraham. "Angeles Crest: The Creation of a Forest Highway System in the San Gabriel Mountains." *Southern California Quarterly* 50, 3 (September 1968): 309–45.

———. "Mountain Resorts and Trail Camps in Southern California's Great Hiking Era, 1884–1938." *Southern California Quarterly* 58, 3 (Fall 1976): 381–406.

Hokanson, Drake. *The Lincoln Highway: Main Street Across America.* Iowa City: University of Iowa Press, 1988.

Holden, Harley P. "The Shirley Influence." *The Living Wilderness* 39, 132 (January/March 1976): 18–23.

Horton, F. V. "Camps in the National Forests Attract Farm Folks Seeking Recreation." In *Yearbook of the Department of Agriculture,* 121–22. Washington, D.C.: GPO, 1932.

Hounshell, David. *From the American System to Mass Production, 1800–1932: The Development of Manufacturing Technology in the United States.* Baltimore: Johns Hopkins University Press, 1984.

Howard, Ebenezer. *Garden Cities of To-morrow.* London: S. Sonnenschein and Co., 1902.

Howes, Robert M. "The Knoxville Years." *The Living Wilderness* 39, 132 (January/March 1976): 23–26.

Hughes, Thomas P. *American Genesis: A Century of Innovation and Technological Enthusiasm, 1870–1970.* New York: Viking Press, 1989.

Huth, Hans. *Nature and the American: Three Centuries of Changing Attitudes.* Berkeley: University of California Press, 1957.

Hyde, Anne Farrar. *An American Vision: Far Western Landscape and National Culture, 1820–1920.* New York: New York University Press, 1990.

———. "William Kent: The Puzzle of Progressive Conservationists." In *California Progressivism Revisited,* edited by William Deverell and Tom Sitton, 34–56. Berkeley: University of California Press, 1994.

Interrante, Joseph. "You Can't Go to Town in a Bath-tub: Automobile Movement and the Reorganization of Rural American Space, 1900–1930." *Radical History Review* 21 (Fall 1979): 151–68.

Ise, John. *The United States Forest Policy: A History.* New Haven: Yale University Press, 1920.

———. *Our National Park Policy: A Critical History.* Baltimore: Johns Hopkins University Press, 1961.

Jacobs, Jane. *The Death and Life of Great American Cities*. New York: Random House, 1961.

Jacoby, Karl. *Crimes Against Nature: Squatters, Poachers, Thieves, and the Hidden History of American Conservation*. Berkeley: University of California Press, 2001.

———. "Class and Environmental History: Lessons from the 'War in the Adirondacks.'" *Environmental History* 2, 3 (July 1997): 324–42.

Jakle, John A. *The Tourist: Travel in Twentieth-Century North America*. Lincoln: University of Nebraska Press, 1985.

James, William. "The Moral Equivalent of War." *Popular Science Monthly* (October 1910), 400–410.

Jennings, Francis. *The Invasion of America: Indians, Colonialism, and the Cant of Conquest*. New York: Norton, 1975.

Jessup, Elon. *The Motor Camping Book*. New York: Putnam, 1921.

———. *Roughing It Smoothly*. New York: Putnam, 1923.

Johnpoll, Bernard K. and Mark R. Yerburgh, eds. *The League for Industrial Democracy: A Documentary History*. 3 vols. Westport, Conn.: Greenwood Press, 1980.

Johnson, Benjamin Heber. "Conservation, Subsistence, and Class at the Birth of Superior National Forest." *Environmental History* 4, 1 (January 1999): 80–99.

Johnson, F. R. "Farmers Numerous in Throng of Motorists that Camp in Forests." In *Yearbook of the Department of Agriculture* (1930), 247–49. Washington, D. C.: GPO, 1930.

Jordan, John M. *Machine-Age Ideology: Social Engineering and American Liberalism, 1911–1939*. Chapel Hill: University of North Carolina Press, 1994.

Kaiser, Harvey H. *Great Camps of the Adirondacks*. Boston: David R. Godine, 1982.

Kasson, John. *Amusing the Million: Coney Island at the Turn of the Century*. New York: Hill and Wang, 1978.

Kellogg, R. S. "As I See It." *Journal of Forestry* 28 (April 1930): 461.

Kelly, Lawrence C. "The Indian Reorganization Act: The Dream and the Reality." *Pacific Historical Review* 44, 3 (August 1975): 291–312.

Kennedy, David. *Over Here: The First World War and American Society*. New York: Oxford University Press, 1980.

Kephart, Horace. *The Book of Camping and Woodcraft; A Guidebook for Those Who Travel in the Wilderness*. New York: The Outing Publishing Company, 1906.

Kevles, Daniel. *The Physicists: The History of a Scientific Community in Modern America*. Cambridge: Harvard University Press, 1971.

Kirby, Jack Temple. *Rural Worlds Lost: The American South, 1920–1960*. Baton Rouge: Louisiana State University Press, 1987.

Klein, Kerwin L. "Frontier Products: Tourism, Consumerism, and the Southwestern Public Lands, 1890–1990." *Pacific Historical Review* 62, 1 (February 1993): 39–71.

Knapp, Richard F. and Charles E. Hartsoe. *Play for America: The National Recreation Association*. Arlington, Virginia: National Recreation and Park Association, 1979.

Kniepp, L. F. "These Tame National Forests." *Service Bulletin* 11, 10 (March 7, 1927): 2–3.

———. "The Place of Wilderness Areas in Our National Life." *Service Bulletin* 12, 11 (March 12, 1928): 2.

———. "What Shall We Call Protected Recreation Areas in the National Forests?" *American Planning and Civic Annual* 1 (1929).

Koch, Elers. *Forty Years a Forester, 1903–1943.* Edited by Peter Koch. Missoula, Mont.: Mountain Press Publishing Company, 1998.

Kolko, Gabriel. *The Triumph of Conservatism: A Reinterpretation of American History, 1900–1916.* New York: The Free Press, 1963.

Koppes Clayton. "From New Deal to Termination: Liberalism and Indian Policy, 1933–1953." *Pacific Historical Review* 46 (November 1977): 543–66.

Krog, Carl E. "'Organizing the Production of Leisure': Herbert Hoover and the Conservation Movement in the 1920s." *Wisconsin Magazine of History* 67 (Spring 1984): 199–218.

Larson, Edward J. *Summer for the Gods: The Scopes Trial and America's Continuing Debate over Science and Religion.* Cambridge: Harvard University Press, 1997.

Layton, Edwin. *The Revolt of the Engineers: Social Responsibility and the Engineering Profession.* Cleveland: The Press of Case Western Reserve University, 1971.

Leach, William. *Land of Desire: Merchants, Power, and the Rise of a New American Culture.* New York: Vintage Books, 1993.

Lears, T. J. Jackson. *No Place of Grace: Antimodernism and the Transformation of American Culture, 1880–1920.* New York: Pantheon, 1981.

———. "From Salvation to Self-Realization: Advertising and the Therapeutic Roots of Consumer Culture." In *The Culture of Consumption,* edited by Richard Wightman Fox and T. J. Jackson Lears, 1–38. New York: Pantheon, 1983.

———. *Fables of Abundance: A Cultural History of Advertising in America.* New York: Basic Books, 1994.

Leopold, Aldo. "The Wilderness and Its Place in Forest Recreational Policy." *Journal of Forestry* 19, 7 (November 1921): 718–21.

———. "Conserving the Covered Wagon." *Sunset Magazine* 54, 3 (March 1925): 21, 56.

———. "Wilderness as a Form of Land Use." *Journal of Land and Public Utility Economics* 1, 4 (October 1925): 398–404.

———. "The Last Stand of the Wilderness." *American Forests and Forest Life* 31 (October 1925): 599–604.

———. "A Plea for Wilderness Hunting Grounds." *Outdoor Life* 56, 5 (November 1925): 348–50.

———. "The Vanishing Wilderness." *Literary Digest* 90, 6 (August 7, 1926): 54, 56–57.

———. "Comment." *American Forests and Forest Life* 32 (1926): 410–22.

———. "Mr. Thompson's Wilderness." *Service Bulletin* 12, 26 (June 25, 1928): 1–2.

———. "Game Management in the National Forests." *American Forests and Forest Life* 36 (July 1930): 412, 414.

———. *Game Management.* New York: Scribner's, 1933.

———. "The Conservation Ethic." *Journal of Forestry* 31, 6 (October 1933): 634–43.

———. "Coon Valley: An Adventure in Cooperative Conservation." *American Forests* 41, 5 (May 1935): 205–8.

———. "Why the Wilderness Society?" *The Living Wilderness* 1, 1 (September 1935): 6.

———. "*Naturschutz* in Germany." *Bird-Lore* 38, 2 (March–April 1936): 102–11.

———. "Deer and *Dauerwald* in Germany: I. History." *Journal of Forestry* 34, 4 (April 1936): 366–75.

———. "Deer and *Dauerwald* in Germany: II. Ecology and Politics." *Journal of Forestry* 34, 5 (May 1936): 460–66.

———. "Conservation Esthetic." *Bird-Lore* 40, 2 (March–April 1938): 101–9.

———. "Wilderness as Land Laboratory." *The Living Wilderness* 6 (July 1941): 6.

———. "Wilderness Values." *The Living Wilderness* 7, 7 (March 1942): 24.

———. *A Sand County Almanac.* New York: Oxford University Press, 1949.

———. *The River of the Mother of God and Other Essays by Aldo Leopold.* Edited by Susan Flader and J. Baird Callicott. Madison: University of Wisconsin Press, 1991.

———. *For the Health of the Land: Previously Unpublished Essays and Other Writings.* Edited by J. Baird Callicott and Eric T. Freyfogle. Washington, D.C.: Island Press, 1999.

Lewis, Ralph H. *Museum Curatorship in the National Park Service 1904–1982.* Washington, D.C.: National Park Service, GPO, 1993.

Lewis, Sinclair. *Main Street.* New York: Harcourt, Brace, Jovanovich, 1929.

Lewis, Tom. *Divided Highways: Building the Interstate Highways, Transforming American Life.* New York: Viking, 1997.

Lillard, Richard G. "The Siege and Conquest of a National Park." *American West* 5 (January 1968): 28–31, 67, 69–71.

Little, Charles E. *Greenways for America.* Baltimore: Johns Hopkins University Press, 1990.

Long, J. C. and John D. Long. *Motor Camping.* New York: Dodd, Mead, and Company, 1923.

Lord, Clifford and Elizabeth H. Lord. *Historical Atlas of the United States.* New York: Henry Holt and Company, 1944.

Lord, Russell, ed. *Forest Outings by Thirty Foresters.* U.S. Forest Service. Washington, D.C.: GPO, 1940.

Lowitt, Richard. *The New Deal and the West.* Bloomington: Indiana University Press, 1984.

Lubove, Roy. *Community Planning in the 1920s: The Contribution of the Regional Planning Association of America.* Pittsburgh: University of Pittsburgh Press, 1963.

Luke, Timothy. "The Wilderness Society: Environmentalism as Environationalism." *Capitalism, Nature, Socialism* 10, 4 (December 1999): 1–35.

Lutts, Ralph. *The Nature Fakers: Wildlife, Science, and Sentiment.* Golden, Col.: Fulcrum, 1990.

Lynd, Robert S. and Helen M. Lynd. *Middletown: A Study in Contemporary American Culture.* New York: Harcourt, Brace, and Co., 1929.

MacCannell, Dean. *The Tourist: A New Theory of the Leisure Class.* New York: Schocken Books, 1976.

MacKaye, Benton. "The Forest Cover on the Watersheds Examined by the Geological Survey in the White Mountains, New Hampshire." In *The Relation of Forests to Stream Flow*, edited by Benton MacKaye, M. O. Leighton, and A. C. Spencer. U.S. Department of the Interior. Washington, D.C.: GPO, 1913.

———. "Recreational Possibilities of Public Forests." *Journal of the New York State Forestry Association* 3 (October 1916): 4–10, 29–30.

———. "Some Social Aspects of Forest Management." *Journal of Forestry* 16, 2 (February 1918): 210–14.

———. "Lessons of Alaska." *The Public* 22 (August 30, 1919): 930–32.

———. *Employment and Natural Resources: Possibilities of Making New Opportunities for Employment through the Settlement and Development of Agricultural and Forest Lands and Other Resources*. U.S. Department of Labor. Washington, D.C.: GPO, 1919.

———. "The First Soldier Colony—Kapuskasing, Canada." *The Public* 22 (November 15, 1919): 1066–68.

———. "A Plan for Cooperation Between the Farmer and the Consumer." *Monthly Labor Review* 11, 2 (August 1920): 213–33.

———. "An Appalachian Trail: A Project in Regional Planning." *Journal of the American Institute of Architects* 9 (October 1921): 325–30.

———. "Progress Toward the Appalachian Trail." *Appalachia* 15, 3 (December 1922): 244–52.

———. "The New Exploration: Charting the Industrial Wilderness." *Survey Graphic* 54, 3 (May 1925): 153–57, 192, 194.

———. "Outdoor Culture—The Philosophy of Through Trails." *Landscape Architecture* 17, 3 (April 1927): 163–71.

———. *The New Exploration: A Philosophy of Regional Planning*. New York: Harcourt, Brace, and Co., 1928.

———. "Wilderness Ways." *Landscape Architecture* 19, 4 (July 1929): 327–49.

———. "A New England Recreation Plan." *Journal of Forestry* 27, 8 (December 1929): 927–30.

———. "The Townless Highway." *New Republic* 62 (March 12, 1930): 93–95.

———. "The Appalachian Trail: A Guide to the Study of Nature." *Scientific Monthly* 34 (April 1932): 330–42.

———. "The Tennessee River Project: First Step in a National Plan." *New York Times* (April 16, 1933).

———. "The Challenge of Muscle Shoals." *The Nation* 136 (April 19, 1933): 445–46.

———. "Tennessee—Seed for a National Plan." *Survey Graphic* 22, 5 (May 1933): 251–54, 293–94.

———. "Flankline vs. Skyline." *Appalachia* 20 (1934): 104–8.

———. "Why the Appalachian Trail?" *The Living Wilderness* 1, 1 (September 1935): 7.

———. "From Homesteads to Valley Authorities." *The Survey* 86, 11 (November 1950): 496–98.

———. *From Geography to Geotechnics*. Edited by Paul T. Bryant. Urbana: University of Illinois Press, 1968.

MacKaye, James. *The Economy of Happiness*. Boston: Little, Brown, and Co., 1906.
————. *Americanized Socialism: A Yankee View of Capitalism*. New York: Boni and Liveright, 1918.

MacKaye, Percy. *Epoch: The Life of Steele MacKaye*. New York: Boni and Liveright, 1927.

Maier, Charles. "The Politics of Productivity: Foundations of American International Economic Policy after World War II." *International Organization* 31, 4 (Autumn 1977): 607–33.

Marchand, Roland. *Advertising the American Dream: Making Way for Modernity, 1920–1940*. Berkeley: University of California Press, 1985.

Marshall, George. "Benton as Wilderness Philosopher." *The Living Wilderness* 39, 132 (January/March 1976): 13.

Marshall, Robert. *The High Peaks of the Adirondacks*. Albany: Adirondack Mountain Club, 1922.

————. "Recreational Limitations to Silviculture in the Adirondacks." *Journal of Forestry* 23 (1925): 173–78.

————. *The Growth of Hemlock Before and After Release from Suppression*. Petersham, Mass.: Harvard Forest Bulletin No. 11, 1927.

————. "The Wilderness as Minority Right." *Service Bulletin* (August 27, 1928), 5–6.

————. "Forest Devastation Must Stop." *The Nation* 129 (August 28, 1929): 218–19.

————. "The Problem of the Wilderness." *Scientific Monthly* 30, 2 (February 1930): 141–48.

————. "A Proposed Remedy for Our Forestry Illness." *Journal of Forestry* 28, 3 (March 1930): 273–80.

————. *The Social Management of American Forests*. New York: League for Industrial Democracy, 1930.

————. "An Experimental Study of the Water Relations of Seedling Conifers with Special Reference to Wilting." *Ecological Monongraphs* 1, 1 (January 1931): 37–98.

————. "The Perilous Plight of the Adirondack Wilderness." *High Spots* 9, 4 (October 1932): 13–15.

————. *Arctic Village*. New York: Harrison Smith and Robert Haas, 1933.

————. *The People's Forests*. New York: Harrison Smith and Robert Haas, 1933.

————. "Should the Journal of Forestry Stand for Forestry?" *Journal of Forestry* 32 (1934): 904–8.

————. "Priorities in Forest Service Recreation." *American Forests* 41, 1 (January 1935): 11–13, 30.

————. "Fallacies in Osborne's Position." *The Living Wilderness* 1, 1 (September 1935): 4–5.

————. "Ecology and the Indians." *Ecology* 18, 1 (January 1937): 159–61.

————. "Impressions from the Wilderness." *Nature Magazine* 44 (November 1951): 481–84.

————. "Mountain Ablaze." *Nature Magazine* 46, 6 (June–July 1953): 289–92, 330.

————. *Arctic Wilderness*. Edited by George Marshall. Berkeley: University of California Press, 1956.

Marshall, Robert and Clarence Averill. "Soil Alkalinity on Recent Burns," *Ecology* 9, 4 (October 1928): 533.

Marshall, Robert, et al. "A Letter to Foresters." *Journal of Forestry* 28 (April 1930): 456–58.

Marshall, Robert, John Collier, and Ward Shephard. "The Indians and Their Lands." *Journal of Forestry* 31, 8 (December 1933): 905–10.

Marshall, Robert and Althea Dobbins. "Largest Roadless Areas in the United States." *The Living Wilderness* 2, 2 (November 1936): 11–13.

Marx, Leo. *The Machine in the Garden: Technology and the Pastoral Ideal in America*. New York: Oxford University Press, 1964.

Mathiessen, F. O. *American Renaissance: Art and Expression in the Age of Emerson and Whitman*. New York: Oxford University Press, 1941.

McArdle, Richard E., with Elwood Maunder. "Wilderness Politics: Legislation and Forest Service Policy." *Journal of Forest History* 10 (1975); 166–79.

McClelland, Linda Flint. *Presenting Nature: The Historic Landscape Design of the National Park Service, 1916–1942*. National Park Service. Washington, D.C.: GPO, 1993.

————. *Building the National Parks: Historic Landscape Design and Construction*. Baltimore: Johns Hopkins University Press, 1998.

McCormick, Richard L. *The Party Period and Public Policy: From the Age of Jackson to the Progressive Era*. New York: Oxford University Press, 1986.

————. "The Discovery that Business Corrupts Politics: A Reappraisal of the Origins of Progressivism." *American Historical Review* 86, 2 (April 1981): 247–74.

McCullough, Robert. *The Landscape of Community: A History of Communal Forests in New England*. Hanover, N.H.: University Press of New England, 1995.

McKibben, Bill. *The End of Nature*. New York: Anchor Books, 1989.

McKinsey, Elizabeth. *Niagara Falls: Icon of the American Sublime*. New York: Cambridge University Press, 1985.

May, Lary. *Screening Out the Past: The Birth of Mass Culture and the Motion Picture Industry*. Chicago: University of Chicago Press, 1983.

Meine, Curt. *Aldo Leopold: His Life and Work*. Madison: University of Wisconsin Press, 1988.

————. "The Farmer as Conservationist: Leopold on Agriculture." In *Aldo Leopold: The Man and His Legacy*, edited by Thomas Tanner, 39–52. Ankeny, Iowa: Soil Conservation Society of America, 1987.

Meine, Curt and Richard L. Knight, eds. *The Essential Aldo Leopold: Quotations and Commentaries*. Madison: University of Wisconsin Press, 1999.

Meinecke, E. P. *Camp Planning and Camp Reconstruction*. California Region, U.S. Forest Service, n.d. [ca. 1934].

Mencken, H. L. "Utopia in Little." *American Mercury* 29 (May 1933): 124–26.

Merriam, John Campbell. *Published Papers and Addresses of John Campbell Merriam*. Baltimore: Waverly Press, 1938.

————. *The Garment of God: Influence of Nature on Human Experience*. New York: Scribner's, 1943.

Merriam, Lawrence. "The Irony of the Bob Marshall Wilderness." *Journal of Forest History* 33, 2 (April 1989): 80–87.

Merrill, O. C. "Opening Up the National Forests by Road Building." In *Yearbook of the Department of Agriculture*, 521–29. Washington, D.C.: GPO, 1917.

Meyer, Stephen. *The Five Dollar Day: Labor Management and Social Control in the Ford Motor Company, 1908–1921*. Albany: State University of New York Press, 1981.

Miles, John. *Guardians of the Parks: A History of the National Parks and Conservation Association*. Washington, D.C.: Taylor and Francis, 1995.

Miller, Sally. *Victor Berger and the Promise of Constructive Socialism, 1910–1920*. Westport, Conn.: Greenwood Press, 1973.

Mumford, Lewis. *The Story of Utopias: The Other Half of the Story of Mankind*. New York: Boni and Liveright, 1922.

———. *The Golden Day: A Study in American Experience and Culture*. New York: Boni and Liveright, 1926.

Nasaw, David. *Going Out: The Rise and Fall of Public Amusements*. New York: Basic Books, 1993.

Nash, Charles Edgar. *Trailer Ahoy!* Lancaster, Penn.: Intelligencer Printing Company, 1937.

Nash, Gerald. *The American West Transformed: The Impact of the Second World War*. Bloomington: Indiana University Press, 1985.

———. *World War Two and the West: Reshaping the Economy*. Lincoln: University of Nebraska Press, 1990.

Nash, Roderick. "The Strenuous Life of Bob Marshall." *Forest History* 10 (1966): 18–25.

———. *Wilderness and the American Mind*. 3d edition. New Haven: Yale University Press, 1982 (first published 1967).

———. *The Rights of Nature: A History of Environmental Ethics*. Madison: University of Wisconsin Press, 1989.

Neumann, Roderick P. *Imposing Wilderness: Struggles over Livelihood and Nature Preservation in Africa*. Berkeley and Los Angeles: University of California Press, 1998.

Nicolson, Marjorie Hope. *Mountain Gloom, Mountain Glory: The Development of an Aesthetics of the Infinite*. Ithaca, N.Y.: Cornell University Press, 1959.

Nixon, Herbert B. *Franklin D. Roosevelt and Conservation, 1911–1945*. 2 volumes. Hyde Park, N.Y.: Franklin D. Roosevelt Library, 1957.

Novak, Barbara. *Nature and Culture: American Landscape Painting, 1825–1875*. New York: Oxford University Press, 1980.

Nye, David. *Electrifying America: Social Meanings of a New Technology*. Cambridge: MIT Press, 1990.

O'Brien, Michael. *The Idea of the American South, 1920–1945*. Baltimore: Johns Hopkins University Press, 1979.

Oelschlaeger, Max. *The Idea of Wilderness: From Prehistory to the Age of Ecology*. New Haven: Yale University Press, 1991.

Oesher, Paul. "A Yankee Traditionalist." *The Living Wilderness* 39, 132 (January/March 1976): 8.

Olmsted, Frederick Law, Jr. and William P. Wharton. "The Florida Everglades:

Where the Mangrove Forests Meet the Storm Waves of a Thousand Miles of Water." *American Forests* 38 (March 1932): 142–47, 192.

Olson, Sig. "Quetico Superior Elegy." *The Living Wilderness* 13, 24 (Spring 1948): 5–12.

Olwig, Kenneth. "Reinventing Common Nature: Yosemite and Mount Rushmore—A Meandering Tale of Double Nature." In *Uncommon Ground: Rethinking the Human Place in Nature,* edited by William Cronon, 379–408. New York: Norton, 1996.

Orvell, Miles. *The Real Thing: Imitation and Authenticity in American Culture, 1880–1940.* Chapel Hill: University of North Carolina Press, 1989.

Outdoor Recreation Resources Review Commission. *Outdoor Recreation for America: A Report to the President and Congress by the Outdoor Recreation Resources Review Commission.* Washington, D.C.: GPO, 1962.

Outdoor Recreation Resources Review Commission. "Wilderness and Recreation—A Report on Resources, Values, and Problems." *ORRRC Study Report No. 3.* Washington, D.C.: GPO, 1962.

Paige, John C. *The Civilian Conservation Corps and the National Park Service, 1933–1942: An Administrative History.* National Park Service. Washington, D.C.: GPO, 1985.

Parsons, Kermit C. "Collaborative Genius: The Regional Planning Association of America." *Journal of the American Planning Association* 60, 4 (Autumn 1994): 462–82.

Patton, Phil. *The Open Road: A Celebration of the American Highway.* New York: Simon and Schuster, 1986.

Paxson, Frederic L. "The Highway Movement, 1916–1935." *American Historical Review* 51 (1946): 236–53.

Pearson, G. A. "Preservation of Natural Areas in the National Forests." *Ecology* 3, 4 (October 1922): 284–87.

Peiss, Kathy. *Cheap Amusements: Working Women and Leisure in Turn-of-the-Century New York.* Philadelphia: Temple University Press, 1986.

Pells, Richard H. *Radical Visions and American Dreams: Culture and Social Thought in the Depression Years.* Urbana and Chicago: University of Illinois Press, 1998 [1973].

Perdue, Charles, Jr. and Nancy Martin Perdue. "Appalachian Fables and Facts: A Case Study of the Shenandoah National Park Removals." *Appalachian Journal* 7, 1–2 (Autumn/Winter 1979–80): 84–104.

———. "'To Build a Wall Around These Mountains': The Displaced People of Shenandoah." *The Magazine of Albemarle County History* 49 (1991): 48–71.

Philp, Kenneth R. "Termination: A Legacy of the Indian New Deal." *Western Historical Quarterly* 14, 4 (April 1983): 165–80.

Pisani, Donald. "Forests and Conservation, 1865–1890." *Journal of American History* 72, 2 (September 1985): 340–59.

———. *Water, Land, and Law in the West: The Limits of Public Policy, 1850–1920.* Lawrence: University Press of Kansas, 1996.

Pomeroy, Earl. *In Search of the Golden West: The Tourist in Western America.* New York: Alfred A. Knopf, 1957.

Post, Emily. *By Motor to the Golden Gate*. New York: D. Appleton and Company, 1916.

Pratt, George D. "The Use of the New York State Forests for Public Recreation." *Proceedings of the Society of American Foresters* 11, 3 (July 1916): 281–85.

Price, Jennifer. "Looking for Nature at the Mall: A Field Guide to the Nature Company." In *Uncommon Ground: Rethinking the Human Place in Nature*, edited by William Cronon, 186–203. New York: Norton, 1996.

Pritchard, James A. *Preserving Yellowstone's Natural Conditions: Science and the Perception of Nature*. Lincoln: University of Nebraska Press, 1999.

Pyne, Stephen. *Fire in America: A Cultural History of Wildland and Rural Fire*. Princeton: Princeton University Press, 1982. Reprint, Seattle: University of Washington Press, 1997.

Quammen, David. *The Song of the Dodo: Island Biogeography in an Age of Extinctions*. New York: Scribner, 1996.

Rae, John B. *The Road and the Car in American Life*. Cambridge: MIT Press, 1971.

———. *The American Automobile Industry*. Boston: Twayne, 1984.

Reich, Justin. "Re-Creating the Wilderness: Shaping Narratives and Landscapes in Shenandoah National Park." *Environmental History* 6, 1 (January 2001): 95–117.

Reich, Leonard S. "Ski-Dogs, Pol-Cats, and the Mechanization of Winter." *Technology and Culture* 40, 3 (1999): 484–516.

Reid, Bill G. "Franklin K. Lane's Idea for Veterans' Colonization, 1918–1921." *Pacific Historical Review* 33, 4 (November 1964): 447–61.

Reiger, John. *American Sportsmen and the Origins of Conservation*. New York: Winchester Press, 1975.

Reisch-Owen, A. L. *Conservation under FDR*. New York: Praeger, 1983.

Reisner, Marc. *Cadillac Desert: The American West and Its Disappearing Water*. New York: Penguin Books, 1986.

Robbins, Roy M. *Our Landed Heritage: The Public Domain, 1776–1936*. Princeton: Princeton University Press, 1942.

Rockland, Michael Aaron. *Homes on Wheels*. New Brunswick: Rutgers University Press, 1980.

Rodgers, Daniel T. *The Work Ethic in Industrial America, 1850–1920*. Chicago: University of Chicago Press, 1974.

———. "In Search of Progressivism." *Reviews in American History* 10, 4 (December 1982): 113–32.

———. *Atlantic Crossings: Social Politics in a Progressive Age*. Cambridge, Mass.: Belknap, 1998.

Rome, Adam Ward. *The Bulldozer in the Countryside: Suburban Sprawl and the Rise of American Environmentalism*. New York: Cambridge University Press, 2001.

Roosevelt, Theodore. "The Strenuous Life." In *The Strenuous Life and Other Essays*, 1–21. New York: The Century Company, 1903.

Rose, Mark H. *Interstate: Express Highway Politics, 1941–1956*. Lawrence: The Regents Press of Kansas, 1979.

Rosenzweig, Roy. *Eight Hours for What We Will: Workers and Leisure in the Industrial City, 1870–1920*. New York: Cambridge University Press, 1983.

Rosenzweig, Roy and Elizabeth Blackmar. *The Park and the People: A History of Central Park.* New York: Henry Holt, 1992.

Ross, John. "Benton MacKaye: The Appalachian Trail." In *The American Planner: Biographies and Reflections,* edited by Donald Kreuckeberg, 196–207. New York: Methuen, 1983.

Roth, Dennis. "The National Forests and the Campaign for Wilderness Legislation." *Journal of Forest History* 28 (July 1984): 112–25.

———. *The Wilderness Movement and the National Forests.* College Station, Tex.: Intaglio Press, 1988.

Rothman, Hal K. "'A Regular Ding-Dong Fight': Agency Culture and Evolution in the NPS-USFS Dispute, 1916–1937." *Western Historical Quarterly* 20, 2 (May 1989): 141–62.

———. *The Greening of a Nation? Environmentalism in the United States since 1945.* Fort Worth: Harcourt Brace, 1998.

———. *Devil's Bargains: Tourism in the Twentieth-Century American West.* Lawrence: University Press of Kansas, 1998.

Runte, Alfred. *National Parks: The American Experience.* 2d edition. Lincoln: University of Nebraska Press, 1987.

———. *Yosemite: The Embattled Wilderness.* Lincoln: University of Nebraska Press, 1990.

Rydell, Robert W. *All The World's a Fair: Visions of Empire at American International Expositions, 1876–1916.* Chicago: University of Chicago Press, 1984.

Sale, Kirkpatrick. *Dwellers in the Land: The Bioregional Vision.* San Francisco: Sierra Club Books, 1985.

Salmond, John. *The Civilian Conservation Corps, 1933–1942: A New Deal Case Study.* Durham: Duke University Press, 1967.

Schaffer, Daniel. *Garden Cities for America: The Radburn Experience.* Philadelphia: Temple University Press, 1982.

———. "Ideal and Reality in 1930s Regional Planning: The Case of the Tennessee Valley Authority." *Planning Perspectives* 1 (1986): 27–44.

———. "Benton MacKaye: The TVA Years." *Planning Perspectives* 5 (1990): 5–21.

Scharff, Virginia. *Taking the Wheel: Women and the Coming of the Motor Age.* New York: The Free Press, 1991.

Schivelbusch, Wolfgang. *The Railway Journey: The Industrialization of Time and Space in the 19th Century.* Berkeley: University of California Press, 1986.

Schmitt, Peter J. *Back to Nature: The Arcadian Myth in Urban America.* New York: Oxford University Press, 1969.

Schrepfer, Susan R. *The Fight to Save the Redwoods: A History of Environmental Reform.* Madison: University of Wisconsin Press, 1983.

Schullery, Paul, ed. *The Grand Canyon: Early Impressions.* Boulder: Colorado Associated University Press, 1981.

Schuyler, David. *The New Urban Landscape: The Redefinition of City Form in Nineteenth-Century America.* Baltimore: Johns Hopkins University Press, 1986.

Schwantes, Carlos. *The Pacific Northwest: An Interpretive History.* Lincoln: University of Nebraska Press, 1989.

Scott, James. *Seeing Like a State: How Certain Schemes to Improve the Human Condition Have Failed.* New Haven: Yale University Press, 1998.

Searle, R. Newell. *Saving Quetico-Superior: A Land Set Apart.* St. Paul: Minnesota Historical Society Press, 1977.

Sears, John. *Sacred Places: American Tourist Attractions in the Nineteenth Century.* New York: Oxford University Press, 1989.

Sears, Paul. *Deserts on the March.* Norman: University of Oklahoma Press, 1935.

Seely, Bruce E. *Building the American Highway System: Engineers as Policy Makers.* Philadelphia: Temple University Press, 1987.

Sellars, Richard West. *Preserving Nature in the National Parks: A History.* New Haven: Yale University Press, 1997.

Sellers, Christopher. "Thoreau's Body: Towards an Embodied Environmental History." *Environmental History* 4, 4 (October 1999): 486–514.

Shaffer, Marguerite. "See America First: Tourism and National Identity." Ph.D. dissertation, Harvard University, 1994.

———. "Negotiating National Identity: Western Tourism and 'See America First.'" In *Reopening the American West,* edited by Hal K. Rothman, 122–51. Tucson: University of Arizona Press, 1998.

———. "'See America First': Re-Envisioning Nation and Region through Western Tourism." *Pacific Historical Review* 65, 4 (November 1996): 559–81.

Shankland, Robert. *Steve Mather of the National Parks.* 3d edition. New York: Alfred A. Knopf, 1970.

Shelford, Victor E. "Preserves of Natural Conditions." *Transactions of the Illinois Academy of Science* 13 (1920): 37–58.

Shelford, Victor E., ed. *Naturalist's Guide to the Americas.* Baltimore: Williams and Wilkins Company, 1926.

Shi, David E. *The Simple Life: Plain Living and High Thinking in American Culture.* New York: Oxford University Press, 1985.

Silverstein, Hannah. "No Parking: Vermont Rejects the Green Mountain Parkway." *Vermont History* 63 (1995): 133–57.

Sims, Blackburn. *The Trailer Home, with Practical Advice on Trailer Life and Travel.* New York: Longmans, Green, and Company, 1937.

Slotkin, Richard. *Regeneration Through Violence: The Mythology of the American Frontier, 1600–1860.* Middletown, Conn.: Wesleyan University Press, 1973.

———. *The Fatal Environment: The Myth of the Frontier in the Age of Industrialization, 1800–1890.* New York: Atheneum, 1985.

Smith, Henry Nash. *Virgin Land: The American West as Symbol and Myth.* Cambridge: Harvard University Press, 1950.

Smith, Michael L. *Pacific Visions: California Scientists and the Environment, 1850–1915.* New Haven: Yale University Press, 1987.

Spann, Edward. *Designing Modern America: The Regional Planning Association of America and Its Members.* Columbus: Ohio State University Press, 1996.

Spence, Mark. "Dispossessing the Wilderness: Yosemite Indians and the Wilderness Ideal, 1864–1930." *Pacific Historical Review* 65, 1 (February 1996): 27–59.

———. "Crown of the Continent, Backbone of the World: The American Wilderness Ideal and Blackfeet Exclusion from Glacier National Park." *Environmental History* 1, 3 (July 1996): 29–49.

———. *Dispossessing the Wilderness: Indian Removal and the Making of the National Parks.* New York: Oxford University Press, 1999.

Spirn, Anne Whiston. "Constructing Nature: The Legacy of Frederick Law Olmsted." In *Uncommon Ground: Rethinking the Human Place in Nature,* edited by William Cronon, 91–113. New York: Norton, 1996.

Steel, Ronald. *Walter Lippmann and the American Century.* Boston: Little, Brown, and Co., 1980.

Steen, Harold K. *The U.S. Forest Service: A History.* Seattle: University of Washington Press, 1991 [1976].

Steiner, Jesse Frederick. *Americans at Play: Recent Trends in Recreation and Leisure Time Activities.* New York: McGraw-Hill, 1933.

———. "Challenge of the New Leisure." *New York Times Magazine* (September 24, 1933), 1–2, 16.

Stocking, George, Jr. "Ideas and Institutions in American Anthropology: Toward a History of the Interwar Period." In *The Ethnographer's Magic and Other Essays in the History of Anthropology,* 114–77. Madison: University of Wisconsin Press, 1992.

Strasser, Susan. *Satisfaction Guaranteed: The Making of the American Mass Market.* Washington, D.C.: Smithsonian Institution Press, 1989.

Strauss, David. "Toward a Consumer Culture: 'Adirondack Murray' and the Wilderness Vacation." *American Quarterly* 39, 2 (Summer 1987): 270–86.

Stricker, Frank. "Affluence for Whom?—Another Look at Prosperity and the Working Classes in the 1920s." *Labor History* 24, 1 (Winter 1983): 5–33.

"A Summons to Save the Wilderness." *The Living Wilderness* 1, 1 (September 1935): 1.

Susman, Warren. *Culture as History: The Transformation of American Society in the Twentieth Century.* New York: Pantheon, 1984.

Sussman, Carl, ed. *Planning the Fourth Migration: The Neglected Vision of the Regional Planning Association of America.* Cambridge: MIT Press, 1976.

Sutter, Paul. "Paved with Good Intentions: Good Roads, the Automobile, and the Rhetoric of Rural Improvement in *Kansas Farmer,* 1890–1914." *Kansas History* 18, 4 (Winter 1995–1996): 284–99.

Swain, Donald. *Federal Conservation Policy, 1921–1933.* Berkeley: University of California Press, 1963.

———. "The Passage of the National Park Service Act of 1916." *Wisconsin Magazine of History* (Autumn 1966), 4–17.

———. *Wilderness Defender: Horace M. Albright and Conservation.* Chicago: University of Chicago, 1970.

———. "The National Park Service and the New Deal, 1933–1940." *Pacific Historical Review* 41, 3 (August 1972): 312–32.

Taylor, Alan. "'Wasty Ways': Stories of American Settlement." *Environmental History* 3, 3 (July 1998): 291–310.

Terrie, Philip. "The Adirondack Forest Preserve: The Irony of Forever Wild." *New York History* 62, 3 (July 1981): 260–88.

———. *Forever Wild: Environmental Aesthetics and the Adirondack Forest Preserve.* Philadelphia: Temple University Press, 1985.

———. *Contested Terrain: A New History of Nature and People in the Adirondacks.* Syracuse, N.Y.: The Adirondack Museum/Syracuse University Press, 1997.

Thomas, John. *Alternative America: Henry George, Edward Bellamy, Henry Demarest Lloyd, and the Adversary Tradition.* Cambridge, Mass.: Belknap Press, 1983.

———. "Lewis Mumford, Benton MacKaye, and the Regional Vision." In *Lewis Mumford: Public Intellectual,* edited by Thomas P. Hughes and Agatha C. Hughes, 66–99. New York: Oxford University Press, 1990.

Thompson, Manly. "A Call from the Wilds." *Service Bulletin* 12, 20 (May 14, 1928): 2–3.

Tjossem, Sara. "Preservation of Nature and Academic Respectability: Tensions in the Ecological Society of America, 1915–1979." Ph.D. dissertation, Cornell University, 1994.

Tobey, Ronald. *Saving the Prairies: The Life Cycle of the Founding School of American Plant Ecology, 1895–1955.* Berkeley: University of California Press, 1981.

———. *Technology and Freedom: The New Deal and Electrical Modernization of the American Home.* Berkeley: University of California Press, 1996.

Trachtenberg, Alan. *The Incorporation of America: Culture and Society in the Gilded Age.* New York: Hill and Wang, 1982.

Truxal, Andrew G. *Outdoor Recreation Legislation and Its Effectiveness: A Summary of American Legislation for Public Outdoor Recreation, 1915–1927.* New York: Columbia University Press, 1929.

Tunnard, Christopher and Boris Pushkarev. *Man-Made America: Chaos or Control?* New Haven: Yale University Press, 1963.

Tweed, William C. *Recreation Site Planning and Improvement in the National Forests, 1891–1942.* U.S. Forest Service. Washington, D.C.: GPO, 1981.

Twelve Southerners. *I'll Take My Stand: The South and the Agrarian Tradition.* New York: Harper and Brothers, 1930.

Tyler, Robert L. *Rebels of the Woods: The I.W.W. and the Pacific Northwest.* Eugene: University of Oregon Books, 1967.

Tyrrell, Ian. *True Gardens of the Gods: Californian-Australian Environmental Reform, 1860–1930.* Berkeley and Los Angeles: University of California Press, 1999.

U.S. Congress. Senate. *Use of Automobiles in National Parks.* 62nd Cong., 2d sess., March 13, 1912. S. Doc. 433.

———. *Proceedings of the National Conference on Outdoor Recreation.* 68th Cong., 1st sess., May 1924. S. Doc. 151.

———. *Proceedings of the Meeting of the Advisory Council of the National Conference on Outdoor Recreation.* 68th Cong., 1st sess., December 1924. S. Doc. 229.

———. *Proceedings of the Second National Conference on Outdoor Recreation.* 69th Cong., 1st sess., January 1926. S. Doc. 117.

———. *A Report Epitomizing the Results of the Major Fact-Finding Surveys and Projects which Have Been Undertaken under the Auspices of the National Conference on Outdoor Recreation.* 70th Cong., 1st sess., 1928. S. Doc. 158.

———. *A National Plan for American Forestry.* 2 vols. 73rd Cong., 1st sess., March 1933. S. Doc. 12.

U.S. Department of Agriculture. Forest Service. "Report[s] of the Forester." In *Annual Reports of the Department of Agriculture.* Washington, D.C.: GPO, 1910–24.

———. *The Use Book: A Manual of Information about the National Forests.* Washington, D.C.: GPO, 1918.

———. *Report[s] of the Chief of the Forest Service.* Washington, D.C.: GPO, 1925–40.

U.S. Department of the Interior. National Park Service. *Annual Report[s] of the Director of the National Park Service.* Washington, D.C.: GPO, 1916–40.

———. *A Study of the Park and Recreation Problem of the United States.* Washington, D.C.: GPO, 1941.

U.S. Department of Labor. *Report of the Secretary of Labor.* Washington, D.C.: GPO, 1915.

Van Name, Willard. *Vanishing Forest Reserves: Problems of the National Forests and National Parks.* Boston: Richard G. Badger, 1929.

Varney, Porter. *Motor Camping.* New York: Leisure League of America, 1935.

Veblen, Thorstein. *The Engineers and the Price System.* 1921. Reprint, New York: Harcourt, Brace, and World, 1963.

Wallis, Allan D. *Wheel Estate: The Rise and Decline of Mobile Homes.* Chicago: University of Chicago Press, 1991.

Warren, Louis. *The Hunter's Game: Poachers and Conservationists in Twentieth-Century America.* New Haven: Yale University Press, 1997.

Waterman, Laura and Guy Waterman. *Forest and Crag: A History of Hiking, Trail Blazing, and Adventure in the Northeast Mountains.* Boston: Appalachian Mountain Club, 1989.

Watkins, T. H. *Righteous Pilgrim: The Life and Times of Harold L. Ickes, 1874–1952.* New York: Henry Holt and Co., 1990.

Waugh, Frank. *Recreation Uses on the National Forests.* Forest Service. Washington, D.C.: GPO, 1918.

Weinstein, James. *The Corporate Ideal and the Liberal State, 1900–1918.* Boston: Beacon Press, 1968.

Westbrook, Robert. "Tribune of the Technostructure: The Popular Economics of Stuart Chase." *American Quarterly* 32, 4 (Fall 1980): 387–408.

———. *John Dewey and American Democracy.* Ithaca: Cornell University Press, 1991.

Wheelright, Jeffrey. *Degrees of Disaster: Prince William Sound: How Nature Reels and Rebounds.* New Haven: Yale University Press, 1996.

White, Richard. "Poor Men on Poor Lands: The Back-to-the-Land Movement of the Early Twentieth Century—A Case Study." *Pacific Historical Review* 49, 1 (February 1980): 105–31.

———. *"It's Your Misfortune and None of My Own": A New History of the American West.* Lincoln: University of Nebraska Press, 1991.

———. *"'Are You an Environmentalist or Do You Work for a Living?': Work and Nature."* In *Uncommon Ground: Rethinking the Human Place in Nature,* edited by William Cronon, 171–85. New York: Norton, 1996.

———. *The Organic Machine: The Remaking of the Columbia River.* New York: Hill and Wang, 1995.

Wiebe, Robert H. *The Search for Order, 1877–1920.* New York: Hill and Wang, 1967.

Wik, Reynold. *Henry Ford and Grass Roots America.* Ann Arbor: University of Michigan Press, 1972.

Wilderness Society. *Protecting America's Wildlands: The 15 Most Endangered Wildlands 2000 Report.* Washington, D.C.: The Wilderness Society, 2000.

Williams, Michael. *Americans and Their Forests: A Historical Geography.* New York: Cambridge University Press, 1989.

Williams, Raymond. *The Country and the City.* New York: Oxford University Press, 1973.

———. "The Idea of Nature." In *Problems in Materialism and Culture: Selected Essays,* 67–85. London: Verso, 1980.

Williams, William J. "Bloody Sunday Revisited." *Pacific Northwest Quarterly* 71 (1980): 50–62.

Wilson, Christopher. "The Rhetoric of Consumption: Mass Market Magazines and the Demise of the Gentle Reader." In *The Culture of Consumption,* edited by Richard Wightman Fox and T. J. Jackson Lears, 39–64. New York: Pantheon, 1983.

Wilson, Richard Guy, Dianne H. Pilgrim, and Dickran Tashjian. *The Machine Age in America, 1918–1941.* New York: Brooklyn Museum, 1986.

Worster, Donald. *Dust Bowl: The Southern Plains in the 1930s.* New York: Oxford University Press, 1979.

———. *Nature's Economy: A History of Ecological Ideas.* 2d edition. New York: Cambridge University Press, 1985.

———. *Rivers of Empire: Water, Aridity, and the Growth of the American West.* New York: Pantheon, 1985.

———. *Under Western Skies: Nature and History in the American West.* New York: Oxford University Press, 1992.

———. *The Wealth of Nature: Environmental History and the Ecological Imagination.* New York: Oxford University Press, 1993.

———. *An Unsettled Country: Changing Landscapes of the American West.* Albuquerque: University of New Mexico Press, 1994.

———. "Wild, Tame, and Free: Comparing Canadian and American Views of Nature." In *On Brotherly Terms: Canadian-American Relations West of the Rockies,* edited by Ken Coates and John Findlay. Seattle: University of Washington Press, 2002.

Wrobel, David M. *The End of American Exceptionalism: Frontier Anxiety from the Old West to the New Deal.* Lawrence: University Press of Kansas, 1993.

Yard, Robert Sterling. *National Parks Portfolio.* Washington, D.C.: GPO, 1916.

———. "Director of the Nation's Playgrounds." *Sunset* 37 (September 1916): 27.

———. *The Book of the National Parks.* New York: Scribner's, 1919.

———. *Glimpses of Our National Parks.* 3d edition. Washington, D.C.: GPO, 1920.

———. "The People and the National Parks." *The Survey* 48, 13 (August 1, 1922): 547–53, 583.

———. "Economic Aspects of Our National Parks Policy." *Scientific Monthly* 16 (April 1923): 380–88.

———. "But We Must Hold Our Heritage." *National Parks Bulletin* 47 (January 1926): 1–2.

———. "Politics in Our National Parks: Shall Standards Be Lowered to Serve Political Expediency?" *American Forests* 32, 392 (August 1926): 485–89.

———. "Needless Road Project Endangers Yosemite." *National Parks Bulletin* 52 (February 1927): 16.

———. "The Motor Tourist and the National Parks." Parts 1 and 2. *National Parks Bulletin* 52 (February 1927): 11–12; 53 (July 1927): 17–19.

———. *Our Federal Lands: A Romance of American Development*. New York: Scribner's, 1928.

Zahniser, Ed. "Walk Softly and Carry a Big Map: Historical Roots of Wildlands Network Planning." *Wild Earth* 10, 2 (Summer 2000): 33–38.

Zimmerman, William, Jr. "Wilderness Areas on Indian Lands." *The Living Wilderness* 5, 5 (July 1940): 10–11.

Zunz, Olivier. *Why the American Century?* Chicago: University of Chicago Press, 1998.